AF506001

Howard Barker: Ecstasy and Death

Also by David Ian Rabey

Criticism:

BRITISH AND IRISH POLITICAL DRAMA IN THE TWENTIETH CENTURY

HOWARD BARKER: POLITICS AND DESIRE: An Expository Study of his Drama and Poetry, 1969–1987

DAVID RUDKIN: Sacred Disobedience

ENGLISH DRAMA SINCE 1940

THEATRE OF CATASTROPHE: New Essays on Howard Barker *(co-edited with Karoline Gritzner)*

Drama:

THE WYE PLAYS

LOVEFURIES

Howard Barker: Ecstasy and Death

An Expository Study of his Drama, Theory and Production Work, 1988–2008

David Ian Rabey

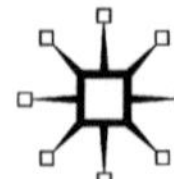

First published 2009 by
PALGRAVE MACMILLAN

Palgrave Macmillan in the UK is an imprint of Macmillan Publishers Limited, registered in England, company number 785998, of Houndmills, Basingstoke, Hampshire RG21 6XS.

Palgrave Macmillan in the US is a division of St Martin's Press LLC, 175 Fifth Avenue, New York, NY 10010.

Palgrave Macmillan is the global academic imprint of the above companies and has companies and representatives throughout the world.

Palgrave® and Macmillan® are registered trademarks in the United States, the United Kingdom, Europe and other countries.

ISBN-13: 978–1–4039–9473–8   hardback
ISBN-10: 0–1–4039–9473–0   hardback

This book is printed on paper suitable for recycling and made from fully managed and sustained forest sources. Logging, pulping and manufacturing processes are expected to conform to the environmental regulations of the country of origin.

A catalogue record for this book is available from the British Library.

A catalogue record for this book is available from the Library of Congress.

10   9   8   7   6   5   4   3   2   1
18   17   16   15   14   13   12   11   10   09

Printed and bound in Great Britain by
CPI Antony Rowe, Chippenham and Eastbourne

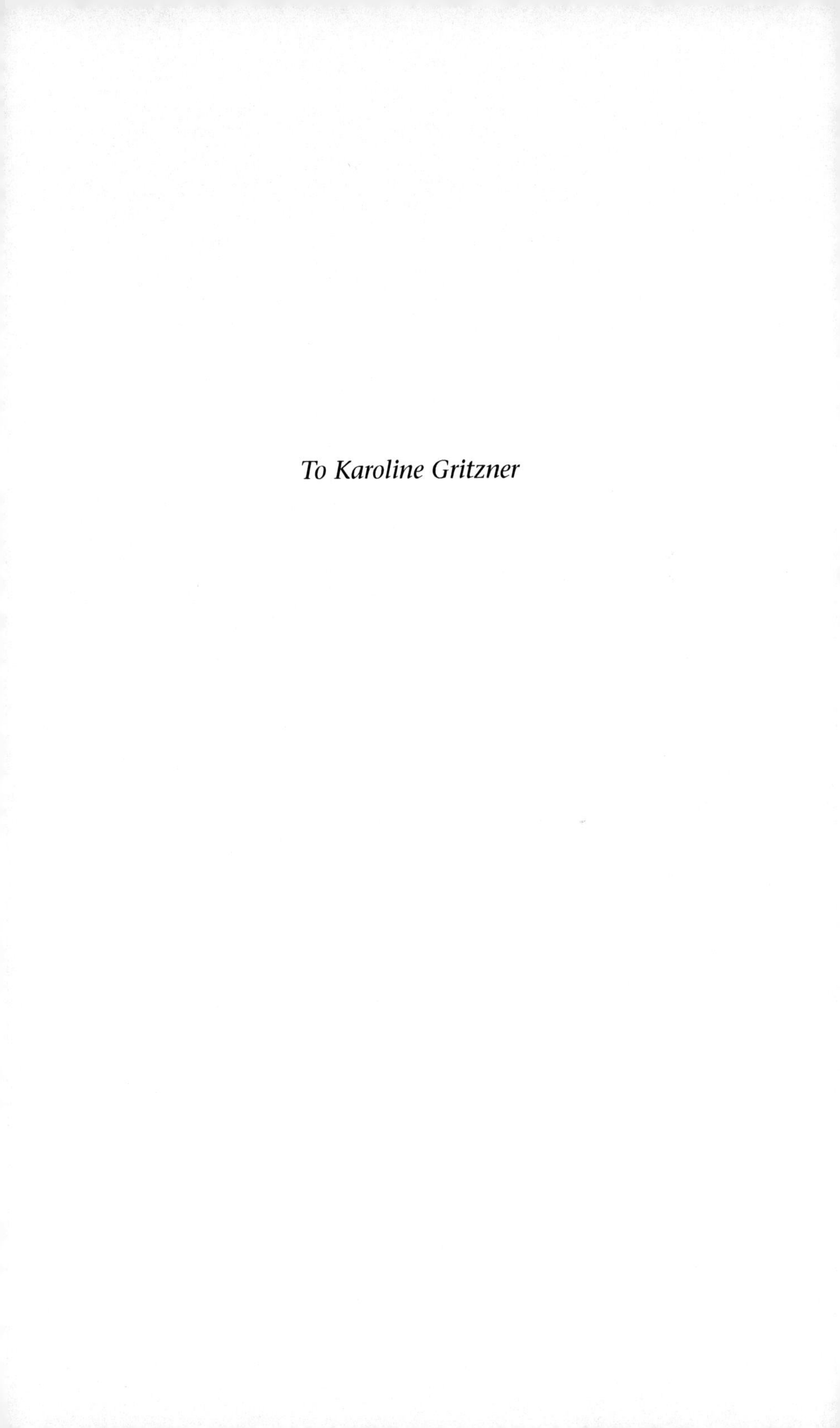

To Karoline Gritzner

'love's garment is pain / And impossible daring'
                            – Gwyneth Lewis, *Parables and Faxes* XXI

                        I have a sense all sorts of things
                        Things we prefer
                        At this juncture
                        Not to contemplate
                        Will be
                        Oh, certainly will be
                        UTILIZED
                                    – *Defilo*

Contents

Acknowledgments

Quotations from Howard Barker's published plays are reprinted with kind permission of the publisher, Oberon Books (James Hogan); quotations from unpublished work are reprinted by kind permission of Judy Daish Associates Ltd. Some material from Chapter Seven appears in extended form in *Theatre Research International* Volume 30 number 2 (July 2005), pp. 175–189 and is reprinted by kind permission of Cambridge University Press. Epigraph from Gwyneth Lewis, *Chaotic Angels: Poems in English* (Bloodaxe Books, 2005) reprinted with kind permission of the publisher on behalf of the author.

This study originally incorporated readings of Barker's three volumes of poetry published during the period under review: *Lullabies for the Impatient* (1988), *The Ascent of Monte Grappa* (1991) and *The Tortmann Diaries* (1996). I have excised my analysis of these in deference to concerns expressed to me about the overall length of this manuscript; but will endeavour to present this in some future context.

Everything impedes your mission if you have a mission. But there were some honourable exceptions, whom I thank: Ioan Williams for his approval of study leave, from my teaching at Aberystwyth University, to permit me to write this; Howard Gooding at Judy Daish Associates for his prompt and thorough provision of unpublished works; Roger Owen, Andy Cornforth, Antoinette Walsh and John O'Brien for their developments of opportunities for, and articulate company on, our adventures in productions of Barker plays; Beth Hoffman (convenor), Daniel A. Sack, Mary Karen Dahl (chair): the Barker panel at ATHE New Orleans, July 2007; David Kilpatrick for sending me his fine essay; George Hunka for support and discussion via his website, 'Superfluities Redux'; Kelly V. Jones for permission to quote from her as yet unpublished thesis; Susan Russell, Thomas J. Barnes and Sarah Rose Evans for permission to quote their observations; Jane Bertish, Edward Petherbridge, Julia Tarnoky, Gerrard McArthur, Justin Avoth, Séan O'Callaghan and Melanie Jessop for contributing their testimonies; Victoria Wicks – in appreciation of her important stage performances for The Wrestling School; the companies of my Aberystwyth productions of *Ursula* (particularly Vicki-Jane Fergus as Ursula, Kate Davies as Placida, Dan Price as Lucas, Heather Stevenson as Leonora) and *Gertrude – The Cry* (particularly Henry Pickett as Claudius, Gemma Crabtree as Cascan, Hannah Jones as Ragusa, Adam Stevens as Albert, Graeme Westgate as Hamlet, Jennifer Woodhouse as Isola, and, again, Sarah Rose Evans as Gertrude, without whom); Sarah Goldingay, bright and indefatigable in her initiatives; my editor at Palgrave, Paula Kennedy, for her support of this project; Glenys Hartnell and Catrin Davies for secretarial assistance; Ken Rabey (who read and commented on the manuscript), Charmian Savill, Isabel Rabey and Ryan Jack Rabey, for enthusiastic company in attending Barker productions and good discussions afterwards; HB for, in many ways, giving me a sense of direction.

DIR, Machynlleth, May 2008

Abbreviations and References

(All places of publication: London except where noted)

AT	*Arguments for a Theatre* (Manchester: Manchester University Press, third edition, 1997)
CP2	*Collected Plays Volume 2* (Calder Publications, 1993)
CP3	*Collected Plays Volume 3* (Calder Publications, 1996)
CP4	*Collected Plays Volume 4* (Calder Publications, 1998)
CP5	*Collected Plays Volume 5* (Calder Publications, 2001)
DH	*Dead Hands* (Oberon Books, 2004)
DTOAT	*Death, The One and the Art of Theatre* (Abingdon: Routledge, 2005)
EB	*The Ecstatic Bible* (Oberon Books, 2004)
FTY	*The Fence in its Thousandth Year* (Oberon Books, 2005)
GTC	*Gertrude – The Cry* and *Knowledge and a Girl* (Calder Publications, 2002)
GTGU	*Gary the Thief/Gary Upright* (John Calder, 1987)
HH	*A Hard Heart* and *The Early Hours of a Reviled Man* (Calder Publications, 1992)
HN	*Hated Nightfall* and *Wounds to the Face* (Calder Publications, 1994)
LS	*The Last Supper* (John Calder, 1988)
OP1	*Plays One* (Oberon Books, 2006)
OP2	*Plays Two* (Oberon Books, 2006)
SAG	*The Seduction of Almighty God* (Oberon Books, 2006)
SAO	*A Style and its Origins* (Oberon Books, 2007)
SL	*Seven Lears* and *Golgo* (John Calder, 1990)
SN	*The Swing at Night* (Calder Publications, 2001)

Editions of Barker's plays, formerly published by John Calder and Calder Publications, are no longer available for order and purchase from this source (but may be available for consultation though library holdings). The plays formerly published by John Calder and Calder Publications are being republished by Oberon books as part of their ongoing programme of making Barker's complete dramatic *oeuvre* available in book form.

Part One

A Style and its Contexts

1
Gifts of Loss: An Introduction to Barker's Writing and Theatre

It is a crime to look, and a crime to avert your eyes...a crime to protest and a crime to enjoy...

– *The Brilliance of the Servant* (CP5, 125)

Howard Barker has been acclaimed as 'England's greatest living dramatist' in *The Times* and as 'the Shakespeare of our age' by the late Sarah Kane. His uniquely stylish and rigorous work as a director for his theatre company, The Wrestling School, has presented his writing in increasingly alluring and startlingly original forms, to create an aesthetic and an experience which he terms Theatre of Catastrophe. He has also emerged as a major philosophical theorist on the possibilities of theatre, and on the theatricality of the human mind, body and soul in situations of extremity, with the writing and publication (since 1989) of the theoretical essays which underpin his practical experimentations, and moreover articulate his philosophic and artistic interrogation of conventional social objectives. This study considers the full range of Barker's theatrical objectives and achievements, and reflects his international recognition and esteem as a major artistic thinker and practitioner. It follows on from my 1989 publication, *Howard Barker: Politics and Desire: An Expository Study of his Drama and Poetry, 1969–87*, and considers Barker's work from 1988 to 2008, including his emergence as a major director and aspects of productions by The Wrestling School (if circumstances permit, I will aim to write a third volume circa 2028). I will briefly characterize some of the conceptual and formal propositions and demands of his theatre in this chapter, and identify major developments in his theoretical writings in the next. Thereafter, this volume offers a broadly chronological mapping of Barker's plays, in a performance context centred on his own productions where possible, taking account of Barker's moves into the direction and design of his own work, in development of a highly specific theatrical aesthetic and methodology.

Barker's theatre brings together classical discipline, scenographic ambition, moral ruthlessness and savage humour, in ways which offer uniquely startling examples of Anne Bogart's intimation that the theatre is 'an art form that can open us to the infinite through complexity'[1]. Indeed, Barker insists that (now as in sixteenth-century England) 'the theatre is a place for poets, it cannot share a discourse with film or television' (and should not need to compete with the effects, means and ends of those media); 'it needs its own voice'[2], its distinctive style and objectives. In the period covered by this study, Barker moves beyond acceptability to the major national subsidized companies (The Royal Court and The Royal Shakespeare Company) which presented his earlier work and into directing and designing productions of the plays he has written, for The Wrestling School, a theatre company dedicated solely to productions of his work.

As a director and scenographer, designing both sets and costumes under aliases such as Tomas Leipzig, Billie Kaiser, and Caroline Shentang, Barker has developed a sense of pictorial stage composition, alongside scrupulously delivered language of operatic reach and scale. However, for all the considerable visual invention of his scenography, Barker ensures that actors are not overwhelmed by their surroundings or subdued by them: spaces remain clear and uncluttered, though flying objects or sliding panels may propose unforeseen visual depths. Barker/Leipzig's sets always draw attention to the actor in space and time (as examples: in *Ursula*, a metallic floor underpinned the final scene in which the virgins were brought on to be slaughtered on hospital trolleys, involving effects of sound, such as trundling, grating and clattering; in *(Uncle) Vanya* a disintegrating ship housed a situation of drifting inertia, in which metal trays fell explosively to the floor, expressing Vanya's horror of domesticity). Frequently actors occupy or move through the space exploring the limits of their balance, yet never falling (except when precisely specified). Characters are strangely familiar rather than exotic; they often have a regal quality or background, and their costumes reflect style and taste (often based on classic *haute couture* of the 1930s, 40s and 50s). But their language, like their experience, is often raw, strongly explicit, surgically incisive and disruptive of what might otherwise be socially harmonious.

Formally, Barker repudiates the immediate recognizability and familiarity which is valorized and claimed by the dominant British dramatic tradition of social realism (as principally identified with the repertoire, "cultural functions" and perceived achievements of the Royal Court Theatre). Rather, his 'first instincts were always to invent and not describe'[3], to imagine what might be (im?)possible, to estrange the familiar, and resist conventional empathy. This has led him to create a series of highly original and well-plotted stories, energized by a suspenseful narrative drive (such as *The Love of a Good Man, Animals in Paradise, Albertina, Acts (Chapter One), The Fence, I Saw Myself*); as well as Shakespearean "subversions", in which events and characters depart startlingly from familiar narrative sources, to question the morality with which the original versions are associated (*Brutopia, Seven*

Lears, (Uncle) Vanya, Minna, Gertrude – The Cry, Knowledge and a Girl); and more "musical" plays in which elements of expressionism and surrealism contribute to linguistically and pictorially poetic forms (*Ten Dilemmas, The Ecstatic Bible, The Swing at Night, Found in the Ground*). In all cases, Barker offers a *speculative* drama, which is distinguished, both theatrically and philosophically, by estrangement and surprising reversal. In the 1992 Barker play *Ego in Arcadia*, Sleen – a recurrent character, and something of a Barkerian 'avatar' who rejoices 'in his status as a manifestation of excess and a refutation of history's urge to order'[4] – presents a song of individuating defiance: 'I exaggerate / It is one of my triumphs / Exaggeration / Nothing in proportion / And illogical in all respects / I don't want truth I only want effects / I exaggerate' (*CP3*, 318). Sleen's artful self-mythologization reflects (somewhat wryly) back on Barker's own search beyond the surfaces of so-called "reality", wanting to discover something more (human?) than the "human", using the imagination to go further than the mirror reflections of a complacently positivist and behaviourist material culture. Indeed, Barker's work seeks to demonstrate how all limitations are originally or fundamentally *imaginative*. This often involves the challenge, or dissolution, of definition as it is constituted from the basis of any single historical (temporal, cultural, legal, moral) point, exposing both the myopia and the restrictive agenda of its ideology.

In Barker's play *The Seduction of Almighty God*, the protagonist muses: 'Possibly our whole existence is contingent on deceit' (*SAG*, 26). The practice of deceit is a major theme of refinement and discovery in Barker's work: it is located in political structures in his drama of the 1960s and 70s, and, since then, more widely, in the depths of interpersonality (the strategic dynamics of which are unsentimentally identified by a character in the 2006 Barker play *A Wounded Knife*: 'this is chess so like chess me versus you'), and even metaphysics. His work explores social, political and personal strategies of mesmerism and seduction; indeed, the work itself achieves its effects through seductive strategies which lay no claim to "the truth", an exclusive verifiability. Barker's work continues to be political and revolutionary, but not in the sense that it has an agenda or politics to prescribe; rather, his work explores vulnerabilities in systems of (self-)fortification, and how forces of control, circumscription and immobilisation will always be (particularly) prey to anarchic upheaval. 'Barker's way as a dramatist is always to expose the public crisis through the personal agony of individuals', as is noted on the back cover of the 2005 Oberon edition of his play *The Fence in its Thousandth Year*. Martin Crimp – a younger British dramatist who is also unusual in his rejection of social realism – demonstrates a point of contact with (and possible inspiration from) Barker, when he similarly asserts the importance of the theatrical objective (both thematic and stylistic) of 'constantly reminding us that human beings are more contradictory and strange than any ideologue could ever imagine'[5].

Barker's theatre, increasingly during this study's period of review, characteristically explores that which we might approach by calling 'the uncanny': that which generates a moment of anxiety about what it might mean to be (in)human in the context of a specific social and moral value system. In the 2007 auto/biographical volume *A Style and its Origins*, Barker/Houth observes 'The uncanny was routine to [Barker] and he included it in his thought as an aspect of things, almost a law' (*SAO*, 82). In the discourse of modern cultural theory, this term is charged by associations with Sigmund Freud's writing on what he terms in German *das Unheimliche*. Michael Mangan, developing from (and finally quoting) Rosemary Jackson, presents a resonant disquisition on this word:

> *Das Heimliche* – the non-negated version of the word – contains a double meaning. It refers on the one hand to that which is familiar ('homely') and comfortable. Its opposite, therefore, *das Unheimliche*, carries a feeling of estrangement, of no longer being 'at home' in the world. However, a second meaning of *das Heimliche* means that which is hidden or concealed. In this sense of the word *das Unheimliche* is that which is revealed. *Das Unheimliche* exposes that which is usually kept out of sight. Thus 'the uncanny combines these two semantic levels. It uncovers what is hidden and, by doing so, effects a disturbing transformation of the familiar into the unfamiliar'.[6]

This last sentence also identifies key effects of Barker's theatre. Mangan notes how Freud's consideration of the uncanny centres on theories of psychoanalysis in which 'something which has been buried or hidden from itself in our conscious mind' is 'triggered by an experience in the present', 'comes to light and reasserts itself'; and how, more widely, 'the uncanny' might be associated with 'boundary work which both marks and problematizes the distinction between the human and non-human'[7] in the prevalent terms.

Barker's plays and characters are compulsively and intensely involved in such 'boundary work'. His characters repudiate received wisdom, seeking out personal knowledge, although this necessitates personal cost which is frequently appalling; but, as Barker observes, 'To deepen one's experience of life is not negativity, not pessimism'[8] (Sleev, the protagonist of *I Saw Myself*, asserts that 'loss is necessary' as part of this searching). Barker situates his eruptions of the uncanny in contexts of historical, political, cultural and moral value; his work is *anti*-domestic, catastrophically disturbing and individuating, opposing any notional consensus that domesticity, sentimentality and populist collectivism may propose as an apologia for dismissal of the *specifics* of humanity. Consider the words of Katrin, survivor of a war atrocity and protagonist of Barker's 1987 play, *The Europeans*, who insists on uncontrollably *reversing* the integratory assumptions of those around her by attacking everything associated with, and rendered tolerable by *das Heimliche*, 'the homely': 'Home is the instrument of reconciliation ... the suffocator of temper, the place where the preposterous becomes the tolerable and hell itself is stacked on shelves, I wish to hold on to my agony, it's all I have' (*OP1*, 99).

I have remarked elsewhere[9] how Barker opposes History – the imposition of ideological and moral narrative form – with Anti-History[10] – the disruptive fragmentation of this form by the testimony and performance of individual pain, articulated by those (like Katrin) marginalized but capable of displacing the labellers and "Historical Authorities" from centre stage. The spatial dynamic of several Barker plays[11] are strong examples of a literally "hell-raising" trope, in that some force conventionally repressed or marginalised as 'abject' (in Julia Kristeva's sense of 'something rejected from which one does not part'[12]) erupts into the centre of the stage action to claim dramatic primacy and wreak havoc. The 'living dead' – in that they are consigned to insignificance within the dominant discourse – turn the tables, and discover language[13].

Barker is very much a modernist, in so far as his work often dramatizes the assertions of the will, strength and dignity of the individual; but the dramatization involves precise and explicit theatrical attention to the falterings of this will, which makes it imaginatively detailed and engagingly poignant. By way of examples, Becker in *Ten Dilemmas* articulates the existential nausea of her indefinite self-in-transition in a searingly and churningly precise couplet: **'I've ceased to be / And yet not yet become'**; and the character Mosca, again in *Ego in Arcadia*, struggles to confront and assimilate the loss of his beloved, and the new, raw senses of time which are exposed, through self-assurances ('I am quite capable of accommodating myself to the new conditions after all what is sophistication but the knowledge of inevitable decay') which are undercut by his exclamations of the name of his beloved, in a confrontation with what he recognizes as the 'rotting of the single passion that gave [him] breath' (*CP3*, 298).

These are notable instances of what the actress Jane Bertish identifies as the way that Barker's characters speak in highly emotionally direct and specific ways: their terms of perspective are surprisingly original, because of their uniquely keen and astutely expressed lack of sentimentality[14]. As Charmian Savill has observed:

> The power of Barker's characters lies in their ability to disrupt the conventional relations of theatre spectating, when the conventional relations are sentimentally romantic associations for the audience with superior power (either of the protagonists or of the notional collective of the audience). Barker's characters are disruptive because they permit the theatre audience neither envy nor power. They deny envy, because audience members may admire their courage, but are unlikely to wish

consistently to be those characters. The characters defy power, because audiences are unlikely to want consistently to despise or pity them, because they are brave, articulate and not self-pitying (except briefly and comically). The characters' resources can bring them to a point of apotheosis or triumph, which is always painful, yet nevertheless represents an individual form of redemption.[15]

Barker's characters resist programmes of ideological coercion in the name of neo-utilitarian "progressive enlightenment", which would ultimately render human nature mechanistic and predictable, because dependent for its definitions on theories of rationality and/or gratification. Barker's Preface to the BBC radio production of his dramatic compendium *The Possibilities*[16] expresses this spirit of wilful resistance: 'We fail to think correctly / There is the hope / We persist in our blindness / There is the hope / We are intransigent at the wrong moment / And capitulate at the wrong moment / There is the sole chance of deliverance'.

Dramatic Context

Thus, Barker, like (and perhaps even more than) Shakespeare, rejoices in contradictions. Oliver Ford Davies notes how Shakespeare's writing is sometimes clear and immediate, sometimes obscure and distant; so are his characters, at different dramatic junctures; and 'how this is one of Shakespeare's great discoveries, that character lies in INCONSISTENCY'; I would add that Barker's work develops this discovery, so that, similarly, the challenge for the actor is 'to learn how to play each moment, each scene, for what it's worth, how to turn on a sixpence and not to try to iron out the inconsistencies' as a television actor may most often be expected to do; rather, as Ford Davies notes, the contradictions 'are there, and we have to play them, because it's in their collision that the central dynamic is to be found'[17], and because the most persistently resonant plays are those with the greatest plurality of meaning.

Barker's characters wrestle with profound experiences which estrange others, their surroundings and even their selves to themselves, and this wrestling is the index of their spiritual and expressive vitality: consider the statement of the emotional revolutions involved for the character Photo, in *The Fence*, when he learns that his lover/"aunt" is in fact his mother, and tells her: 'it is impossible to like you and impossible to dislike you ... I ADORE YOUR TERRIBLE LYING I ADORE YOUR TERRIBLE TRUTH' (*FTY*, 50).

These words require the actor not just to speak but to *embody* the complexity of tensions between opposites: to incarnate, not merely describe. They demand that, in the words of Anne Bogart, the actor create 'a body and a state that would say the particular words of the play and then speak'; so that what is at stake in the words is manifested to the audience by 'the condition of the body', its specific and opposing visceral energies, and (im)balance. Thus, Barker's theatre, addressed with proper seriousness, is not only intensively linguistic but intensively physical: as Jane Bertish observes, the work demands that the actor 'experiences the language bravely'[18]. This demand for an emotional athleticism in the performer which is both vocalized and physicalized is one of several ways in which it is both appropriate and precise to speak of Barker as a 'Shakespearean dramatist'; others include his imaginative instinct to interrogate and subvert the closures of previously existent and familiar dramatic narratives (considered in Chapter Four and subsequently), and his overt theatrical address to *the limits of language* (considered later in this chapter and subsequently). Mangan notes how Shakespeare responded to an anxiety of his quickly-changing social world and time, that there were 'more things than there are words to express them by', with an 'unconventional and resourceful' sense of wordplay which 'had the effect of extending the meaning of individual words, and of generating suggestions which go far beyond the paraphrasable meaning of his sentences'[19]. In Barker's plays, like Shakespeare's, definition, language, and other forms of self-expression, such as art, are an issue and a problem: the characters' experiences of extreme situations bring them into collision with 'the limits of language – the extent to which words are adequate to "capture reality", and the way in which meaning continually slips away'[20]. This collision may be tragic, as when plays dramatize how 'the meanings which men construct for themselves through their words, their language and their social conventions are subject to collapse'; or something even further, when a verbal and visual poetry coalesce into a distinctive theatricality which suggests and demonstrates 'a possible way of coming to terms with, and perhaps overcoming, some of the limits of language'[21]. This brings us to approach the characteristic effect of **ellipsis**: of which, more later.

It is notoriously difficult to identify Barker within the convenient confines of a more recent dramaturgical and theatrical lineage; he has acknowledged his admiration of Charles Wood (b. 1932)[22] and has written formal ripostes to Edward Bond (*No One Was Saved*) and Bertolt Brecht (*The Europeans*, and subversions of the conventions of Brecht's *lehrstücke* in *The Power of the Dog, The Possibilities* and other short play 'compendia', and *The Last Supper*). Like Samuel Beckett, Barker presents remorselessly concentrated verbal and visual stage imagery which depict experientially the consequences of repressed expression, where a bleakly humorous negativity jostles dualistically with a wry honouring of isolation; however, Beckett frequently depicts a consciousness turning inwards and against itself, whereas Barker's characters more usually respond to their pain by coming back on to the offensive against its social causes.

In his explorations of separation, sacrifice and sainthood, Barker has one further important theatrical forerunner (but not conscious influence) in the

English dramatist John Whiting (1917–63), whose plays are increasingly existential in an anti-ideological way, depicting characters who are surprisingly (if tragically) decisive, articulate and deliberately ennoblingly self-destructive. Whiting is, as I have remarked, 'opposed both to Osborne's fundamental if vague empathetic veneration of emotion and to Wesker's faith in social improvement'[23], and his work's increasing deliberate offensiveness to these twin liberal criteria (which might more recently be associated with the drama of David Hare and David Edgar) provoked the baffled fury of liberal critics and of audiences who accepted and internalised their terms of analysis. E. R. Wood observes that even Shaw, an iconoclastic writer 'who had once shocked and annoyed audiences by his paradoxical and somewhat heretical ideas', 'left them in no doubt about which character was the author's spokesman or about what conclusions he was putting before them at the end'; however, Whiting's major plays presage 'a different age, when the very solutions of the old problems seem to have produced new and more difficult ones; when the playwright does not claim to "know the answers"'[24]. Whiting's extrapolation of Shaw's expressionistic tendencies into an unforgiving bleakness is a major riposte to the British theatre traditions, its criteria and objectives of social realist redemptive drama and popular entertainment, which both make (conventional) sense by confirming (predictable) expectations. Instead, Whiting calls controllability and calculation into question, anticipating (and perhaps inspiring) effects in the work of Rudkin, Arden, and Kane[25]; and his major plays – *Saint's Day* (1951) and *The Devils* (1961), as well as the more linear and limited *Marching Song* (1954) – investigate how a character's actions might make and unmake them into something more and less than human, a compulsive theme comparable to Barker's; as are the revelations in his principal dramas of how 'post-war decay and populist masquerade prove unsatisfying and ultimately hostile' towards the protagonist 'who partly reflects its edicts in unacceptable terms'[26], but who demands witness as s/he insists 'upon their own crucifixion, on crosses they have fashioned from the broken jetsam of conventional heroism'[27] and gender identity. I am not proposing an equivalent or derivative relationship between Whiting and Barker; Whiting's major accomplishments are limited to two plays, *Saint's Day* and *The Devils*; but I offer an overlooked analogy between their purposeful formal anti-predictability, their dramatisations of self-consciously perverse saints who defy digestion by and reconciliation with the forces of social unity, and the mutually compounding offensiveness of these objectives for the popular and liberal critical apparatus of their respective ages; a less populous, more fragile, but no less significant seam in English theatre[28].

  More recently, Barker's work is a visible major influence on what Graham Saunders identifies as 'the key distinguishing feature of the dramatic strategy employed in Sarah Kane's work', which is the rejection and subversion of the conventions of realism so that characters 'constantly elude psychological verisimilitude and do not allow us with any certainty to pin down their moral standpoint'[29], so as to remove 'the psychological signposts and social geography'[30] associated with traditional British instructive social realist drama. However, the classicism of Barker's work means that it cannot be claimed as a convincing forerunner or adjunct to the so-called "In-Yer-Face" wave of 1990s British theatre, of which Kane was the vanguard writer, except perhaps in Barker's emphasis on the experiential aspect of theatre: a furrow which this wave ploughed in more sensationalistic (and less speculative) brutal, contemporary and materialist terms[31].

In the field of visual art, the major twentieth-century existential artist Francis Bacon demonstrates an essentially dramatic imagination (like Rembrandt and Goya). Indeed, Bacon's paintings depict images of anxiety, human distortion and breakdown, straining against frames of containment, with what Grey Gowrie calls an 'implied theatricality' suggesting 'humankind trapped by history and its own sensibility'[32], and 'the world of Godot or King Lear'[33]. However, Louis Le Brocquy recalls how Bacon 'constantly regretted the absence of a dramatic mythology in our time', finding the inertia in Beckett's writing insufficient to his own 'exhilerated despair'[34]. Bacon might have found this 'dramatic mythology' of nervous stress in Barker's drama. Barker is himself an acclaimed and exhibited painter, proclaiming himself 'inevitably drawn towards the body and its gestures and the vocabulary of the gesture, of how the body "speaks" its pain, even when, as in the picture plane, words are absent'[35] (as in the wordless, or barely worded, plays which comprise his dramatic compendium *The Forty*); 'Nothing in my plays or my pictures ever pleads for recognition, they exist in that territory of tragedy which creates its own emotional and physical *milieux*, where the cry or the howl comes to the relief of a language silenced by pain...'[36].

Barker describes himself as influenced by very few dramatists, but rather by some poets (Apollinaire, Paul Célan, Rilke, Attila Jozsef, François Villon, George Oppen) and some musical composers (including Bartók, Ligeti and Stockhausen), and observes: 'Bartók's way with a string quartet is not unlike the way I approach a dramatic text: that continual variation on a theme almost to the point of a neurotic re-examination of it, an explosion of it, a re-assembling of it'[37]. Indeed, Barker as a director uses musical "samples" from these composers as what he calls theatrical 'punctuation', to support not only dramatic transitions but also images of heightened states or extreme actions.

Language and Limits

> Kiss me but in such a way I
>
> *– A House of Correction (CP4, 379)*

Barker imbues his characters with an articulacy and a fluency which is fully and poetically expressive, rather than concerned with the reproduction of contemporary everyday speech. They are characterisingly explicit in their distinctive anxieties, as when one character in *A Wounded Knife* (literally) finds himself speaking the words: 'I am saying everything that comes into my head polite vulgar sense stupidity'; elsewhere in the same play, another admits: 'I talk when I am anxious … happiness lives in the eyes not in the mouth when you are happy how few words you need possibly none'[38].

The actor Juliet Stevenson proclaims

> the unrivalled beauty of speaking words Barker has written: the rhythm of his writing is exquisite, he has a poet's and a musician's understanding of the rhythms of language, and how these are linked to our internal lives: the ways that language flows, or is chipped and broken, staccato or fluent, [reflect] how the rhythms of our speech are linked directly to how we think, feel and breathe.[39]

Yet, having identified the modernist aspects of Barker as an artist, and the expressiveness of his language, one should equally note how his work rises to the challenge of (at least one version of a) postmodern experimentation, which Jean-François Lyotard suggests should be characterized by indeterminacy, 'exploring things unsayable and things invisible' (a form of 'experimentation' which is therefore 'poles apart' from literal 'experience' – but which instead offers a poetic and speculative experience)[40]. Rather than familiarity or immediate recognition, Barker's work offers the contradictory intensifications of surprise (pleasure, terror) in admitting (welcoming, contemplating) the unknown (and even the terror of the failure of language). In his writing, this often takes the form of ellipsis: the deliberate omission of words from speech, which invites the audience to attempt imaginative completion (*The Last Supper* yields a splendid example: 'Pity the man who has no wall. Pity his limitless choice of.'); and the set of dots which extend the resonances and ramifications of a phrase beyond formal conclusion, inviting the audience to pursue and entertain further possibilities. This offers a syntactic means of presenting the unpresentable rather than imitating nature, and may disrupt the usual way we experience time: the philosopher Alphonso Lingis notes how 'At the moment when none of the anticipated words are there, we are held in the present'[41]. Lyotard notes how, by hinting at 'the actualization of a figure potentially there in language', surrealist painting 'tries to get around this inadequacy by including the infinite in its compositions'[42]. There are indeed surreal elements in Barker's drama: not just in its delimiting effects of poetic syntax, but in the surprising formal and pictorial juxtapositions which it arranges; these disrupt and question habitual notions of order in space and time, presenting elements which 'are at least defined if not always recognizable', but arranging them together 'in paradoxical fashion', as in a dream or nightmare[43]. This develops imaginatively Theodor Adorno's proposition that, in the wake of Auschwitz, it is inappropriate to pretend that 'all that is real is rational, and all that is rational is real'[44]. The actor and lecturer Susan Russell reflects analytically from her experience of performance: 'Through text and setting, Barker insists that the actor exist within discontinuities and disruptions of time and place; therefore, a Barker actor must constantly define and re-define the text and the performing self through fearless engagements with displacement'[45].

Barker's poem 'Infinite Resentment' (in his 1996 collection *The Tortmann Diaries*) imagines the dead mutinying against God and the injustice of mortality: a cosmic repudiation of what might be identified as the ultimate divine "confidence trick", the poverty of the world and the restrictions of decay, both of which serve crucially to paralyse the imagination and will. The result is that it is often only belatedly ('with a thickening senility') that one begins to glimpse 'Alternatives', such as 'The infinite reversibility of things / The seduction of opposites / And how to part the colours of the night': alternatives which permit the transcendence of conventional human limitations, perceptions and choices[46].

Barker's drama and *mise-en-scène*, as well as his poetry, aim to demonstrate these alternatives, these possibilities, in action. Yet these glimpses of alternatives occur – significantly for a writer who is primarily (though not exclusively) a *dramatist* – through experiences of, or impulses to, passionate contact (where passion is the pain of wanting), and/or joint or mutual action (based on emotions which circumvent conventional appeals to social obedience based on fear and assurances of safety and privacy: emotions such as eroticism, pity, anger, courage, grief and awe). Barker is a writer-director who uses an intense verbal text to approach images and tableaux which challenge, subvert or surpass conventional terms of definition, and precision of verbal expression remains one of the prime modes by which his characters seek to elaborate and refine (the morality of) their selves in action ('if one cannot distinguish between words how might one distinguish between values' – *Lot and his God*). However, action may take the self into moments which move beyond the immediate encapsulations of language – notwithstanding the compelling challenges which they provide for language (and other capacities, including the impulses to political organization) to encompass them. Barker's development as a director and scenographer may be construed as a bid to privilege and orchestrate (in time and space) a 'silent component', the apparently indefinable moment which words permit one to approach, as scaffolding permits the construction of an edifice or a plunge into a void: the experience of what Lingis terms Catastrophic Time, discussed later in this chapter; the experience of indeterminacy, as described from the actor's perspective by Gerrard McArthur, in the Appendix.

In these respects, Barker is addressing what Lyotard identifies as 'The avant-gardist task ... of undoing the presumption of the mind with regard to time'[47]. The surprising challenge to definitions (sociopolitical, individual, perceptual, imaginative) can turn out to be strangely expansive, informative, haunting, even pleasurable...

Ecstasy

> She looks for pleasure in the strangest places ... where others might draw back from shame she ... discovers inspiration ... happiness and pleasure seem to me far from synonymous one might know ecstasy and still
>
> *– I Saw Myself*

Barker's vision of sexuality is best approached via Karoline Gritzner's observations:

> In Howard Barker's plays, sexual desire necessarily *complicates* life; it signifies a tragic encounter with the Other and catapults individuals into an awareness of their own limitations and possibilities ... In Barker's 'promiscuous theatre', erotic sexuality, pain and death are construed as exclusively personal, solitary experiences, which affirm 'the individual's right to chaos, extremity and self-description' (Barker) ... The notion of ecstasy is of crucial significance to the ways sexuality is theatricalised in Barker's work. In addition to the Oxford English Dictionary's definition of ecstasy as 'being beside oneself; being thrown into a frenzy or stupor, with anxiety, astonishment, fear or passion', ecstasy also implies the idea of losing one's centre. In performance of Barker's work this is physicalised in the characters' unbalanced postures and uneven walks...[48]

This links to Susan Russell's observation of how the Barker actor is required to 'exist within discontinuities and disruptions of time and place' and constantly re-define both 'the text and the performing self' through engagements with displacement. Juliet Stevenson has observed from a performer's perspective:

> [Barker's] plays are increasingly about sexual energy between men and women in quite extreme and socially unacceptable forms ... where men and women are engaging with each other at their deepest and most inexpressible, but also most true and complicated, levels ... [including situations] where women are allowing themselves to be quite overtaken by their sexual needs, to be overwhelmed, or to allow themselves to be overwhelmed, by the sexual needs of the man they are with... Even if difficult, this is relevant to explore.[49]

I agree, but would add that Barker's works also frequently depict male characters seduced to the point where the self-defined limitations of their selves are violated and exploded (partly in search of what might particularly appropriately be associated with '*le petit mort*'), in a pursuit of risk which proves unstoppable and uncontrollable (The Priest in *The Ecstatic Bible*, Holofernes in *Judith*, Eff in *Dead Hands*, Doja in *He Stumbled*, amongst others). In Barker's plays, characters of varying genders and sexualities seek or achieve the state of ecstasy. Ecstasy should be distinguished here from simpler notions of 'pleasure' or 'gratification'; rather it refers to the most intensely convulsive drama of the body and the self experienced between life and death, where rapture mingles with ordeal, involving the possible sense of disintegration from a former bodily centre and sensory relocation in other constituent parts, perhaps involving the sense of being outside oneself, looking in; where (as Gritzner says) the defining limitations of the participants, and their world, are startlingly re-visioned (Bernini's sculpture, *The Ecstasy of Saint Teresa*, for example, depicts this experience, within a frame of religious ascription). Barker identifies the compulsive re-presentations of the uncontainable force of sexuality, and its cresting and falling tidal rhythms, in his work:

> Sexuality is the only thing that draws us completely out of ourselves and exposes the extreme, the will to conquer and submit on both sides of the partnership, the desire to conquer, own, possess, and simultaneously to surrender and be possessed – such a delicate oscillation and quite the most extraordinary facet of existence as far as I'm concerned; and that is why so many of my characters eventually, not "take refuge in it", but find it the only place where they can be themselves ... in the passionate embrace with another if she, he, permits it...[50]

Death and Catastrophic Time

Lingis argues that human perception is intrinsically based on time: that society makes time intelligible through work, and an envisioned future of possibilities. However, when death strikes suddenly, it 'immediately drives its shock wave into the future', 'rendering not only inoperative but senseless, the order and system that had come to pass, whose endurance and momentum was shaping the emergence of goals and results'[51]. Death strikes, annihilates the future of possibilities that was being extended from the present, and 'makes the time in which it occurs appear as the empty endurance of the void'[52]: this is the devastation of catastrophe.

However, Heidegger suggests that the heightened sense of imminent and inevitable catastrophe – the once-in-a-lifetime situation which demands all our passions and forces (which may prove wanting) – may create a sense of what he calls 'authentic (authentifying) time', a life that lives and works in, despite, and because of, the imminence of catastrophe[53], a veritable state of emergency which drives people to marshal all their resources and forces for release in action. 'We live in crushed time' says Sleev, the protagonist of Barker's *I Saw Myself*, rallying her seamstresses to complete a tapestry even as enemy armies approach, unavoidable harbingers of irresistible destruction. Yet this is a form and experience of time that is not only devastating but possibly galvanising: 'it can also hold us, and even draw us into it'[54], just as Sleev charismatically succeeds in commanding an irrational loyalty in her women, to abandon their separateness in hopes and values, and nevertheless complete the tapestry. Lingis delineates the force of that which he prefers to designate as Catastrophic Time:

> When the dreaded comes to pass, our futile fear, anguish, and panic often give place to something else. The striving and anticipation of our reason give place to observing the advance of the catastrophe with attention and something like fascination ... producing not panic or desperation but rather a kind of intellectual lucidity, almost a curiosity ... moments when we abandon the serene pleasure of accomplishment in the world of work and reason ... [to] sink into an anxiety that unfolds as an inhuman exhilaration.[55]

I propose that this reflects the uncanny (*unheimlich*) appeal of Barker's Theatre of Catastrophe: its dramatic situations, its theatrical effects and its compulsive speculation into the excessive, which challenges prevailing habits and rules of perception: seeing, thinking, imagining what exceeds the usual tolerable possibilities (for example, in *Lot and his God*, when Lot's wife urges transgression on an angel, to become her lover: 'in my own home let us violate all those things conventionally described as homely'). Lingis again:

> To be sure it is strange to speak of valuing catastrophe. But something in us impels us to look beyond the present, future, and past deployed in our work, to let our eyes drift into the abyss beyond. Something in us convinces us that to do so is of value, even though there can be no question of anything being maintained, secured or acquired.[56]

And in an age which privileges maintenance, security and acquisition, Barker's Theatre of Catastrophe offers the strange gift of *loss*: the loss of every consoling 'net / of / habit / fiction / or / relief' (to quote Barker's poem *Gary the Thief*, GTGU 13); an experience which breaks our commitment to the value of familiarity in forms, and unusually (or even uniquely) resists seizure, measurement, categorization and appropriation. This is not only or merely a devastation; the breakup of equilibrium offers release. Compare Derrida's sense of the gift, which is a gift to the extent that it '*gives time*'; it does not give itself as an object, 'to be possessed, or consumed', rather the gift gives 'a possible future'[57], a restructured and 'delimited time'[58]. Barker has observed how, in his theatre:

> something is always *lost*. In a world of relentless aggregation, to lose might be a lightness, an intoxication. What is it that is lost? The burden of your moral convictions. This is a loss that might be experienced as a *privilege*... (DTOAT, 41)

Similarly, the tragic character experiences a *'taking-away'* of common consciousness, values and hope that parallels 'the *moral condition of dying'* (DTOAT, 99).

Lingis also notes how sexual excitement 'surges in the shattering of built-up structures', inviting what Marguerite Duras terms the 'unleashing of whole, deadly passions'[59]. If one follows Lingis's argument, the invitation and promise of catastrophe might involve and find form in eroticism: 'the release of excess forces, craving extreme experiences in extreme torments and extreme pleasures'[60] (and I pursue connections between concepts of Catastrophic Time, eroticism and nakedness in Chapter Seven). Barker's theatre offers a domain of dreamlike rapture and nightmarish anxiety where the repugnant and exultant meet: a disquietingly almost-familiar space of created and creative disruption, where, in 'the decomposition of the world of work and reason, transgressive and ruinous passions catch sight of the sacred'[61].

2

'The ecstasy of vanishing meaning'

Arguments for a Theatre; Death, The One and the Art of Theatre

> How should we enter death? Is this not the subject of all philo-
> sophy and all theatre...?
>
> *– Death, The One and the Art of Theatre, 28*

Barker's instincts for the discovery and interrogation of the fundamental social and moral terms of existence lead him into theories and speculations on the role of aesthetics – specifically, the aesthetics of a practically and morally experimental theatre – in the challenging of their *received* (and often implicit) terms. The first 91-page Calder edition of *Arguments for a Theatre* appeared in 1989, collecting Barker's articles, conference papers, lectures and reflections on practice, developing the articulation of his belief that 'in a society of increasingly restricted options ... a creative mind owes it to his fellow human beings to stretch himself and them, to give others the right to be amazed, the right even to be taken to the limits of tolerance and to strain and test morality at source' (*AT*, 37).

As Lyotard has observed, 'the rule of the philosopher's discourse has always been to find the rule of his/her own discourse'; and from this basis the philosopher is thus 'someone who speaks in order to find the rule of what s/he wishes to say'[1]. Similarly, Barker is striving in his theoretical writing to discover his own critical discourse *emerging* from his drama and theatre, impatient with journalistic misperceptions and dismissals which would evaluate his work in terms of inappropriately conventional criteria and objectives, and miss the point(s). Rather, Barker's theoretical essays provide a space and opportunity for him to articulate *his own terms*: the crucially different aesthetics and challenging propositions of his unusually ambitious endeavours.

Arguments for a Theatre might initially be identified also as arguments *against* a theatre: a theatre which presents the audience (only) with what theatre managements, theatre journalists and government funding agencies want, and believe their audiences require or deserve. This imaginative deadlock (and potential self-fulfilling prophecy) usually appeals to the notion that the audience is itself a static, monolithic thing, requiring appeasement, social improvement and "good value". This is founded on the myth of *access(ibility)*: the pre-eminent importance of social inclusion, emotional unison, familiarity and predictability; and is based on the moral certainty (often expressed as ironic comedy) which characterizes most conventional offerings in film and television.

Barker's theories suggest that the theatre should leave these expectations to those media, rather than ape, or compete with, them:

> The notion of the stable audience is a reactionary one, a blunt weapon used against the revolutionary text. It is a notion of the people used against the people, and the sing-song of the populist state. The cultural managers will demonstrate the frivolity, the absence of concentration, the impatience, the dictatorship of television habits over the minds of the audience, but never the appetite for challenge, truth or discrimination. The Public, as an invention, becomes the enemy of the artist, a solid block of immovable entertainment-seekers whose numbers and subsequent economic power forbid intelligence. I refused this as a description of the audience and [found] the audience for the work, sometimes half-hostile, but wanting. The tension between the audience and the play became for me an aesthetic, the nature of experience. This involved challenges to common morality, common socialism, even what passes for common humanity. (*AT*, 22)

Indeed – despite their occasional fleeting attractions – familiarity, predictability and irony become, if dominant, the cornerstones of a personal (and cultural) depression: they take no account of the human curiosities about, and appetites for, surprise, strangeness, passion, risk, mystery, excess and transgression. A diet of constant familiarity, predictability and irony restricts theatre (and any medium) and infantilises its audience (they may indeed consequently develop a fearful distrust of the unfamiliar and complex). Barker proposes that such staple ingredients, expectations and criteria be relinquished, to film, television and the conventionally anodyne 'theatre', by an art form he practises and identifies successively as Tragedy, Theatre of Catastrophe and The Art of Theatre (this is one manifestation of how *Arguments* progresses through a series of interrogations, variously developed identifications of conflicts, rather than through a single consistently developed argument in search of a single empiricist "sense" or "truth", which is after all what Barker is opposing).

Barker associates his 1983 play *Victory* with his breakthrough into the form of anti-reconciliatory tragedy he calls Theatre of Catastrophe:

> At the end there is no restatement of collective principles such as justify Greek and Shakespearean work; the individuals who break the social

code, in the essential moment of tragic action, do not apologise, are not brought back into the polis, nor do they necessarily die; they possibly wander off into solitude, into a very empty landscape, and disappear into that...[2]

The audience participate in the struggle to make sense of the journey, which becomes their journey also. Consequently, what is achieved by them is achieved individually and not collectively. There is no official interpretation. (*AT*, 46)

This brings the Theatre of Catastrophe into 'a process of illegality' which 'exists not on the surface, but in the interstices of culture': 'whilst it does not humiliate its audience or offend it by intention ... it overwhelms the audience by the *plethora* of what it experiences', 'a vastness and disjointed character which defies all clarity':

The sense of witnessing *too much*, of being *out of control* is essential to the Catastrophic play, which as in all other respects, rejects the socially-acceptable subject matter and the socially-acceptable duration of experience ... This quality of plethora invites new ways of seeing and witnessing since lineal narrative is regarded as reactionary and the experience is one of constant digression and a multitude of spectacles which are related not by theme but by tension, psychology or atmosphere. (*AT*, 147)

Theatrical Context

On its first appearance, *Arguments* was the first book of British theatre studies to identify and attempt to move beyond the clichés of contemporary theatre since John McGrath's 1981 polemical work, *A Good Night Out*. McGrath's book, at its best, was critical of the self-satisfied circularity of 1970s British theatrical strucures, and briefly exhorted the valuable danger of the unpredictable self-inventing performer. However, McGrath sited the potential power of that performer within a self-consciously and determinedly *popular* theatre, wherein 'popular means: intelligible to the broad masses, adopting and enriching their forms of expression, assuming their standpoint, confirming and correcting it'[3]: a significantly elastic variety of manoeuvres to be covered by a single word. It also contained a neo-religious, potentially authoritarian awe before 'the people', who remained conveniently undefined except through the mediations of McGrath's own instincts and rhetoric. His call for a theatre which celebrates 'the people' is challenged by Barker's observation: 'The left's insistent cry for celebration and optimism in art – sinister in its populist echo of the right – implies fixed continuity in the public, whereas morality needs to be tested and re-invented by successive generations' (*AT*, 49). Celebratory art rejoices in what is *already known*: the collective reiteration of shared meanings, which is essentially conservative. Barker's theatre creates a crucial disjunction from the self-reassurance of cultural solidarity advocated by Brecht, McGrath or Fo and Rame. The strategic rhetorical advantage of 1980s Thatcherism was its acknowledging and embracing of the breakdown in consensus, promising self-determination whilst in fact instituting a fundamentally populist narrowing of social meanings, interests and choices. This was the cultural climate within which Barker opposed 'popular theatre' with his consciously provocative objective of a radically elitist theatre; consciously provocative because Barker located his elitism, not in economic opportunity, but in imagination, which is available to all (classes), but defies generalization. Barker shares one very basic starting point with McGrath, and Brecht before him, and Boal subsequently: these writers perceive a social alienation which the dominant forms of theatre are failing to address or transmute into an interrogatory energy; and propose (in their diametrically opposing ways) how the theatre might be reinvented to do so. Barker, however, does not try to second-guess the audience's life experiences and address them in Marxist terms of class politics: rather he proposes that the audience attend his theatre on *his imaginative terms* (or, if they cannot, or do not wish to, that they stay away), in full knowledge and admission of the fact that this is currently a doorway only some will want to enter (because of the ways in which society is currently organized, not least in terms of education in art and culture, taste and expectations). Barker does not propose an artistic experience which prescribes, or aims to issue in, an identifiable social action or outcome, like these Marxist-revolutionary writers. However there is a small element of common ground in terms of objectives, between Barker's depictions of liberation in primarily individual terms, and Boal's more recent bids to analyse and dismantle the internalization of repression (in Boal's Rainbow of Desire techniques, which seek to confront and overpower pre-emptive reflexes of shame and determinism which demand the individual's forfeit of power and rights).

British theatre from the 1980s onwards venerated and reiterated a particularly deterministic reading of Chekhov's drama[4]: this British neo-Chekhovian theatre sought reassurance through atonement, located maturity in self-restriction and was effectively complicit in reduction to order. Barker's *(Uncle) Vanya* (written 1992) is an appropriately, thoroughly theatrical cultural and emotional riposte, which, like Barker's other work, offers an expansion of the vocabulary, both theatrical and existential: an expansion of terms of language, experience and being, in defiance of the prevalent restrictions and diminutions of options presented. The structures and funding of British theatre, increasingly dominated by the figure of the director, also increasingly resisted (and disparaged) the prospect of theatre artists speaking, writing, and acting for themselves. In most cases, the mediation of the dramatist by the director and a major subsidised company (and the dramatist's acceptability to the long-term economic prospects of these) was enshrined as crucial and necessary; hence the offensiveness of a self-determining artist who proclaimed the artistic, political and ethical redundancy of most theatrical practice and proposed a new aesthetics. Moreover, Barker's increasingly manifest ability to direct his own writing better than any other professional director challenged theatrical structures for the creation of reputations, their traditional hierarchies (and vested interests) of authority and economic dependency alike.

Barker's theatrical impulses were (and are), here as subsequently, more identifiably European than British. Artaud had similarly suggested that the theatre is a place to 'discover not so much what is best, but what is most exceptional and *vulnerable* in you'[5]; he compared theatre and plague as harbingers of catastrophic revelations and extreme gestures, which disrupted received notions of "the social" and "the natural". Barker's essays, like Artaud's, offer provocations rather than contexts, amplifications of antagonism which propose theatre as something both sacred and dissonant, revelatory and irrational, classical and visionary.

Eugenio Barba is an exploratory director and theorist who shares Barker's objective in creating theatre productions which the audience member does not "consume" (or even understand or know how to evaluate), but with which they continue to have a burgeoning imaginative dialogue. Barba's distinctions, in a programme essay for his Odin Teatret company's 1990 production, *Talabot*, demonstrate a similar concern to unlock the strangulatory myth of the homogenized audience as sacrificial totem:

What does it mean to work keeping the spectators in mind but not the Public? The public ordains success or failure, that is, something which has to do with breadth. The spectators, in their uniqueness, determine that which has to do with depth: to what extent the performance has taken root in certain individual memories.

... The necessity of distinguishing between public and spectators derives from the will to consciously exploit an inevitable condition: even though some or many reactions can be unanimous and common (these are "the public's" reactions), communion is impossible. Intense relationships can be established, but based on reciprocal estrangement.

This estrangement is not only a source of difficulties but can be exploited as a precious source of theatrical energy. Instead of trying to construct an organism which speaks to all spectators with the same voice, one can think of it as being composed of many voices which speak together without each voice necessarily speaking to all spectators.

One might compare the idea and experience of a surprising and estranging but compelling *dis-integration* – sought and sensed both publicly and individually in Barker and Barba's theatres – with Lingis's sense of Desire, as 'a desire for the absent, for infinity ... [which] may function to maintain a nonconfrontational coexistence of different sectors of oneself' because 'One may want the enigmas and want the discomfiture within oneself[6].

Thus Barba, like Barker, envisions theatre as an 'interweaving of one solitude with another', exploring estrangements of physicality and power; and for specific spectators, as for specific performers, this theatre can become a necessity, 'a relationship which neither establishes a union nor creates a communion, but ritualizes the reciprocal strangeness and the laceration of the social body hidden beneath the uniform skin of dead myths and values'[7]. Barba's direction of his own theatre company aspires to what he terms *asociality*[8]: an expressive relationship which does not renounce individual differences of experience, perception, knowledge and needs; not an escape from society but a divergence from its norms; a conscious refinement through which to test an aspirational life.

Barker's theatre also has some contact points with the visceral poetics and fractured re-visionings of earlier narratives which we find in the work of German dramatist Heiner Müller; and I will later argue that Barker's unique orchestration of all facets of *mise-en-scène*, as well as his willingness to imagine large-scale works (such as *The Bite of the Night*, *Rome* and *The Ecstatic Bible*), provide some kinship, in terms of imagination and ambition, with the (otherwise differently individual) director-dramatists Ariane Mnouchkine and Robert Wilson. I would even venture to suggest that Barker's work fulfils, from a European perspective, some of the initiatives demanded by the Indian writer and theorist Rustom Bharucha, who is critical of 'the essentially decorative use of "tradition"' in the creation of so-called 'political spectacles'[9]. Bharucha rather proposes theatrical confrontations with, and sharing of, 'the dark corners of our consciousness'[10]: which aptly describes the aims of The Wrestling School. Bharucha argues that theatre is 'neither a text nor a commodity' but an activity which needs to be in ceaseless contact with both the pragmatics of possibility and choice and 'the inner necessities of our lives': he suggests 'If theatre changes the world, nothing could be better, but let us also admit that this has not happened so far. It would be wiser (and less euphoric) if we accepted that it is possible to change our own lives through theatre'[11]. It would also be less ideologically prescriptive, and more existentially anarchic.

Secrets of Death

Barker produced an expanded second edition of *Arguments* in 1993, *en route* to the fully developed and completed third edition, published by Manchester University Press in 1997. His observations do not constitute, nor should be mistaken for, an instruction manual on directing or theatre-making; an observation which Mangan makes of Edward Bond is even more pertinent to Barker, that his theoretical writings 'should be considered – the same way as the directing of his own plays should be considered – not as attempts to tie down the meanings of his plays, but as further explorations in themselves of the questions and issues which the plays raise'[12]. Rather, Barker's essays constitute a speculative and increasingly philosophical (yet always theatrical) delineation of imaginative objectives.

The third edition of *Arguments* introduces Barker's use of *the secret* as a key concept and quality: in defiance of the commercial principles of maximum audience figures (such as might be identified, invoked and sought through the commercial equations of "market research") and 'Absolute Access', and the voguish fashion and underdeveloped novelty which prompts journalists and managements to issue brief hyperbolic claims of "relevance" and "immediacy" ('What's New...?'), Barker asserts a 'permanently fugitive' art (*AT*, 158) which is elusive rather than directly or reductively communicative: obsessive, ecstatic, enraging; illogical, anti-utilitarian and self-inventive; a theatre for an audience who knowingly and willingly expose themselves to an estrangement and destabilization, even an anxiety. The 1996 lecture 'Love in the Museum' posits a theatre which offers an addictive encounter that is as secretive, surprising, deliberate, self-threatening and potentially destructive as the encounter with one's next lover. There is a contact point between Barker's vision of the possibilities of the secret and that posited by the French philosopher Jean Baudrillard, who proposes the secret as a supremely artificial, seductive, initiatory force, 'never an economy of sex or speech, but an escalation of violence and grace'[13].

Developing Artaud's plague simile, Barker envisions theatre as a potential 'House of Infection': defying a culture of social hygiene and mutual surveillance, which would make each woman her own panopticon, each man vindicated by policing self and others to a point where anaesthetization and immunity from danger are the ultimate social promises (rather than any disruptive, because differently meaningful, contact with others). Barker's observations, in 'Love in the Museum', have proved particularly prescient: '"Accountability", "access", "the right to know", "information technology", "the public domain", "scrutiny" ... are apparently unobjectionable elements in the construction of an open society'; yet repeated invocations of these constructs masks 'their tendency to become the instruments of an inquisition in a society that is not merely open but transparent', a totalitarian obsession 'about the need to eliminate all forms of secrecy' and a moral authoritarianism 'assuming the mantle of the popular will' (*AT*, 171). In refutation of the edict that 'all must be known, all must be accessible, unambiguous, essentially functional', Barker intimates: 'The spectre of tragedy – its inviolable secret – its terrible power of dislocation lies in the forbidden knowledge that these citizens have a fatal susceptibility to instincts which are perfectly incompatible with collective discipline'. This marks a dissociation from classical models of tragedy, and its purgative and harmonizing functions: 'For all that Aristotle attempted to create an aesthetic which placed tragedy safely within the bounds of social regulation, its seductive authority reaches always to the self-determining, the self-describing and the erotic' (*AT*, 173). Moreover, Barker claims that even those contemporary theatre artists 'blithely posturing in the garments of dissidence in the contexts of the national theatre system' are permitted to do so, because 'incorporated into the culture of transparency, propped into position by aesthetic prejudices like clarity and lucidity'; contrastingly, he proposes a truly counter-cultural, anti-utilitarian dramatic and theatrical practice which marshals '*knowledge* against information, *speculation* against journalism, *imagination* against evidence': a theatre which identifies with 'the darkness from which it emerged', an 'absence of illumination' (*AT*, 185; compare Claudius's ecstatic proclamation in *Gertrude – The Cry*: 'FAITH COMES FROM SECRET PLACES AND IN THE DAZZLING CATHEDRALS OF LIGHT IT DIES', *GTC* 66).

What, then, might be the audience for such a theatre? In another context of writing (on transvestite theatre), Lingis senses the possibility of a congregation which is not absorbed into but 'troubles the rational community, as its double or shadow'[14]; and his observations find resonances in Barker's objectives:

> Theatre, which represents transactions with representatives of self, represents society to itself, but also opens up a space of its own outside society. In its absolute form ... does it travesty, parody, undermine, consume all our representatives of self, in an implosion of all our simulacra, leaving, as Baudrillard says, the absolute of death alone on the stage? Or darkening a space for the naked surgings and relapses of lust? ... Is it transposing or releasing, subverting or trumpeting lust? That is its secret. The power to keep its secrets is the secret of its power.[15]

Derrida also shares this sense of the secret as constituting a problem, not only for philosophy, science and technology, but also religion, morality, politics and the law, suggesting something which remains out of reach and control; he proposes different concepts of time: lived time and secret time, and the anti-totalitarian 'temptation' that "secret time" will determine what is to be remembered[16].

In the deceptively slight volume, *Death, The One and the Art of Theatre* (2005), Barker writes with an unprecedented emotional immediacy and arc of conceptual ambition. Its starting point might be the apparently fatalistic litany which occured in Barker's *(Uncle) Vanya*: 'The theatre is a contract / Between the living and the dead / The dead inform the living of their fate / A requirement / A necessity' (*CP2*, 329). As part of his characteristic explorations of reversibility, Barker now reads out of this intimation, not an oppressive determinism, but a strange (*unheimlich*) liberation from the prioritization of values as elaborated by the *living*.

Firstly, Barker looks unflinchingly at the mixture of anxiety and curiosity that the prospect and nature of Death necessarily strikes into the human consciousness. Unlike other animals, humans glimpse the inevitability of their own death, and, through imagination, are able to picture a world in which they are not present, seeking personal meaning in the face of the incomprehensible.

> Nothing *said* about death by the living can possibly relate to death as it will be experienced by the dying. Nothing *known* about death by the dead can be communicated to the living. Over this appalling chasm tragedy throws a frail bridge of imagination. (*DTOAT*, 1)

Barker then traces how this unavoidable sense of human *specificity* is reflected, and in many ways elaborated, in the emotional arena of sexual longing, adoration and seduction; and he identifies The Art of Theatre as the appropriate form for the delineating and incarnating of Death and Eroticism; and indeed how imaginative confrontations with the demands of Death and Love are, themselves, intrinsically *theatrical* (as expressed by the poetic refrain of the book, 'All I describe is theatre even where theatre is not the subject'). Arranged as a collection of speculations, deductions, prose poems and poetic *aperçus*, *Death, The One...* perceives human vitality in the instability which is simultaneously its frailty and its power, linking exclusivity with allure: as in the elementally forceful figure of 'The One', who is beloved in and for her irreducible *uniqueness* (however infuriating some of its forms). *Death, The One...* proposes a vision of Tragedy, and its central compulsion to make dignities of one's obsessions, which takes its cue from Nietzsche and Baudrillard, but projects their explorations into more tightly focused, profoundly individual and unflinching admissions of encounters (both mortal and erotic) which await all, yet are ultimately particular to each sensibility:

> All cultures are enslaved by idealism – they are defined by their servitude to the ideal. Only tragedy locates the ideal in death, but because death is the first enemy of political systems, tragedy is caricatured as *negativity*. The bravery of tragedy – where not even sexual love is sufficient to abolish the fascination of death – lies in its refutation of pleasure as an organizing principle of existence. Who would deny that this contempt for pleasure is also an ecstasy? (*DTOAT*, 3)

The Art of Theatre is characterized as the compulsive negotiation of the Imagined with the Dead (imagined as clustering on the opposite side of the Styx, compulsively re-presenting something inexpressible): a hymn to the fascination of the boundary, and an insistent personification of limit-lessness, which alters conventionally sensed possibilities:

> The play of *the theatre* asks *how shall we live?* The tragedy asks *how should we die?* But where is the antithesis, for the tragedy answers the question *how shall we live* in the very act of exposing the way into death. It draws death back into life, and consequently *alters life...* (DTOAT, 94)

Thus *the art of theatre* is both inherently tragic (a term Barker prefers in this volume to the 'catastrophic') and distinguishes itself from the conventional 'theatre' ('*The theatre* purports to give pleasure to the many. *The art of theatre* lends anxiety to the few. Which is the greater gift?', DTOAT, 1). This brings Barker to the question: 'To *admit death* ... Is this political?' (*DTOAT*, 8) – as when The Art of Theatre confronts and rehearses death, in 'a confession of ignorance, of the limits of knowledge' (*DTOAT*, 31):

> Profound emotional experiences even where they fill the audience with despair (perceived 'negativity') serve to increase resistance to social coercion (the more so since all social coercion justifies itself on the pleasure principle). The pessimistic work of art – who dares talk of death as pessimism? – strengthens the observer by obliging him to include death in his categories of thought. (*DTOAT*, 13–14)

To make a cross-reference to the world of visual art, this 'inclusion of death' is artfully achieved in the 1533 painting by Hans Holbein the Younger, *The Ambassadors*: Holbein's incorporation of an anamorphic skull directs the viewer to re-view the image from an unconventional point of view, which distorts the centralized claims to authority of two fat diplomats and their resplendent trophies, and demonstrates, through an almost surreal effect, how the legislative confidence of human consciousness depends on the forgetting of its own contextual (temporal and social) limitations: in the painting, 'the apparently wrong perspective' allows the spectator to 'understand death as the truth behind all the triumphant equipment displayed in it'[17].

Here, as Stephen Greenblatt observes: 'The skull expresses the death that the viewer has, in effect, himself brought about by changing his perspective, by withdrawing his gaze from the figures of the painting', when 'that gaze is, the skull implies, reality-conferring': without it, the surrounding objects carefully represented in their apparent substantiality become hallucinatory, or even vanish like an illusionist's trick; thus, a fractional change of perspective brings 'death into the world'[18]. Greenblatt deduces:

> The effect of these paradoxes is to resist any clear location of reality in the painting, to question the very concept of locatable reality upon

which we conventionally rely in our mappings of the world, to subordinate the sign systems we so confidently use to a larger doubt. Holbein fuses a radical questioning of the status of the world with a radical questioning of the status of art[19].

Barker's intention in *Death, The One...* recalls Holbein's; both artists seek not only to remind us *that*, but demonstrate *how*, to enter the 'non-place' constituted by the sustained contemplation of death, which their work directs, is 'to render impossible a simple return to normal vision'; rather, as Greenblatt claims, 'we do return and reassume that perspective that seems to "give" us the world, but we do so in a state of estrangement'[20].

Death, The One... is Barker's most profound address to the recurrent theme and dynamic in his work, the practice of (self-)deceit that permits the confidence trick. *Death, The One...* has the effect of placing the 'anamorphic skull' of Barker's tragic Art of Theatre amidst the 'confidence trick' constituted by conventional manifestations of theatre, and perspectives on the world; like Holbein's painting, Barker's treatise fuses a radical questioning of the status of the world with a radical questioning of the status of art. The recognition and admission of the immanence and inevitability of death becomes, not fatalistic or debilitating, but paradoxically vivifying, discovering an 'ecstasy' in the 'vanishing' of a conventionally dominant totalitarian 'meaning' (*DTOAT*, 14):

> What was, is, and forever must be, cannot be *depressing*. Depression is a failure of the spirit. Who are the most depressed? The comedians. Fear is their territory. Tragedy fears nothing, it enters in, it must enter in, it senses this entering as an *ecstatic obligation...* (*DTOAT*, 81)

> Death renders irrelevant all that was relevant. Whilst this injures the still-living (the still-loving) might this not be one of death's *generosities*? (*DTOAT*, 37)

Barker observes how the conventional theatre gratifies when it gives no more than is expected of it, offering performance as accountability and predictability (in accordance with the contemporary British municipal social discourse of utility, production and purpose, and its consequent repressive instrumentalization of human behaviour, scornfully observed by Iain Sinclair: 'commitment to duty of Best Value, ensuring clarity of objectives and customer and performance focus at the heart of our cultural and organisational exchange'[21]). In contrast and opposition, '*the art of theatre* gratifies when it violates the tolerance of its public (is this not a contradiction? But the effects of the work of art are characterised by *delay...*)' (*DTOAT*, 103). Thus, The Art of Theatre deliberately rejects the 'commitment to the duty of Best Value' in the immediate and quantifiable terms beloved by arts management and funding, of "customer services"; rather, it summons the unforeseen, and the unforeseeable. Like the dead, it ceases to exchange and prefers to transform, suggesting alchemy rather than equivalence; just as 'The One' may transform (herself) but cannot be *exchanged* for another, because of her fateful (perhaps even fatal) uniqueness. The Art of Theatre gratifies more, the longer its *duration* of effects, the longer its images linger in dialogue with previously held pieties; it is impious in its 'determination to imagine beyond the perimeters of experience' (*DTOAT*, 49). Like death, it ceases 'the elaboration of values, the prioritization of values, the valuing of values' (*DTOAT*, 21); and thereby the valorizing of the normative 'Best Value' (and the satisfaction of the associated bureaucratized "culture" of evaluation, association, preference, hopes of reward, hopes of security). The Art of Theatre purposefully avoids conventional resolution, and disrupts the predominantly privileged and prioritized forms of time (compare Céline's wry observation: 'We can't get together while we're alive. There are too many colours to distract us and too many people moving around us. We can only get together in silence, like the dead'[22] – or, one might add, like the audience of The Art of Theatre). In this respect, The Art of Theatre is importantly *anticipatory*, reaching beyond the current forms of the known and accepted:

> *The theatre* thus: '*let us describe what the audience feels. It will thank us for it*'. *The art of theatre* thus: '*let us describe what the audience does not know it feels. So what if it reviles us for it?*' (*DTOAT*, 39)

Importantly, Barker locates the process and experience of dying elsewhere than in the state of death. However, the confrontation of death, upon which The Art of Theatre insists (albeit fictionally), involves the audience in a congregated focus on individuals struggling at their limits; as when one attends a dying relative, there is a simultaneous focus on the untransferable and inexchangeable aspects of terminal experience, and a forceful confrontation with the (ultimately inexpressible) boundary all must one day witness, then cross (inexpressibly). Lingis suggests the 'experience of death-boundedness isolates and unites us in a community of those who have nothing in common'[23]; and this may be a more pertinent angle by which to approach the audience's experience of The Art of Theatre than more conventional and vacuous invocations (such as 'immediacy', 'relevance', 'collectivism'). Here and elsewhere, Barker recognizes and exposes how the conventional forms and terms of understanding are insufficient, and increasingly predicated on consumerist expectations and paradigms: what Lingis terms 'identifying, measuring, coordinating work' which 'does not give rise to evaluation' but informs 'a systematic enterprise of devaluation', a recoil into reductivity; in contrast, Barker's work offers an expansive address to the strong emotions which characteristically 'seek out what is incoherent, inconsistent, contradictory, countersensical; they endorse what is unpredictable, unworkable, insurmountable, unfathomable'[24].

Death... represents an important advance on *Arguments*, surpassing even the range and depths of the earlier collection, not only because of the brilliant identification and delineation of its provocatively linked central conceptual triptych, but because of its pursuit of the attendant impulses, anxieties and questions at the heart of *that which distinguishes mortal aspiration* (as opposed to hope) in the human (the drive to go *beyond* human precedent, *beyond* the human, even at the cost and point of death). Barker's speculations poeticize a juncture between aesthetics and metaphysics, and argue for the art of theatre as the proper crucible of philosophy: the analysis and bittersweet appreciation of the paradoxes in life's melancholies and ecstasies.

Part Two

Plays and Productions

3
Intimacy with the Unforgivable

The Last Supper, The Early Hours of a Reviled Man, Golgo, Judith, Rome, Ten Dilemmas

> what makes us…unforgivable…this ability…**I won't say appetite**…for continuance
>
> *Ten Dilemmas (CP2, 367)*

The Royal Shakespeare Company had shown commitment to Barker's late 1970s plays, which dealt with the failures of British forms of social democracy, with productions at their London Warehouse studio (*That Good Between Us*, 1977; *The Hang of the Gaol*, 1978; *The Loud Boy's Life*, 1980); however, the RSC's transfer of operations to the Barbican theatres was characterized by a more populist era of programming. In a near-final flourish, a 1985 RSC retrospective season, 'Barker at the Pit', presented the new play *The Castle* alongside the professionally unproduced older plays *Downchild* (written 1977) and *Crimes in Hot Countries* (written 1980) in strongly acted productions, but the RSC's attention and marketing priorities were more focused on the main house production of *Les Misérables*, limiting the Barker productions to a small number of performances, despite critical acclaim for *The Castle* in particular. In the late 1980s, Barker's work was proving increasingly unacceptable to his former institutional allies. The RSC rejected the play they had commissioned, *The Europeans*, when Barker submitted it in 1987; they did however stage *The Bite of the Night* in 1988, Barker's most ambitious play up to that time; this had been commissioned by The Royal Court and written 1985-6, but the Court would not proceed to a full staging. The artistic leadership of the Court was nonplussed by the play's extreme challenge to social realist convention. Their concession was to stage extracts in a private rehearsed reading at their Theatre Upstairs before an invited audience, with a formidable cast including Julie Covington as Helen, Bill Paterson as Savage and Tilda Swinton as Gay (Ian McDiarmid presented another one-off reading of scenes from the play at the Almeida in 1988, with Frances Tomelty as Helen and McDiarmid as MacLuby). Barker's innovative epic seemed doomed to languish without the Royal Court's full commitment.

Nevertheless, Spring 1988 saw two productions of Barker's work playing simultaneously in London: Ian McDiarmid's production of *The Possibilities* at the Almeida Theatre; and Kenny Ireland's production of *The Last Supper*, arriving at the Royal Court after opening at Leicester Haymarket. *The Possibilities* was presented by 'Not the RSC', a loose collective of disenchanted actors formerly involved with the Royal Shakespeare Company now seeking to undertake challenging work. *The Last Supper* was the first play conceived for, and presented by, The Wrestling School, the new theatre company specifically formed to examine Barker's work.

McDiarmid's production of *The Possibilities* was sharp and forceful, demonstrating an appropriately surgical cruel precision in physical movement and interaction, and fluency in movement from one short play to the next. It presaged McDiarmid's long-term involvement with the Almeida: he and Jonathan Kent took over the running of the theatre for ten years, opening their 1990 inaugural season with the stage premiere of Barker's *Scenes from an Execution* (originally broadcast by BBC Radio in 1985 and winner of the Prix Italia Radio Drama award). McDiarmid's production of *Scenes* featured Glenda Jackson, who had so impressively played the central role of Galactia on radio, reprising her performance in what would be her last major stage work before her professional self-dedication to politics. The production hinged on the central idea of having the principal unseen image of the play, Galactia's controversially anti-heroic painting of the sea battle of Lepanto, located on the "fourth wall" of the theatre auditorium – or, more specifically, with audience members spatially "standing in for" the human details of the painting to which characters referred, its unconventionally terrified, cruel or butchered human faces and bodies. Jonathan Hyde as the Doge – like the best of the performers in McDiarmid's *The Possibilities* – demonstrated an appropriately precise grasp of his director's own distinctive performing style, in which the breathing patterns of Barker's language are permitted to inspire physical posture and momentum of movement, involving a choreography of fully physically manifested transitions from the full reach of one impulse to the next.

Later in 1988, Danny Boyle's production of *The Bite* for the RSC at The Barbican Centre's small Pit theatre gave a lucid exposition of Barker's play. Boyle had directed *Victory* for Joint Stock in 1983; his four and a half hour production of *The Bite* (its running time publicized apologetically by the RSC) valuably presented some sense of the later play's ambitious mythic reach, cumulative force and fierce excoriation of contemporary British political and moral edicts. The production's clean design, predominantly bright lighting and confinement to a small studio space tended to limit the force of the play's darkly poetic imagery and experiential range, but it nevertheless featured laudable performances. *The Bite* was to be the RSC's last Barker production; the conciliatory spirit associated with Adrian Noble's tenure at the Company was – like the Royal Court's sense of fashion, novelty, immediacy and social realism – increasingly antithetical to Barker's expanding ambitions.

The Wrestling School's production of *The Last Supper* manifested a new visual boldness in *mise-en-scène* which would become a continuing hallmark of the company. Costumes and make-up were eclectic and deliberately extreme, literally fantastic rather than re-presentations of notional familiarities and period consistencies which might provide imaginative footholds for an audience seeking to locate the play in the specific limitations of a given "world". Angular rhythmic music and other non-naturalistic sound cues (such as disembodied laughter) erupted into the play, frequently destabilizing a mood rather than sustaining or generalizing it: sometimes, these sonic forces provoked an almost spasmodic choreographic reaction, akin to an electric shock, from a chorus of characters; at others, provided a stark rhythmic underpinning for an individual character to deliver a heightened utterance in the manner of a chant or incantation.

The two Prologues to *The Last Supper* both, in different ways, establish the play's decisive break with the various forms of ingratiating populism, which are significantly established as the "climate" of the play: it drifts across, and erupts into, the action as a *'terrible storm'* of depersonalised laughter. The First Prologue establishes the enthronement of 'The Public' in lieu of God for those whose shame requires subordination and atonement; The Second emphasizes *The Last Supper*'s deliberate disappointment of conventional neo-journalistic presumptions (**'The play contains no information'**) and avoidance of the reductively instant single meaning (rather, it constitutes an invitation to hang up the **'Suffocating overcoat of communication'**), even as the speaker disarmingly acknowledges his pedagogical posture ('Oh, I lecture you ... Forgive!', *LS* 2). The formal models of the Prologues recall those in *The Bite of the Night*, but more pithily establish the regressive hope of an infantilizing cultural *milieu*, and the enticing promise of a less predetermined alternative.

Charles Lamb aptly identifies two recurrent Barker motifs which are at the heart of *The Last Supper*: firstly, the seductive tactic of refusing shame, which 'challenges by transgressing the limits of the (socially defined) self and attempting to lure the other into a complicity, a pact'[1]; and the common scenario of the 'magisterial' relation, 'teacher and pupil, a classic example of a duel where the participants attempt to drive each other mad', here developed to the point where 'each relationship is a vertiginous and unrelenting duel between master and disciple'[2]. In some ways *The Last Supper* grows out of Barker's poetic monologues for performance, *Don't Exaggerate* (1984), *The Breath of the Crowd* (1986) and *Gary the Thief/Gary Upright* (1987), all of which develop the focus on the individual foregrounded against the crowd, and the compulsion of instinct (for death in the first, for shamelessly continued life in the last) to break through socially imposed form and received wisdom. The itinerant preacher Gary Upright is the Barker character who first proposes that each man and woman might discover and become their own divinity, forgiving and refining themselves through rehearsals of charismatic persistent existential performance:

In this effect of seizure
I have the attribute of a god unnamed
Which you are also
But since you won't affirm your deity
Since you will smother
Since you will obscure the light with your hand
Observe the practices of one who
Stands before you unashamed (*GTGU*, 26)

The Last Supper is also punctuated by a series of "parables" which provide scenes of three soldiers' disingenuous abjection, which (like *The Possibilities*) invite, only to disrupt, comparisons with Brecht's *Lehrstücke* and Edward Bond's socially deterministic fables[3]. The parables – from which it is impossible to distil a generalized meaning – provide a contrast to the main theme's atmosphere of despair, twilight, contraction, degeneration and self-conscious finality, through the literally wilful eccentricity of the soldiers who 'do not kill': their irrepressible optimism, philosophical wonder and childlike curiosity.

Lvov attempts to exemplify for his followers the breaking with 'trivial conscience', '**doing the undoable**' in ways he proclaims 'the essence of love' (*LS* 11; compare Bradshaw's discovery of the self-overcoming terms of Victory in a catastrophic political landscape, in Barker's 1983 play of that name[4]). He senses their frittering of resolve in competitive and resentful 'wit and banter' as they witness and experience the advance of a totalitarian political culture, presaging 'the death of complexity' (*LS*, 18), which often accompanies the end of a war. His ultimately predictable followers are outstripped only by Judith, who can remind him 'you taught me that information is nothing, and expression, all' (*LS*, 12). Her relentless appetite for knowledge and skill in performance is required to rescue even her mentor Lvov from his residual impulses towards the mundane satisfactions of common life. Lvov secretly acknowledges 'I am not afraid of death. I am afraid of being revealed' (*LS*, 26), and of oblivion. His intimations of disintegration are confirmed by his disciples' increasing indignation: one notes his own drift into imitation, 'That can't be freedom, can it? Freedom must be ceasing to be you'; another observes: 'The beauty of Lvov was that he arrived complete ... The question is, however, whether we admire completeness. Whether we might not rather admire – flexibility, growth, deterioration, alteration'(*LS*, 35); another confesses the lure of happiness, the promise of how 'The aching for the absent thing might die...and in its place there might be...a continuing fascination with the existent' (*LS*, 36); another subjects him to the accusation of the pseudo-empiricist criteria of popular success, in language reminiscent of an Arts Council form: 'Can you explain how you intend to move from the minority to the majority?' (*LS*, 38). Whilst consciously 'turning their love into hate', Lvov's problem is to drive them to betray him in a way which seals his myth and charisma rather than diminishes him (or them, despite their impulses to self-diminishment). He tries to do so by admitting sexual contact with Gisela, an intellectual who cannot resist his preference of her, but who then recoils from the revelation of his 'authentic **commonness**'. However, even this (self-)debasement does not muster the required pitch of fervour in the disciples. Only Judith 'prefers knowledge to peace' (knowing 'The charismatic have no friends' or equals); she intends to prove herself, momentarily but crucially, greater than him, challenging and forcing him into death in order to seal both his and her immaculate artificiality of will: 'We have to kill Lvov, because he is ceasing to be Lvov' (*LS*, 51).

The disciples kill and eat Lvov, placing him beyond arrest or incorporation by the new authorities who have come to 'claim' him, forces opposed to any form of 'certainty' not defined by the restored ideology. The disciples are united in their secret experience of Lvov, but also publicly identified and arrested because of it. Lamb reads the final image as a vindication of the disciples: 'Lvov seduces them beyond the transcendental law, beyond the prohibition, and this, of course, becomes the "secret" that bids them and will be the source of their power'[5]. I find the image more ambivalent: their "secret" is not powerful enough, nor secretive enough, to prevent their identification and immobilisation, and thus their "power" might be that of a muted (if willing) totemic martyr, like Scrope in *Victory*. As Lvov knew, their *distinction* threatens to become their crucial *limitation*, unless artfully managed. Even the soldiers who bind them reveal a darker colouring to the appealingly resolute consistency in their Schweykian antics of survival, by proclaiming all death and disappearance which accompanied their wake as '**accidental**'; perhaps they have proved themselves better performers than the disciples. The final aimless 'drift', of the knot of disciples, as '*The cloud passes overhead*', makes for a melancholy (rather than triumphant) image of complicity in continued subordination, the finitude which Lvov recognised as 'the investment the servant makes in servitude' (*LS*, 48). Like many irreducible, mythic moments, the final tableau is like a prism poised to reflect the mutual interpenetration of victory and defeat, gain and loss, love and death.

In *The Last Supper*, time seems to pull in two opposite directions simultaneously: on the one hand, there is a sense of imminence, momentum, pressure, impatience and finality; on the other, the characters frequently express a sense of strain, deceleration and resentment, both tendencies being reflected in Lvov's complaint 'You announce the last supper and they don't hurry themselves, do they?' (*LS*, 9). The character of Judith is the principal counterforce to the sense of everything running down, which even Lvov has to depend on others to transmute; and it is an artful strategy of Barker's to make the Judas-figure a splendidly artificial female aesthete who shares the name of the slayer of Holofernes, and who insists on discovering the one thing Lvov cannot teach her: what it is to live without him. In some ways, *The Last Supper* brings to full dramatization what is glimpsed by the priest Orphuls in Barker's play *The Europeans* (written 1987): 'Judas was cruel for knowledge, and without Judas there could be no resurrection, Beauty, Cruelty and Knowledge, these are the triple order of the Groaning God... (*OP1*, 146). Orphuls's sermon is immediately judged to be magnificent and yet fatally unforgivable. However, becoming unforgivable is a hallmark of existential apotheosis for Barker's protagonists: in *The Bite of the Night*, the compulsively transgressive Helen maintains 'Could I ever forgive myself if I were forgivable?' (*CP4*, 105). There is even a suggestion of the insight glimpsed by Riddler, protagonist of the later play, *A Hard Heart*: 'Gods are born in pain, not pleasure. They are the product of extremity, a manifestation of the will of peoples ... which flows into their mortal bodies and inspires them, enabling them to breathe in unfamiliar airs' (*HH*, 23–4; however, Riddler's self-identification proves a hubristic delusion). *The Last Supper* is, as its subtitle proclaims, 'A New Testament' on the *difficulty* of becoming unforgivable and attaining existential grace. Even the Parables, with their wryly abstract titles, *slow down* a sense of progressive understanding; the wilful reversals they depict serve to complicate, rather than simplify, the application of any generalized deduction to life, and so they deliberately disappoint the promise of the conventional expectations of the Parable form. The increasing sense of characters irrupting through and 'bleeding' across the parameters of Play and Parable, further accelerating the audience's sense that there are no bearings left to get, will be amplified by Barker in Isaac's appearances in the main theme of *Rome*.

Barker's fourth volume of poetry, *Lullabies for the Impatient*, was published in 1988. Unlike its predecessors, this collection features no extended piece intended for performance by an actor; it is suffused with a more meditative atmosphere, which is itself consciously dramatized, even interrogated. One poem, 'The Early Hours of a Reviled Man', imagines the French writer Louis-Ferdinand Céline awaiting the nocturnal atmosphere of illegality to cease practising medicine and to write in mockery of 'the catastrophe of cultures / Wishing it was over', like a gargoyle on the edifice of post-war Europe, burning to escape the damnation of continued life, and mocking the corruption in the surrounding 'violent symmetry' of reconstruction and reconciliation. This poem provides the seed of germination for a full-length play of the same title. The play *The Early Hours of a Reviled Man* (broadcast by BBC Radio 1988; directed for the stage by Roger Owen, Aberystwyth

Theatr y Castell 1990) develops the theatrical potential of the figure of Céline, whose writing depicted life fatalistically as a series of catastrophes, in which only death and the process of dying have dominion. For Céline, only the energy of hatred promises relief from feelings of personal emptiness and persecution (by forces he personified in the figure of the Jew)[6]. The play explores the will and power of the unforgivable, with dark seductive wit and nimbleness. In Barker's most haunting evocation of nocturnal cityscape, his protagonist Sleen takes his regular perambulation through the streets where he has loved, triumphed and suffered, exposing the futility of all vested interests which are cloaked by the conventionally social: 'If I do not walk the streets I cannot write. Why write, you say, it would be better if you didn't. Better for whom?' (*HH*, 54). He is hijacked by old friends and new enemies, all invoking a sense of justice he is determined never to satisfy. Barker's Sleen is a masterful dramatization of the compulsive drive and stealthy ability to defy all forms of moral superiority; Sleen defines himself through doing the 'Unfashionable / And / Untolerated / Thing' (*HH*, 51) and, like other charismatic Barker protagonists (such as Starhemberg, Gary Upright, Lvov, Judith, Vanya, Gertrude, Sleev), his potency resides in his command of imaginative (self-)invention which evacuates social forms and rituals of reintegration, and relocates the sacred in the individual imaginative will; these characters achieve what Greenblatt identifies as the quintessential sign of power, 'the ability to impose one's own fictions on the world'[7]. A brief Preface to the radio production aptly establishes a context of contemporary political correctness, where a student objects, ideologically rather than aesthetically, 'I do not think this author should be on the course … I do not think a bad man can write good books'. The persistence of Céline/Sleen's offensiveness, and the resentment which his paradoxical qualities inspire, makes him an appropriate figure for Barker's reanimation: he is another unforgivable prophet, practising medicine amongst the poor in his twilight years (using the grateful poor as 'a mirror' in which to reflect his own 'wit and unhappiness' by playing 'the genius and master' to them), self-armoured in pessimistic fatalism which does not inhibit his wilful refusal to submit or be 'smothered' by conventional ideals of happiness, rationality, longevity or utilitarianism. For example, he perceives and announces 'the ecstasy' of war:

> It refused discrimination. It ridiculed our struggle for variety, cutting the sensitive and the vile in equal quantities. It reduced the ugly and the beautiful to a single bloody shape. How democratic it was! There was democracy for you! That and disease! But who dares say so? (*HH*, 52)

He warns his ex-lover Jane that her charity is misdirected, if she expects him to profess gratitude, as he feels none. He acknowledges the lure of sycophancy when a student, presumed acolyte, approaches; however, the student is one of a succession of characters each of whom blames Sleen for their own lack of fulfilment. The play becomes more hauntingly dreamlike as the city discloses its labyrinthine and subterranean scenes of strange reversibility, through which Sleen leading his outraged but mesmerized followers as in a *danse macabre*, luring them with the challenge 'Come with me … Because you hate me' (*HH*, 61); this strange entourage includes the Jewish surgeon whom he seduces with the promise of 'The ecstasy of breaking [her] oath'. Sleen is scandalously at home with the concept of moral relativity: 'We are not sham. Nothing we say endures, that's all'; 'I like the police. They understand the insignificance of truth. Or rather, there is a greater truth to which they owe a secret loyalty. This is the arbitrary nature of existence which disdains selection' (*HH*, 56); though this is offensively inadmissible to 'those who cannot come to terms with pain'. His necromantic philosophy permits him to slip the knot of his escorts' accusations and expose the insecurity underlying their insistence on separating themselves from him: 'I deserve everything I get, but so do you' (*HH*, 65). Indeed, their outrage is revealed as a hysterical compulsion to distinguish or redeem themselves by insisting on their own superiority to him; this nevertheless gives him an irrational power, to call into question, and even dissolve, their terms of identity and morality.

As noted in Chapter One, Julia Kristeva, in her book of theoretical poetics *Powers of Horror*, identifies the Abject as 'something rejected from which one does not part', a figure which 'does not respect borders, positions, rules'[8] – such as the brilliant novelist and doctor to the poor who is also anti-semitic. She observes:

> The one by whom the abject exists is thus a *deject* who places (himself), separates (himself), situates (himself), and therefore *strays* instead of getting his bearings, desiring, belonging or refusing … the space that engrosses the deject, the excluded, is never *one*, nor *homogenous* nor *totalizable*, but essentially divisible, foldable and catastrophic. A deviser of territories, languages, works, the *deject* never stops demarcating his universe whose fluid confines … constantly question his solidity and impel him to start afresh.[9]

Sleen's progress through the city exemplifies and dramatizes the very effects of which Kristeva writes; appropriately, since her study is partly concerned to identify the dynamics of mutual degradation in the writings of Céline, which constitute his assault on the dichotomy of the sacred and the profane. She notes how the deject calls all borders into question, because 'abjection is above all ambiguity, because, 'while releasing a hold, it does not radically cut off the subject from what threatens it – on the contrary, abjection acknowledges it to be in perpetual danger'[10]. In this sense, *The Early Hours* is a jet-black comedy of abjection and contamination, in which

Sleen's pursuers represent the subjects who wish to separate themselves from his unforgivably questioning presence, but who find their boundaries continually problematized by his sheer persistence in existence, and their inability to murder him.

The figure of the self-parodying, self-compounding, unforgivable and necessary betrayer is alternatively but similarly characterized by Brendan Kennelly in his Introduction to his epic poem, *The Book of Judas*, which lends 'the outcast scapegoat' Judas a 'protean virtuosity' and opportunity 'to speak back and out from his icy black corner of history':

> The Judas-voice is … freakish and free, severed and pertinent, twisting what it glimpses of reality into parodies of what is taken for granted, convinced (if it's convinced of anything) that we live in an age almost helplessly devoted to ugliness, that the poisoned world we have created is simply *what we are*, and cannot be justified or explained away by science or industry or money or education or progress. To this extent, Judas knows nothing is external: where we are is who we are, and what we create is merely the symmetry of our dreams[11].

As I have written elsewhere, the power of the Judas-voice resides in its ability to convince that it makes a virtue of necessity, and a necessity of evil[12]. With parodic extravagance, Sleen assumes the Judas-voice and the role of deject, proclaiming himself a sacrificial totem of corruption: 'The vilest murderer would shine beside me in celestial light' (*HH*, 70); 'How fortunate tonight is my adieu. How sweet the world will be by morning!' (*HH*, 68). However, he also constitutes a constant intimation of the impotence of those who denounce him: 'she senses…the inadequacy of revenge … the chorus of the dead whose howl cannot be satisfied…what's my diseased old corpse to them?' (*HH*, 72). However, on encountering his birthplace, in the slums, he is brought to a highly personal admission: 'Love…which could not bear to show itself…hides in pain…and welcomes every evidence of malignity' (*HH*, 77). When Jane seems closest to murdering him, the surgeon objects 'I think your action is entirely personal' and therefore squalid; Jane retorts in turn to the surgeon, 'I am tired of your purity! It hangs like a suffocating banner on my face' (*HH*, 79). Sleen's actions and opinions are *entirely* personal: unfathomably, outrageously, unforgivably so. His very presence conducts the others through a series of revelatory scenes which demonstrate to them how their dependency on postures of correctness and self-identifications with moral superiority *debar* them from his unshakeable purchase on the imagination – even on theirs. As I have observed elsewhere:

> The Judas-voice might thus involve its speaker in playing roles in a scenario written by others; even in scapegoating him/herself, perhaps fatally.

Another possibility lies in the subversive power of performance, in achieving a level of poetic expression and action which transcends fear, incarnating 'the electricity in the air that burns the labels and restores the spirit of investigative uncertainty' (Kennelly, 9). The speaker of the Judasvoice may move from assuming a role in the prewritten scenario, on others' terms, to playing out the role with an invention and conviction, on his or her own terms, and thus dislocate the labellers – those voyeuristic, deterministic would-be controllers – from centre stage, repositioning them as witnesses to a demonstration of their own comparative half-heartedness.[13]

This is what Sleen – and Barker – achieve in *The Early Hours*. It is one of Barker's most profoundly comic disruptions of the terms of moral identity through what Greenblatt terms 'absolute play', the pursuit of the unforgivable through shameless performance:

The will to play flaunts society's cherished orthodoxies, embraces what the culture finds loathsome or frightening, transforms the serious into the joke and unsettles the category of the joke by taking it seriously, courts self-destruction in the interest of anarchic discharge of its energy. This is play on the brink of an abyss, *absolute* play.[14]

Barker's short play *Golgo*, performed by The Wrestling School and directed by Nicholas Le Prevost in 1989 as a companion piece to *Seven Lears*, provides an even stronger example of 'absolute play'. In *Golgo* this is literally the activity of another artist-protagonist, whose native climate is the 'antithesis of kindness' and who is unstained by humiliation. Conceived for the bicentennial of the French Revolution, *Golgo* deliberately repudiates the celebratory populism of most contemporary responses to the anniversary by centring on Whatto – melancholy frivolist of a courtly world, loosely based on the painter Watteau[15] – even as a Chorus denounces him to the audience, invoking their moral outrage and warning against his charisma (**'He'll twist your horror into pity'**, *SL* 62). Characteristically, the events of the play complicate and problematize the Chorus's doctrinal invocation of morality and History.

Whatto presides over a garden party of consciously doomed aristocrats, the central event at which is a dramatic reinterpretation of the death of Christ: this constitutes the players' final spasm of performance and (self-) dramatization, dancing on the edge of the abyss which promises their own fatal redundancy. Like Sleen, Whatto proclaims a philosophic resignation to prospective death, at the hands of the mob whose 'barmy' ecstasy he otherwise might enjoy: 'who am I to criticize, who never loved life from the beginning, who was born to enjoy nothing and to see the horror even in a kiss' (*SL*, 55–6). A Priest attempts to censor their performance with desperate populist slogans ('Complexity is a conspiracy against the people!'), but is himself drawn into the performance to give a fine portrayal of Barabbas, as the Chorus reminds the audience '**We trust you to discriminate / Between the manipulators / And the genuinely contrite**' (*SL*, 62). Whatto's central dramatic moment is his imputation of 'shame' to Christ, because 'There were others of Golgotha! ... There are those recorded, and those who fail to be recorded, obviously there is **other testament!** ... **Always other testament!**' (*SL*, 71). In the terms of Barker's *The Power of the Dog* (and like *The Europeans*), *Golgo* opposes History (the imposition of ideological and moral narrative form) with Anti-History (the disruptive fragmentation of this form by the testimony and performance of individual pain, demonstrating the insecurity of all promised reconciliation and the instability of all order). *Golgo* achieves this with particular pithiness, in its haunting tableaux which represent flourishes from a last ditch. At the end, even Whatto's outraged peasant ex-lover cannot kill him efficiently, and concludes the play with the exhortation 'God give us the courage of our cruelty ... Or / We *(In the darkness a massive exhalation of breath)*' (*SL*, 81). Of course, her terms of self-overcoming are as applicable to her enemy Whatto (his skill as a mesmeric, disarming performer demonstrates this courage) as to her proletarian allies, the "collective" with which the audience are (increasingly) imperfectly associated. This further demonstrates how (theatrical, like erotic) intimacy with the "unforgivable" serves to disrupt ideological exclusion and moral security in superiority.

This is also the theme of another short play, *Judith* (written and published 1990, directed by Barker for The Wrestling School 1995), based on the Apocryphal story of strategic seduction and ideologically justifiable murder. Barker portrays the Assyrian general Holofernes as surprisingly contemplative, but in a way which informs rather than mitigates his performance of power. He reflects upon, and identifies himself with, the force of death, 'its arbitrary selections', the persistence and proximity of which justifies the military profession, 'for while victory is the object of the battle, death is its subject, and the melancholy of soldiers is the peculiar silence of a profound love' (*CP3*, 243). The soldiers' 'wilful suspension of all logic' and 'collaboration in chaos' is what permits the battle, 'the mad life licensed' in experience and study of pain and death of those around, which constitutes the 'ecstasy' of warfare. This delineation of the philosophical scaffolding for his successful campaigns is, of course, the refining of his self-performance, for self and others, which distinguishes him as a leader who is charismatic in his subtlety and certainty.

However, Holofernes is challenged by another performance, the "double act" represented by the Jewish widow Judith and her Servant. They offer him the promise of skilful erotic performance, with Judith proclaiming 'Naked I must be ... Naked and unashamed', with the Servant assisting because 'undressing is an art' (*CP3*, 247). However, Holofernes interrupts and counters her performance, persisting in his fixation on Death as the ultimate force, capable of overthrowing her invocations and offer of Love. They play out their strategies: Holofernes partly undressing like Judith, Judith provocatively articulating some terms and stakes: 'You might wish to imagine me rather than know me. That is the source of desire, in my view. Not what we are, but the possibilities we allow others to create us' (*CP*3, 250). Holofernes discloses 'a terrible need', that he is not 'the definition of another's life': 'I think we only live in the howl of others. The howl is love' (*CP*3, 253); nevertheless, he admits his refinement of speech is 'a trap' to compensate for other weaknesses. Judith performs the acceptance of this, his apparent self-disarmament, which can presage her own: 'now we have abandoned the search for truth, really, we can love each other!' (*CP*3, 254). But, as the Servant notes in an aside, this is another raising of the stakes of the game: 'One of them is lying, or both of them' (*CP*3, 255); moreover, Judith must be careful, because, in this as in the mystery of any compelling performance, 'sometimes, the idea, though faked, can discover an appeal' (*CP*3, 256). In a moment of (strategically disarming?) relaxation, Holofernes suggests that they both lied, but have enfused the tropes of their performances with a genuine essential abandon, and experienced surprising disclosures: 'But in the lies we. Through the lies we. Underneath the lies we' (*CP*3, 258). The Servant rightly perceives that Judith must intervene decisively if she is to overturn this tactic: when her slogan, demonizing Holofernes, fails, the Servant '*enrages JUDITH with a lie*' that he is smiling, and smug rather than destabilised.

As Lamb observes, 'The notion that Holofernes is grinning in confident anticipation of another easy victory is enough momentarily to abolish his performance in Judith's eyes'[16]. However, rather than ideological satisfaction, Judith experiences a torrent of emotions released by the full, and profound, reach of the duel. First she attempts copulation with his headless corpse (and further scandalizes and physically disrupts the Servant's Manicheism by asking: 'How can he be an enemy? His head is off'); she then experiences a paralysis which prevents their escape. Now the Servant has to raise the stakes of her own game. Just when Judith, like the audience, might be prepared to reduce her dismissively to the status of a somewhat comically consistent ideologue, the Servant unfolds and deploys an existential credo which encapsulates three propositions which are central to, and recurrent in, Barker's exploratory dramatisations:

> Firstly, remember we create ourselves. We do not come made. If we came made, how facile life would be, worm-like, crustacean, invertebrate. Facile and futile. Neither love nor murder would be possible. Secondly, whilst shame was given us to balance will, shame is not a wall ... Come near it and you will see how thin it is, you could part it with your fingers. Thirdly, it is a facility of the common human, but a talent

in the specially human, to recognise no act is reprehensible but only the circumstances make it so, for the reprehensible attaches to the unnecessary, but with the necessary, the same act bears the nature of obligation, honour and esteem ... they are the specially human who drained the act of meaning and filled it again from sources fresher... (*CP3*, 263)

This appears to inspire Judith, who accepts a vision of herself as unforgivable and therefore charismatic, a divine monster and unstoppable golem. Judith is capable of not only subjugating but humiliating the Servant, who is nevertheless impossible to pity because of her ideological faith in the redemption of all pain and death. Judith has been placed beyond The Law, and becomes The Law. It is true not only that, as Lamb observes, 'The moment of seduction detaches the individual from both personal and political history'[17], but also that, as in this instance, personal and political history can prove reversible for the sublimely charismatic: 'Israel claims Judith, but she claims Israel'[18]. Her transfiguration is both triumphantly and appallingly absolute. Barker's 1995 production for The Wrestling School tautly focused exemplary performances by Melanie Jessop (Judith), William Chubb (Holofernes) and Jane Bertish (the Servant); the staging of its pivotal moment is well summarized by Zimmermann[19].

Formally, *Rome* (written 1989, published 1993, and still awaiting a full professional production) is clearly related to *The Bite of the Night*, both in the two plays' extreme range of experience and in their re-visioning of the defining foundation narratives of the classical world: Homer's heroic epic account of the Trojan wars (characteristically, in *Bite* Barker shows Homer himself losing control over, and being enveloped by, the political extrapolations of the events and impulses he celebrated); and the fall of the Roman empire, culture and religious authority to the encroaching 'barbarians' (in *Rome*, the god Benz likewise finds himself surprisingly embroiled in his own creation). *The Bite* presented, in forms analogous to dream and nightmare, relentlessly political explorations of various contemporary cultural fetishes and social prescriptions (including populist celebration, intellectuals' shame, demonization of the body and the restitution of the family), with the feverish rising and falling of Trojan political regimes reflecting the successive manifestations and legislations of a given character's psyche and sexuality. Helen, an embodiment of the unlimited appeal of shamelessness, is mutilated as criminally offensive by each successive regime, yet she still retains her disruptive magnetism. The sub-title of *Rome*, *'On Being Divine'*, indicates how, in this later play, Barker is concerned to investigate how the spiritual authority of church and individuals challenges or colludes with the political authority of state.

Pius, a pontiff, finds his consciousness extended even beyond death by his undimmed fascination with the physical beauty of his shameless mistress, Beatrice, as she copulates with other men. Pius persists as a presence in the action, discovering that Death is endless ('First blow to the optimists'), constituting rather a succession of 'ante-rooms all the way'. The stage directions, stipulating Pius *'revealed, descending'* while Beatrice and her lover *'study'* each other and *'revolve'*, indicate Barker's increasing scenographic ambition. *The Bite* increased the simultaneity of various competing rival points of focus as one character tried to upstage or undercut the performance of another; in *Rome* Barker experiments with scenic composition in further ways which necessitate a large stage, high levels and the "flying in" of details and characters on wires, as well as the 'passage' or irruption of (groups of) characters across and into scenes (for example, the Regiment, who urge onlookers to mediate, reclaim and redeem ideologically their destructive frenzy: 'Don't lend us the power of language / **Or we shall sing the ecstasy of murder!** ... Don't let us be revealed / **As the condition of your laughter!**', *CP2*, 202). The action then moves into one of several periodic interludes, 'First Abraham', in which the Biblical parable of servitude is given savagely humorous dramatization. Instructed to sacrifice his son, Abraham struggles to proceed from the deduction, that 'To love is to inflict impossible pain' on both self and others. When Benz interrupts the sacrifice and says Abraham can kill a ram instead, Abraham objects to this apparently whimsical mitigation which belittles his own will and resolve: 'You ask for everything. And I want to give you everything ... And now you say, not everything, after all' (*CP2*, 207).

Doreen, 'A Cultured Woman of Rome', places her faith in the preservation of objects, oblivious to the political ramifications of her own relative devaluation of the human[20]. Beatrice's widowed daughter Smith is as absolute as her mother is obliging; Smith finds herself attracted to the new pontiff, Park, for his 'perfect' ignorance of common priorities; however, Beatrice rejects the sexual overtures of the paradoxically confounded god, Benz, who, even as he deems destruction necessary, proclaims 'the inadequacy of love' – love which, unlike fear, he finds himself incapable of commanding. Park can identify the presence of Christ even in the wartime atrocity: the pain of its perpetrator as well as the pain of its victim; however, he is embarrassed by Smith's subjugation of her self before him, which she points out is also an expression of (her) will.

The interlude, 'Second Abraham', shows Abraham struggling to subdue Isaac by argument rather than by physical force, by emphasising 'the wholly random nature of existence'. Benz interrupts, objecting that if life is vile, death is a liberation. Rather, Benz wanted Abraham to kill his son 'without the benefit of philosophy', to 'make no sense of the deed, but endure the purest pain', for Benz's sake. Abraham kills Benz in rebellion against the insulting imposition of a diminishing imbecility; Abraham then realizes 'now things will be hard...Now we will have only ourselves to blame' (*CP2*, 230), in a Nietzschean cosmology dominated by human will, without recourse to divine entreaty or apologia.

Park's will to perfection significantly provokes the antagonism of both human and divine sensibilities. Barker:

> *On Being Divine* means to overcome the meanness of man. Again, so much social propaganda asserts the common ordinariness of human beings, the solidarity of the frail, as if we were destined to love one another. Park's assumption of divinity is his overcoming of common humanity ... His divinity is repudiation.[21]

In fact, in this respect Park surpasses the peevishly demanding and increasingly jealous Benz, something which Park moreover *asserts* to Benz: 'Because I am not ashamed, and when the unashamed encounter you, the shame reverts to Him who in the beginning, invented it. **There. I dare to address you as an equal.** So I've sinned' (*CP2*, 234). Benz acknowledges that His own lack of morality may lead Park (and others) to consider Him 'imperfect and conclude He is not better than me this God'. Park senses that, without a sense of spiritual authority, Rome will fall; but his personal claim to represent a Divine Love of Rome makes him the target of an attack, in which assassins blind him. Beatrice finds herself with the '**staggering and unwelcome** treasure ... **precious catastrophe**' of a child, but Benz is outraged by her lack of conventionally obedient maternity, or indeed any belief, whilst jealous of anyone's intimacy with her; Beatrice recognizes 'You even...Ache to submit' as '*he kneels at her feet, hanging his head*' (*CP2*, 237).

But, as he cannot submit, Benz feels driven to spoil and hurt. Park takes his ordeal to give him further licence to identify himself with Rome (like Judith in her self-identification with Israel), and Smith testifies to his charisma, and her own equal resolve, by blinding herself to be like him. A Torturer works on Park further, infantilizing him, but even this makes him an object of physical pity to Beatrice in a way which Benz envies further.

The new pope, Lascar, seeks to abolish discrimination between the Roman and the Barbarian to permit reconciliation; his conscious abandonment of principle and taste is careerist expediency, even as (like the Doge in *Scenes from an Execution*) he insists on attempting to absorb everyone, including Beatrice, in his ingratiating totalitarianism. Lascar claims not to be ashamed of any act or thought in his life; but this is accommodation rather than will, a limiting of self and others. Park, like the mutilated but persistently magnetic Helen in *The Bite*, remains an object of devotion even in his wreckage, hoist to the top of a pillar by interpretive acolytes proclaiming 'He has no words / And therefore is no liar'[22]. Park's elevation and Smith's repudiation of His gift of sight makes Benz increasingly vicious in his demands: He rapes Smith against the pillar. The housemaid Beknown displaces Doreen from her social authority, and mocks her vagrancy;

Beknown's 'unashamed' superiority nevertheless prohibits Isaac's questioning, indicating the fragility of her power, like Lascar's. Life under the Barbarians is defined by the crowd as reassuringly similar in its terms of permission, even charmingly novel ('They let me open my shop on Sundays' / 'And fishing in the reservoir was not banned' / **'A few did'** / 'My girlfriend was still beautiful' / **'A few'** / 'Her hairstyle was if anything improved' / **'Individuals'** / 'And new music with a hypnotic beat' / **'Disappeared so what they were such snobs'**, *CP2*, 268).

In this climate of appeasement, Beatrice remains unaccommodating: she appears naked, inviting worship even as she proclaims herself 'incapable of love', having ceased to apologize to others and even to herself, and becoming therefore 'Lonelier and greater' than Benz, even when he tries to shatter her divinity with the death of their child. He acknowledges, 'There is no tenderness in you that I can injure'; she knows, **'There is the fact / And there is the emotion** ... They come apart and in the gap's divinity' (*CP2*, 272). Smith refuses apology in her own way, persisting as Chief Interpreter of Park despite repeated casual abuse, and, as Park suggests, she personally constitutes the ideal of Rome because of rather than in spite of this adversity: Park characterizes Rome as the act and spirit of **'Wanting'**. The Marquis of Dreux-Breze enters the main action, having punctuated the play with a series of entr'acte appearances as a consciously absurd figure whose 'nimbleness' (or sheer outrageous personal sense of definition) has spared him the 'consequences of **enlightenment justice progress and equality'** and permitted him to elude extermination by their agents. Dreux-Breze resembles a comically resilient version of Whatto in *Golgo*, who is tellingly unreconciled to the necessity of his redundancy and doom in a new age. When Benz is distraught by Beatrice's death, it is Dreux-Breze who can recognize 'It's you [Benz], isn't it, who longs to be human?', with cruel laughter breaking through his apparent sympathy. Though Smith admits 'I fear nothing / But my own divinity...', the final tableau at the ocean's edge suggests a fuller self-acceptance; when Benz has pushed Park out into the sea, Smith is left *'perfectly alone'*, both in her spiritual and political authority, and in her personal integrity.

In *Rome*, therefore, Barker characteristically 'does the unfashionable thing', not only in the duration, extent and form of his epic play; he re-characterizes the empire and city as an ideal of 'wanting' which is opposed and continually resistant to a degenerative accommodation to the barbarian regime. Park and Smith are mutilated but enduring antitheses to Lascar's reconciliation. They and Beatrice also oppose Benz's jealous bids to limit the terms of humanity. Thus the resistance to inclusive ideology is fought out on both the social and political level, and on the divine one, with the deliberately and insistently unfashionable and offensive possibility of personal spiritual authority acting as a conduit between the two. Here, as elsewhere in Barker's work, the terms of opposition to totalitarianism – even or especially when neo-liberal and apparently benign in its rhetorical promises – are the expansion of the terms and vocabulary of **what it might be to be human**. Becoming and being 'unforgivable' is the hallmark and method of such opposition. Intimacy, and/or its refusal, is the solvent of prescribed identities. And, in an insistently collective neo-utilitarian society, the shameless achievements of solitude and beauty may create spaces and constitute terms for a truly wider definition of all things human.

Barker's *Ten Dilemmas in the Life of a God (The Incarceration Text)* (written 1991, published 1993, as yet professionally unperformed), unusually for Barker, dramatizes, as the subtitle suggests, an apparent and thoroughly tragic impossibility of breakthrough. Released from prison, Draper discovers his mistress, Becker, ensconced with another man, Sharp. Draper reels between the pain of things disclosed and the fascination with things imagined, between abject and shameless possibilities which he inflicts on himself and others: he publicly acknowledges himself 'Impotent / And always was with her'; 'I exist only to make love to this woman which / I never have yet and .../ May never' (*CP2*, 354)[23].

A flashback recounts Draper's existential determination on crime through his murder of two chimingly contemptuous musicologist acquaintances of Becker's. This infliction begets another, the harrowing separation of his imprisonment, which compounds an intolerable love with intolerable anticipations. The musicologists, like Draper's sisters, want to rescue Becker from 'the maze' of her unhappiness; however, Becker is 'divine' in her rejection of pity and exhortations to share, her discarding of wit and perhaps even in her lack of sexual facility in 'a love that never was made proper ... Whatever that may be'. She has taken up with Sharp, in the knowledge of her 'unforgivable' 'ability...**I won't say appetite**...for continuance' (*CP2*, 367). She is also distinguished by her skill at performance, specifically of the hand gestures she has learnt from classical paintings on religious themes (annunciation, crucifixion, deposition, grief, abandon).

Estranged from all configurations of former life, Draper exhibits his vicious clarity in pain to passers-by, appealing to them hopelessly 'Help me, I must enter her body'. As if in mocking response, he is forced to witness a (real or imagined) scene of Sharp publicly penetrating Becker, yielding in Becker a pleasure which she attributes and dedicates to Draper. Draper reels on, 'sustained by the very thing that murders' him, flayed into a raw restlessness which defies any possibility of accommodation or relief. At times, he can achieve a Satanic repudiation ('**keep your facile copulations**'); at times, he fractures on the conviction: 'If I can't enter her, I can't live' (*CP2*, 388). His compulsively reperformative pain attracts the murderous pity of his sisters and Sharp. When even his Servant tells Draper that Becker's barrenness is because of him, he relinquishes his life. Becker absorbs her grief and outrage in an effort of aesthetic formality which permits her '**A fictional life**', survival and persistence through the *gestus* of performance.

Ten Dilemmas dramatizes how the madness and offence of intolerable love demand that one attempt to discard the fictions of the human and the natural, in favour of imaginative self-recreation, whilst acknowledging how the attempt may fail. It contains some of Barker's most moving expressions of pain at the nausea of existential dislocation through love. The play is unusually accessible, though uncompromising, in that it is immensely cruel in its explorations and expressions of emotions which are nevertheless available and recognizable to many, if not all. Draper and Becker's attempt to fashion a transgressive sexuality *beyond* conventional goals of orgasm and propagation may fall *beneath* those ideals; when Becker claims that a look or an utterance can penetrate and govern the very soul, she inevitably calls forth the contrary figurative possibility that in some instances it can be an imperfect substitute for the phallus. *Ten Dilemmas* is a thoroughly tragic (rather than catastrophist) play in its demonstration that some passions permit neither redemption and relief nor breakthrough, and that incarceration can occur on every conceivable level of being. But, even then, the play and its protagonists drive beyond the essential restrictions of shame and insist '**Don't look away, this is my beauty**' (*CP2*, 380); and suggest, like the other work considered in this chapter, that it is only at the broken boundaries of social inclusivity that one may discover an intimacy and an isolation which can challenge and expand the readily available terms of human existence, action and expression.

4
Cultural Re-Fashionings and Shakespearean Negotiations

Brutopia, Seven Lears, (Uncle) Vanya, Minna

> I found another living underneath my skin. Also me. I found a
> different sheltering inside my bones. Also me.
>
> *Rome* (*CP2*, 262)

This chapter slightly bends my generally chronological approach to consider
as a group Barker's re-visionings of existent dramatic narratives which he
pursued in the late 1980s and early 90s, which make full dramatic flowerings
from the intimation in *Golgo* that 'There are those recorded, and those who
fail to be recorded', and so there is '**Always other testament!**' to be excavated.
The interrogation of received cultural forms, familiar versions and associated
moral wisdoms is an important, and often particularly subversive, strategy
available to all artists, beyond the potentially sterile literary fashion for (identifying) intertextuality. Sarah Kane's knowledge of Barker's work may well
have informed her observation: 'Art isn't about the shock of something new.
It's about arranging the old in such a way that you see it afresh'[1].

I have noted elsewhere, not only how the work of Shakespeare provides
richly fertile ground for such re-evaluative imaginings, but how his work is
itself the most vibrant classical exemplification of this very strategy, and
how we might speak with proper accuracy and seriousness of someone
'Being a Shakespearian Dramatist':

The genius of Shakespeare's drama might aptly be said to reside in the
incompleteness of its *prescriptions*: hence its challenging power and its
infinitely renewing fascination. I would also add that the 'necessary
choice' of the dramatist, like that of the performer, also excludes, but
simultaneously illuminates, others ... It is worth reminding ourselves in
this context that this is how Shakespeare himself usually worked. With
the exceptions of the three apparently 'sourceless' plays (*Love's Labour's
Lost, A Midsummer Night's Dream* and *The Tempest*), every Shakespeare

play is a consciously surprising re-emphasizing re-animation of some pre-existing story or play, and the explosive power of *King Lear* is amplified by its startling final departure from the happy ending of its chrysalis play, *King Leir*.[2]

Barker's own model of practice might be as least as much influenced by classical traditions of painting, wherein numerous artists present their distinctive and distinguishing compositions of and perspectives on subjects, such as the Crucifixion, narratives from the Apocrypha and images from classical epics and myths, as part of a continuum of both *hommage* and making new argument from old matter.

In 1989, The Wrestling School presented the premiere of Barker's *Seven Lears* in a co-production which started out at The Haymarket Theatre in Leicester – appropriately, as this city is reported to be the underground burial site of the historical king, Leir/Lear – before touring to Sheffield Crucible and London's Royal Court. Kenny Ireland's production of *Seven Lears* was spearheaded by Nicholas Le Prevost's whirlingly pained, nimbly agitated, portrayal of Lear. Dermot Hayes's memorably eclectic design delimited the play's chronology with a bricolage of historical references (Renaissance, nineteenth and early twentieth century). The effective casting of Jemma Redgrave (as Clarissa) and Julie-Kate Olivier (as Cordelia) further emphasized the classicism of the project through the application of young representatives of notable English acting dynasties to the disciplines and challenges of Barker's work. At the Royal Court, *Seven Lears* was accompanied by three performances of Le Prevost's intent and uncompromising production of *Golgo*, an adeptly irritating choice of play for the company to premiere at a venue traditionally associated with the conscience and consensus of social realism in English drama and theatre.

1990 saw The Wrestling School's first reassessment of a Barker play: a confident and polished revival of his 1983 play *Victory*, again beginning its national tour at Leicester and headed by a precise and complex performance by Le Prevost as Charles, supported by notable performances from Philip Franks as Scrope and Tricia Kelly as Bradshaw, and an ambitious set featuring a miniature railway track shoehorned into the Haymarket's cramping studio space. However, Ireland's increasingly conciliatory production values were elsewhere noticeably ill at ease with Barker's aesthetic, as demonstrated most unfortunately in a final sentimental tableau of Bradshaw and Ball sheltering beneath an umbrella. This reflected Ireland's belief that Barker's work could be acceptable to a wider public if one softened its edges, but threatened to resolve the play's powerfully disruptive dynamic into a final palliative image of liberal humanist reconciliation. Significantly, Barker himself chose to assume a major active role in the direction of the company's work from this point. In tandem with *Victory*, the company presented rehearsed readings of the as yet unperformed 1987 play *The Europeans*, with Melanie

Jessop lethally precise and poised in the central role of Katrin; this play would subsequently be directed by John O'Brien and myself at the University of Aberystwyth in 1991 and professionally premiered in 1993 in Ireland's uneven production for The Wrestling School, best served by Philip Franks's portrayal of Orphuls. Elsewhere, Geoff Moore's 1990 production of *The Castle* for his company Moving Being in Cardiff St. Stephen's (now The Point) had the resonant setting of a converted church, but the characterization of the female characters in his modern-dress production unequivocally suggested post-Greenham eco-warriors and a liberal simplification of the moral odds and issues, notwithstanding a spiky, attentive, unsentimentalized performance by Ri Richards as Skinner.

But now for closer consideration of four of what Barker calls his 'conversations with dead authors': the re-visioning of classic texts commenced with the 1986 invitation to modernize Thomas Middleton's *Women Beware Women* for the Royal Court, which led Barker to 'redistribute the moral authority within the text, excavate the minor characters and abolish a transparently inauthentic ending'[3] (for full discussion of this project, see Rabey, 1989, pp. 173–83). Barker has acknowledged how his play *Seven Lears* sprang from a curiosity about Shakespeare's play *King Lear*: 'What made me question the absence of Lear's wife from Shakespeare's tragedy was a writer's feeling for the architecture of a text, and we have slowly re-learned that architecture is about emptiness as well as substance, void as well as materiality'[4]. However, *Seven Lears* is importantly pre-dated by *Brutopia*, written in 1989, originally as a two-part television play, commissioned but unproduced (but first professionally staged in a production by Guillaume Dujardin, Nouveau Théâtre de Besançon, 2002). In *Brutopia*, Barker subversively evokes Robert Bolt's popular "classic" of 1960s and 70s liberal English theatre and 'O' level set text, *A Man for All Seasons* (1960). Like John Osborne's *Luther* (1961), Bolt's play is, in some aspects of style and matter, a significantly English response to the international theatrical impact of Bertolt Brecht. In the late 1950s and early 1960s, native commentators such as Peter Brook, Kenneth Tynan, Harold Hobson and Martin Esslin popularized the name of Brecht as a reference point for historically conscious and re-evaluative "committed" theatre, whilst being wary if not dismissive of his political theories and definite objectives. Bolt and Osborne's plays fall short of Brecht's *Verfremdungseffekt*, being ultimately historically situated empathetic dramas of individual temperament: Bolt's Preface testifies to his imaginative attractions to his protagonist, the figure of Sir Thomas More, for his exemplification of an 'adamantine sense of his own self'[5] who ultimately and unshakably trusts in the 'patterned and orderly' promises of the Law, without which society crumbles[6]. This innate, even compulsive, conservativism is some way from the post-catastrophic existentialist pursuit of meaning through activity which Bolt claims to admire in Camus[7]; rather, Bolt's play (like John Fowles's 1969 novel *The*

French Lieutenant's Woman in relation to Sartre, and David Storey's 1970 play *Home* in relation to Beckett) exemplifies how many English authors seek to register and accommodate the potentially unsettling gaunter perspectives of European existential and stylistic propositions, through reducing them to particularly English domestic and humanistic terms. Bolt himself acknowledged that the formal neo-Brechtian alienation effects which did occur in his play were used to deepen the eventual emotional engagement with the protagonist[8].

I identify Bolt's intentions and style in order to specify the conventionally accepted moral terms, and a significantly venerated and culturally promulgated dramatic form, of the English nomination of Thomas More as 'Christian Saint' and 'hero of selfhood'[9], which indicates the weight of invested associations which Barker's subversion may have to overturn. But it is informative to compare the stubborn simplicity of Bolt's dramatic character More with the more complexly dramatized account of the historical figure who emerges from Greenblatt's study in New Historicism, *Renaissance Self-Fashioning: From More to Shakespeare* (1980).

Greenblatt takes Thomas More's ideas as his starting point in the identification of the effects of *'power,* whose quintessential sign is the ability to impose one's fictions upon the world'[10]. This analysis also provides an apt description of various Barker characters, who "project" – geometrically, psychically, politically, strategically, seductively – images of their individual will, frequently in outrageous terms, to shape the inner lives of others and fashion their behaviour. The most literal and expansive – architectural and governmental – examples might include Stucley, Krak and Skinner in *The Castle*; the various and successive ideologues who define Troy in *The Bite of the Night*, and Rome in *Rome*; Riddler in *A Hard Heart*; even Pharoah in the earliest Barker play, *One Afternoon....* Characters who do this on a level of interpersonal strategic combat, in awareness of the wider social ramifications of victory, are legion, but include Lvov, Sleen, Judith, Vanya, Isonzo, Gertrude, Algeria, Sleev. Indeed, this is, in some chosen or imposed sphere and to some subtle or expansive extent, the defining activity of all principal Barker characters. One of the ways in which Greenblatt's characterization of More most profoundly differs from Bolt's, is crystallized in Greenblatt's assertion that More's insights were capable of *'cancelling,* but not *clarifying,* human politics ... by rendering political life essentially *absurd'*[11]. Bolt's traditionally English liberal artistic project is one of simplification and clarification: no wonder *A Man for All Seasons* is at its most theatrically adept on the frequent instances when it plays the tropes of domestic comedy. Greenblatt contrastingly suggests that More's major artistic and philosophical work *Utopia* operates in terms comparable to the distorted skull in Holbein's *The Ambassadors* by introducing a parallel presence – that of the imagined Utopia to England – which is neither entirely separable nor capable of being brought into accord with what normally, principally exists[12]. In this context, Greenblatt posits More as a renowned ironist, who lived his life 'as a character thrust into a play, constantly renewing oneself extemporaneously and forever aware of one's own unreality – such was More's condition, such, one might say, his project'.[13]

So far, so Barkerian. This vision/version of More also partakes of Baudrillard's seducer, who has the power to make her/his signs flicker, to be simultaneously there/not there[14]. However, More's *Utopia* does not, cannot, sustain this taunting defiance of reconcilability: Lawrence Stone argues, *Utopia* gives idealized expression to the 'rise of the nuclear family in modern England' and the decline of other, competing affective bonds: 'Where Plato's ideal involved the destruction of the family, that of More involved the destruction of all other social units'[15]; in More's *Utopia*, 'family strategies are entirely subsumed under state strategies'[16], with the society resembling a giant panopticon of mutual surveillance, constructed so that all involved are regulated by being under constant observation. More's vision of society operates through the citizens' (and readers') internalization of the ethos of shame to enforce conformity; and 'in emphasizing shame rather than guilt as a social force', *Utopia* would diminish the possibility of the autonomy promised by Protestantism 'by reducing the inner life and strengthening communal consciousness'[17]. It is worth remembering here that the conscience of which More speaks in his writing (and Bolt in his play) is fundamentally his participation in the visible manifestations of the power structures of Catholicism. Whilst transgression is not entirely eliminated in More's *Utopia*, 'the coercive power of public opinion – the collective judgment of the community, perceived as objective, external fact – diminishes the logical necessity for a mechanism of social control' by operating within the inner recesses of individual consciousness[18].

It will be evident how Barker's instinct and appetite for exposing the impulses, strategic fabrications and mechanisms of professed Utopias would lead him to engage with More, the English literary and historical figure centrally associated with both the ideal and the very name/word. However, Barker's play *Brutopia* effectively divides the self-consciously performative, politically cancelling, Nietzsche-anticipating More of Greenblatt's characterization from the determinedly witty but ultimately patriarchal, ideologically "clarifying" (that is, control-oriented) More who emerges from *Utopia*. Barker renders the first, strongly nihilistic, energy in the dramatic form of More's less favoured daughter, Cecilia, who expresses her alienation through her secret composition of a counter-text to her father's central work. She develops the vision of her imaginary alternative:

> In Brutopia there was neither lie nor truth. Everyone believed everything
> ... And the Brutopians, believing everything, sometimes laughed and
> sometimes wept at the same spectacle. They were, luckily, bereft of

> tenderness. This made them perfect citizens. In Brutopia, you cannot be
> unkind. So much hypocrisy is spared by this! (*CP2*, 132–3)

Here and elsewhere, in their liberation from humanist stereotypes and
conventional configurations of pity, Cecilia's Brutopians sound very close
to the ideal Barker theatre audience. At other times, their actions expose
the totalitarian subtext of liberal humanism: 'The most common word in
Brutopia is **we** ... This produces such a climate of mutual celebration ...
until your ears are singing!' (*CP2*, 156).

The subtitle of *Brutopia*, *Secret Life in Old Chelsea*, appropriately captures
this subtextual and subterranean aspect of Barker's play, which reads like
an excavation into the depths and corners of the theatrical terrain of Bolt's
project of clarification and enlightenment. The relationships between characters are mercilessly crystallized and expressed: Henry says to More, 'You
don't like girls, do you?'; More tells Alice that the function of all wives is
'to fear'; the civilized garden banter and courtly witticisms of Henry and
More are literally grounded upon, and excited by, the imprisonment of a
heretic, a simultaneously demonized and domesticated 'other' whom they
interpret, goad and occasionally indulge. But, as Charles Lamb observes,
More (and even Cecilia) replicate a situation common to other Barker plays
in which 'the author of a world ... becomes embroiled by contradictions
within his own creation'[19]. A garden maze permits Cecilia's first encounter
with a Servant, who tutors her in the philosophical intimations associated
with Brutopia, and permits her first glimpse of these imaginary beings, in a
splendid spatial demonstration of the plasticity and reversibility of seductive landscapes (of both maze and stage). The Servant explains how the
poor of this alternatively-perspective world periodically erupt and kill not
only the rich, but each other, their only pleasure and one which the rich
encourage: 'It assures them they are correct. Because in Brutopia nothing is
seen to be good unless it is opposed' (*CP2*, 138). More correspondingly
encounters a Doctor who claims to come from Utopia ('If a thing is imagined, it is born ...Utopia is all consequences', *CP2*, 142), and a Common
Man, who insists on the abolition of all privacy (an extension of the
'nothing to hide, nothing to fear' surveillance ethos which has even more
immediacy in the twenty-first century). Cecilia characterizes Brutopia in
terms diametrically opposed to such neo-utilitarian functional transparency, in ways which identify her as one of Barker's (unusually literally)
hopeless romantics: 'In Brutopia, nothing is what it seems to be. This is
universal and a source of comfort. Where nothing is expected, disappointment is unknown, and hope entirely redundant' (*CP2*, 152).

One of the major issues in *Brutopia* is the existential choice, refinement
and performance of the self. When More is aghast that he is fated to die
because he is determined to be honest, Cecilia replies 'Be dishonest, then'.
The Servant is similarly forthright, insisting that if More is 'sick' with the

self he has created, that he should create another; but More is adamant that no such choice exists. Cecilia's hopelessness is a means of imaginative atonement with the salutary imminence of death, as when she hauntingly offers 'Come and practise death with me' (*CP2*, 175). In this sense she is representative of many Barker characters and plays in that she offers a desentimentalizing opportunity to 'practise death' imaginatively: a strange, unconventional and therefore barely recognizable kind-ness in the face of an experience which is perhaps, after all, our ultimate but least spoken feature of kinship. She blocks More's pardon, realizing that he needs to make his investment in his chosen apotheosis of death; so does the painter Holbein, who observes More to be no longer 'swollen with power', instead 'swollen with the absence of power'. When Cecilia purposefully forces the manuscript 'Brutopia' into printed book form, she demonstrates an existential commitment to chance, or fate, by throwing copies over a wall, trusting to chance as the only force by which 'one copy, and only one perhaps, might reach its loved one and enrich his life'; thus the book crystallizes her existential project: 'I know what kindness is. It is something done to the self. But when this self is made, give it to others' (*CP2*, 193). Cecilia is finally saved from institutionalized (and institutionalizing) wrath by Henry, one of Barker's disarmingly frank autocrats (compare Charles in *Victory*, Leopold in *The Europeans*) whose personality is built on such a riot of impulses that it defies the stability intrinsic to absolute tyranny. Finally, he saves Cecilia from the madhouse, but recognizes, not without compassion or self-awareness, why his courtiers suggest that 'I am the madhouse'.

Brutopia demonstrates a philosophical brilliance and a memorable sense of surprisingly disclosing dramatic space, and is a purposeful subversion of both Bolt's *A Man for All Seasons* and the aesthetically related conventions of the television historical costume drama against which it was originally conceived to play. Even with these strengths, and the incisive exchanges quoted above, it is not a major Barker play, partly because it replays some of the features of *Victory*: an overshadowed female relative of a philosophical literary genius embarks on a quest of self-determination, in demonstration of both spiritual loyalty and desecrating autonomy, exulting in the splintering of former duties but tenuously committing to the future via the written word, unromantically impregnated but saved from imprisonment on the respectful whim of a riotously physical monarch. Thus, it is not a *dramatic* breakthrough for Barker, but contains *philosophical* refinements which will inform and prepare for further advances, and I have dwelt on its context and details in order to identify its status as the first step of an enquiry; and the logical progression from this to Barker's next re-examination of self-fashioning and renegotiation of a cultural 'creation myth' core text on identity and autonomy: no less than Shakespeare's *King Lear*.

King Lear might be identified as Shakespeare's ultimate step in what Greenblatt suggests is that dramatist's 'principal concern': 'the representation of a self-undermining authority'[20]. This description in itself indicates the potential pertinence of the figure of Lear to Barker's theatrical enquiries; as noted, Barker's drama contains numerous self-consciously unstable monarchs. Moreover, the dramatization of a 'self-undermining authority' is central, not only to Barker's principal concerns, but also to his dramatic method, in which characters articulate one impulse or aspect of themselves brilliantly only in order to undercut their own performance from the perspective of another impulse or aspect of themselves. Similarly, *King Lear* presents its theatre audience with experiences of *indefinition* and *delimitation* which are recurrent objectives in Barker's theatre. Greenblatt notes how *King Lear* works 'to unsettle all official lines', 'empties out the centre that it represents and in its cruelty ... paradoxically creates in us the intimation of a fullness that we can savour only in the conviction of its irremediable loss: "we that are so young / Shall never see so much nor live so long"'[21]. Stephen Booth observes how *King Lear* contains and offers a normally unimaginable degree of uncontainability, compulsively shattering dramatic, generic and human forms and hopes of definite confinement[22]: these are also distinctly Barkerian effects (and I have previously identified Booth's terms as particularly applicable to Barker's *The Castle*[23]).

It is not surprising that Barker is drawn towards sustained exploration and study of Shakespeare's (and, according to literary consensus, any dramatist's) most catastrophic tragedy. As with *Brutopia*, Barker begins his enquiries, in an epigrammatic Introduction to *Seven Lears*, with speculation about a pregnantly significant (significantly pregnant?) female presence, not just marginalized but absent: Lear's wife: 'This extinction can only be interpreted as repression. / She was therefore the subject of an unjust hatred. / This hatred was shared by Lear and all his daughters. / This hatred, while unjust, may have been necessary'. This, of course, raises the question, 'Necessary for whom?'. This starting point might lead a naïve reader or audience member to expect an ideologically feminist and anti-patriarchal re-visioning of Shakespeare's play, but Barker offers something less inherently predictable.

From the outset of *Seven Lears: The Pursuit of the Good*, Prince Lear is distinguished from his two brothers by his (heroic? ostentatious? inordinate? unenviable?) sensitivity. Just as the refinement and civilization of More's garden in *Brutopia* literally rested upon the prison of the heretic, the kingdom of Lear's father is founded on a prison, containing 'the dead who aren't dead yet'. The princes Lud and Arthur see the gaol as confirmation of the prisoners' abject evil, and of their own separate(d) good; Lear is contrastingly haunted by the intimation that 'something bad is happening here' (rather than *contained* here), feels pity for the prisoners in their deprivation, and concludes 'I shan't be king, because I am not the eldest, but...if I were king ... I'd stop this!'. Initially, Lear may thus appear to the audience as a prospective reforming liberal humanist. He pursues his musings – 'The function of all government must be ... The definition of, and subsequent encouragement, of goodness, surely?' – sounding rather like a precocious student of Thomas More. But then he is *'inspired'* to associate goodness with the distinctive exercise of the will: 'You make goodness difficult, if anything. You make it apparently impossible to achieve! It then becomes compelling, it becomes a victory, rather as acts of badness seem a triumph now!' (*SL*, 2). When his brothers commit suicide on a whim, the Bishop is appointed as Lear's (particularly Barkerian) educator, with prime responsibility for eradicating sentimentality and instilling a clear-eyed hopelessness. The structure of the play follows Lear through the discarding and making of separate 'Lears', selves which represent stations on his journey of self-fashioning, whilst the Bishop refocuses Lear's attention on the gaol, and expounds on the insignificance of truth and the arbitrariness of power, which underlies the façade of Lear's humanism and permits his deliberations ('If some are to be free, others must be unfree, or they could not know freedom', *SL*, 7).

Entering adolescence, Lear finds himself simultaneously 'barmy for a skinny girl' – Clarissa – 'and infatuated with her mother' – Prudentia. He experiences success in imposing his fictions on others notwithstanding, or even because of, their outrageousness, which overcomes resistance in a way which conventional logic would not. Thus, Prudentia is convinced to permit Lear's association with her daughter (as 'yet another manifestation' of their intimacy); the studious minister Horbling is appointed Fool (and will be driven to less than hilarious proclamations of humour, 'the grating of impertinence upon catastrophe' and 'the consolation of impotence'); and Clarissa is persuaded to acquiesce to Lear, not despite, but because of the 'pain', 'grinding' and 'punishment' which their union will entail.

'Third Lear' shows Lear, now monarch, in battle, responsible for great loss of life, shaken and vacillating ('Time to unlock the gaol! Or maybe not!') until Clarissa rallies reinforcements. However, whereas Clarissa repeatedly invokes the 'good', Lear knows pragmatic objectives such as victory are more complex; his inquisitive sense of near-infinite possibility may be denigrated as childlike, but it also lends him the power of an unbroken will (a recurrent Barkerian strength; compare Isonzo's successfully compelling description of himself as 'an inquisitive child' at the end of *The Twelfth Battle*). Lear disarmingly excuses himself: 'if I am a child, it is because a child must know. Its ravings are the protests of the uninstructed. It thinks the sky is a false barrier, and the floor, pretence. It raves less, year by year, as the barriers are demonstrated to be real, and insurmountable' (*SL*, 17). However, he significantly succeeds in armed combat with a mutineer by appealing 'to all that was good in him', which 'weakened' him by making him 'anybody's fool' (*SL*, 21). His very successes (such as his seduction of Prudentia to the utterance 'truth's a thing you can grow out of') make him recoil (**'What is to be done with me? I think I am evil'**), but accelerate rather than slow his appalling insights ('The nature of beauty, as of goodness, rests in its power to substantiate the self...Which is not goodness at all, is it?', *SL*, 23–4).

'Fourth Lear' pits Lear's debates about the arguable necessity of deaths (with Kent, the Chorus of the Gaoled and the lascivious vagrant who is made Earl of Gloucester) against Lear's consciously provocative experiments with mechanical flight (a passion of the original King Leir reported in chronicles but omitted from Shakespeare's play). In an interlude, Kent and Clarissa kill the Bishop, on the charge that 'Any thought that Lear produced, this man legitimized', though the Bishop manages a resonant retort against these homicidal idealists: **'Kent you have no interior sight and you pass that off as goodness'** (*SL*, 32). 'Fifth Lear' shows Lear in retreat from the world, pursuing sainthood, whilst his impoverished servants fly kites around his tower for every child and suicide who dies: Clarissa, self-proclaimed agent of conscience, arraigns her mother Prudentia for her complicity with Lear, and insists on her death (The Gaol acting as a wryly approving chorus of moral consensus, like that in *Golgo*). 'Sixth Lear' emphasises the rift between Lear and the joined forces of Clarissa and Kent: Lear tries to drown their baby, Cordelia, but even their opposition is momentarily arrested by the appearance of a foreign power, headed by the Emperor of Endlessly Expanding Territory: a religious as well as political potentate, who expresses a challengingly different form of ethical and metaphysical conviction: that 'there are not enough bodies in the world', and that the solution to the weariness of false knowledge lies 'after death'. This is one of Barker's most resonant subversions of Shakespeare's *King Lear*, which after all depends for both its ultimate horror and its residual relief on the conviction that there is no afterlife. The Emperor (a particularly chilling incarnation of all murderous forms of religious fundamentalism) challenges the limits of material definition and the primacy of life. Even as the reeling and bewildered Lear retrieves Cordelia from death, Clarissa indicates the surrounding emotional chaos to the Emperor, 'You see how terrible we find life? How it maddens us? Don't offer us another' (*SL*, 46). Later, Clarissa introduces Cordelia and her sisters to the presence of the Gaol, unforgivably; Lear now calls this place his 'garden', 'the one place I can discover sanity'. Even Clarissa's children cannot summon pity for her in her unassailable attempted goodness, nor do they prevent her extermination, though Lear and all his daughters sense that Cordelia will be the one to correct Lear when his wife is gone. By the final 'Seventh Lear', the Chorus of the Gaol have also proved unforgivable, and the former flowers of Lear's 'garden' lie heaped and dead, in the wry knowledge **'We knew How else could we be free? But knowing How could we be allowed to live?'** (*SL*, 49).

The questions raised by *Seven Lears* are reverberative: to what extent can one fit 'the hand of intelligence into the glove of government'? Is goodness inevitably and necessarily politically unforgivable? Does an acuteness of sensibility mitigate or increase impulses to tyranny? In The Wrestling School premiere production, the visual imagery was also startlingly realized, particularly the seething chorus of the Gaol (finally hanged, suspended from the theatre flies, above Lear and Kent's chess game); Lear's ludicrous yet grandiose flying machine, the unison interjections of the unborn wraiths of Regan and Goneril clamouring for their conception[24]; the strange pathos of the blindfolded figure of the imprisoned Prudentia surrounded increasingly by variegated kites, nodding on wires[25]; the Emperor, imposing, stooped and faceless, festooned in black robes and wraps and supported by twin walking sticks. *Seven Lears* also offers some witty speculations about the obscure preconditions to Shakespeare's play: not only the nature and identity of Lear's wife, but reasons for Cordelia's compulsive dissociation from her sisters and difference in attitude to Lear; and for Kent's sustained watchfulness, Gloucester's libertinism, the Fool's tolerated status as irritant and his repeated failure to amuse. However, it is often the aporias and enigmas of *King Lear* (and other tragedies) which imbue it with enigmatic fascination and infinite imaginative resonance, and Barker's 'prequel' has the odd, and particularly uncharacteristic, effect of tending to adumbrate and apparently forensically rationalize another, ultimately even more impressive, play, seeking to mitigate Shakespeare's confrontation of *indefinition* with effects of *comprehension*: *Seven Lears* can seem to be trying to account for the explosions of *King Lear*, rather than open them up further, which may be why the most vivid stage images listed above are all Barker's inventions entirely independent of the ur-text.

As Barker observed in 1994 about his imaginative reconsiderations of 'voids and absences':

> Sometimes these resonant spaces, when investigated, have led me to an interrogation of the function of the text – and the cultural status of the author – in contemporary society, an elision which is perhaps inevitable in confronting authors who are now much more significant than the sum total of their works. I am here thinking not so much of Shakespeare, now a negligible influence on the tone of contemporary writing in Britain – itself a tragedy for the theatre of our time – as Chekhov, whose uncontested authority in British theatrical and cultural circles has made of him a more luminous icon in this part of Europe even than in his country of origin[26].

I have similarly observed in my critical history *English Drama Since 1940* (2003) how, in British theatre of the 1980s and 90s, 'Chekhov's sad humanist comedies of resignation to "inevitabilities" became second only to Shakespeare as favourite theatrical choices for classical revival'[27]. Barker argues in his 'Notes on the Necessity for a Version of Chekhov's *Uncle Vanya*', which preface his own *(Uncle) Vanya*: 'We love Vanya, but it is a love born of contempt. It is Chekhov's bad faith to induce in his audience an adoration of the broken will. In this he invites us to collude in our own despair'; 'It is necessary therefore to demonstrate the existence of will in a world where will is relegated to the comic and the inept' (*CP2*, 292).

Barker's *(Uncle) Vanya* begins with a savagely comic condensation of Chekhov's world (characterized by the pain and 'odour of deep-seated harmlessness', **'not too much pain however not tragedy'**), developing the technique initially visible in *Brutopia* where characters perform extreme versions of their involution or else express their (conventionally sub-textual) intimations with extravagant ruthlessness. The first half of Barker's *Women Beware Women* presents an exercise in highly skilful, but idiomatically faithful, compression before Middleton's characters erupt from repression into the poetic and sexual relief expressed by Barker's language. Contrastingly, in *(Uncle) Vanya* Barker reduces (in a culinary sense, perhaps comparable only to Heiner Müller's drama) the attitudes of Chekhov's characters to neurotic essences of exasperation, monotony and reflex (for example, Astrov's morally superior testimonies of deference to an ecological hereafter, Telyeghin's futile guitar strumming, Sonya's refuge from personal pain through parroting the clichés of class-based historical analysis insisting on the imaginative constraints of 'economic impotence'). Only Vanya has the painful emotional and imaginative mobility to step outside the frame of the action, to express his agonized passions to the audience, and even though he can pause and comment on the action of the play, he seems powerless to alter its course. At this point of the play, as Cornforth observes, whilst Vanya 'rarely, if ever, speaks alone on stage, much of what he does say is apparently not heard or ignored by other characters', hence, 'his relationship with the audience becomes at least as important as that which exists with the other characters, if not more so'[28]. In his articulate despair, Vanya plays the role of the savagely comic unreliable Chorus, inviting audience judgment on other characters and his own motives whilst artfully challenging their terms of definition and reaches of imagination ('Do you agree the professor is vile or do I slander him am I correct or extreme **He goes to bed with such a young woman** am I poisoned by sexual jealousy or **He puts his flesh inside her body** or or or', *CP2*, 299). Thus, Barker's Vanya combines from the outset the theatrical qualities of both Whatto and the Chorus in *Golgo*, with a startling speed and reach of performance of precise but thwarted (and precisely thwarted) emotions. Even the object of his adoration Helena tells (taunts?) him 'You should stop grumbling and reconcile people to one another', and his beloved niece Sonya drones the mantras of conventional futile hope, that is, suspension in anaesthesia: 'Uncle Vanya will be happy one day, won't you Uncle', whilst Vanya struggles to insist to his uncomprehending onstage audience **'Humans sin. It is the essence of their beauty'**(*CP2*, 303). However, Barker's Vanya successfully shoots Serebryakov (whereas Chekhov's Vanya repeatedly misses), and is thereby committed to a life outside conventional morality, renaming himself 'Ivan' instead of the diminutive endearment, and promptly commanding Helena 'Get undressed'. She readily commences, but Vanya stops her, possibly indicating that even he is as yet somewhat nonplussed and uncertain where to go after his drastic action, lacking, and/or intimidated by, her confidence in sexual directness. Even Helena, his nearest equal in this new-found imaginative release and reach, expects him to turn himself in to the police. However, Vanya is defiantly, exhileratingly unrepentant, even towards the posthumous choric spectre of his victim: he repudiates Chekhov's apparent 'regret' of the 'melodramatic interlude' which disturbs 'the fragile balance of characters': '**Too late too bad too everything**' (*CP2*, 308).

Helena proceeds to proclaim her own forms of sexual 'ache' and 'need', including her appetite for the respect inherent in the sexual command (which Astrov cannot formulate). If 'Vanya's journey through the play could be said to be an exploration of the self which is conceived in desire'[29], so too is Helena's, and she is apparently more practised. Vanya discovers that Helena can express a hunger for confrontation and disclosure, and ferocity of dignity and demand, that at times exceeds his own, dislocating him from the certainty of his own presumption of imaginative superiority and initial somewhat adolescent vengefulness (under the terms of which Helena sometimes seems more of a trophy, or accoutrement, of male power), into the renegotiative dangers of love. Vanya seems prepared to break from the terrible force of inconsequence which dominates the Chekhovian world, a confined space where 'the melancholy of unlived life is exquisitely redeemed' in an inertia which surrenders freedom for a sensed gratification of the need to be 'forgiven' for the crimes of 'self-murder', 'self-betrayal' and 'self-disgust'. These are the terms which constitute its supposed "humanism", but Vanya crucially and defiantly prefers to accept consequences of his action, including the forfeit of sympathy:

> **I don't require sympathy tell him.** It is possible I am not human. I was comic and now I am inhuman. The comic, the pathetic, the impotence, made me lovable, but underneath I was not human. And nor is anyone. Underneath, Human. **Tell Chekhov!** (*CP2*, 309)

As I have commented elsewhere, 'Barker's Vanya reveals that Chekhov's play is only nominally play: it is more accurately an essay in predetermination [and] limitation'; whereas Vanya bursts the walls of Chekhov's play by pursuing the unforgivable, in the more dangerous play of shameless performance which disrupts the containment of conventional morality[30]. The usurping of Chekhovian melancholic inconsequence and comic resentment ('Pain because we do not act ... And civility because we do not act') extends to Sonya's abruptly active, explicit sexual pursuit of Astrov.

However, whilst Vanya has exposed the limitations of Helena's definition as a wife to Serebryakov, she has exposed his limitations as a lover. When Vanya uses all his powers of eloquence to recharacterize his experience of impotence as an apotheosis, Helena's mute but 'uncharitable' squeeze of his hand undermines his linguistic self-elevation, by reminding him, and the audience, of her wants, and the event's shortcomings. This seems to provide a cue for Astrov, who seeks to make his own phallocentric inscription on Helena (exposing Astrov's "new man" posture as the disingenuous bid for ingratiation that this posture often conceals). Helena's degree of co-operation in the ensuing offstage copulation with Astrov is never precisely specified (it raises the possibilities that she may have demonstrated curiosity and he may have seized control and dominated, or that the encounter may have been outright rape). Whatever the case, the envy of Maryia and Sonya impels them to blame Helena for the incident as much as, or more than, Astrov; Helena maintains her dignity against their spite, and Sonya throttles Astrov. Vanya unwisely tries to pass off Helena's experience with Astrov as 'meaningless' ('It's nothing to me') and then as further grounds for his immature idealisation of her (in terms of 'intangible perfection' and 'innocence'); he then finds the emotional intelligence to take a new initiative – 'Describe it to me' – insisting on their imaginative ability to confront, reclaim and enhance contingency (which may even extend to her consequent pregnancy), a further step towards their being 'artificial', the creation of their own wills, 'possibly entirely false' and yet 'quite unashamed'. However, Telyeghin profoundly regrets his spontaneity, and the ghost of Astrov proclaims the supposedly consoling platitude that whole hearts can flow 'only in despair', joining the ghost of Serebryakov to form twin Eumenides of circumscriptive mediocrity, and preparing the way for Chekhov's direct re-entry into the play as a character and force seeking to reassert control. Chekhov maintains that the characters' attempted revolution was only an adolescent rebellion, 'A mutiny is merely the affirmation of things after all'; 'Clearly, I am necessary. In me, there lies a terrible significance' (*CP2*, 327).

In the confrontation of Act Two, Helena initially defies Chekhov more resourcefully and effectively than the sobbing or merely petulant Vanya; she elaborates wilfully on her sexual achievements with Vanya in order to encourage him with her commitment, defy Chekhov's irony and withstand Maryia's envious scorn. The dynamic shifts when Chekhov makes a disarming admission to Vanya, how he abandoned the dream of telling himself, pouring himself 'like a liquid from a jug into the void of another', and discovered a freedom in the realization that 'We are entirely untransferable' (*CP2*, 334). Vanya may be struck by this suggestion, that one must ultimately acknowledge limits to mutuality (one can appreciate, even love, others in and for their separateness, but any attempt to escape into them, or even know them fully, must be doomed to failure), even as he releases

Chekhov's hand in rejection of the dramatist's wish for empty gestures of complicity. But when Chekhov dies, the other characters seem to find it necessary to reinvent and incorporate him in their habitual reflexes (such as Maryia's 'I think we all need a cup of tea') and fears of freedom; even Sonya backslides into fearful determinism and resignation to despair. Helena's threatening demand, 'If you betray me, I will kill you', suggests her own fissure of fear; Vanya may be seeking departure from the Chekhovian world, but he has said or done nothing to suggest that he is contemplating deserting her in the process; she seems to feel that acknowledgment of their separateness would be a betrayal of their love, specifically of her commitment.

Helena's fearful insecurities about the loss of her general desirability become more apparent in Act Three: her opening speech to the mirror reveals her reliance on an admiring audience in order to define and confirm herself (if she is indeed pregnant, her insistence 'I'm thinner' and her identification of power with her own 'thin but pliant' beauty may suggest a further insecurity, that maternity will irrevocably destroy her essentially important statuesque attractiveness). Vanya is silent in response – indeed, her increasingly baroque monologue may partly be intended to deny further comment – as she identifies her body as 'the end of thought the terminus of rationality and instinct both ... **I am the point and purpose of the world'** (*CP2*, 336). Having sensed Serebryakov's resistance of this possibility, she now looks to Vanya for neo-religious adoration; formerly the unsentimental instigator of imaginative initiatives, Helena may be wishing to turn away from further (shared and individual) exploration by escaping into a smothering, oblivious mutuality; she is now afraid of the wind, the world outside the ruins of the house, and of any possibility beyond this desperate clinging to the 'undemocratic' privilege of her beauty. In the period between Acts Two and Three, Helena and Vanya may even have achieved conventional sexual satisfaction, but this seems to have increased rather than quelled Vanya's restlessness as he rebels against the all-important ultimate end-stop that "climax" is supposed to represent: 'Is this happiness? It doesn't feel like happiness, unless happiness is fog ... **Who wants fog, not me!'** (*CP2*, 337–8). For Vanya, the narcotic of happiness is no substitute for passion, the pain of wanting, or desire, which resists definition (Cornforth suggests that Helena has here lost her sense of desire, and so 'ceases to be, for Vanya, desirable'[31]). Vanya tells her that he is afraid only of her – correctly, when she represented a challenge to further action and appetite – but that now their love is 'rinsing the life out of' him; significantly she proposes sex as a possible means to drown his restlessness and her fear (using a phrase, 'Make love now', which Cornforth identifies as being drawn from 'the lexicon of kitsch'[32]), and he refuses, striving to remember the serial number of the gun which, for him, objectifies desire.

Helena's instigation of the suicide pact is resonantly problematic, not least in its permission of more than one interpretation, so first I present the readings which informed the production of the play directed by Andy Cornforth and myself (at the Emlyn Williams Theatre, Theatr Clwyd, November 1994, and Theatr y Castell, Aberystwyth, January 1995), in which I played Vanya[33]; and also the comments made in a letter from Barker (10 May 1995), disagreeing with aspects of our reading. *Cornforth*:

> Though it is Helena who initiates the sequence, we have a sense that she does not truly want, or even expect, Vanya to fire, or at least that once she has entered into the pact she is bound by its unwritten rules ... Whilst Vanya accepts the inevitability of indefinition in life, Helena does not ... In effect, Helena seeks a reinscription of the self: she may be thrown in the air, but she must always be caught. The imperfections and indefinitions of life are to Helena 'hideous', yet to Vanya they are that which makes life livable ...He shoots Helena, partly because the rules of the game demand it, and partly to free both Helena and himself from the Chekhovian domesticity which is contracting around them, [an assertion of will in] an environment once again becoming evacuated of will.[34]

Rabey: Helena proposes the suicide pact as an ostentatious gesture of defiance which is, more fundamentally, the last refuge of an impotent idealism. She prefers escape from the demands of will, and admiration as a beautiful corpse, over the risk, untidiness and discovery of more life. Whilst Vanya is mesmerised by her into an intoxicated complicity in the pact, he cannot shed the sense that this is regressive, that they 'did succeed' in eluding the Chekhovian world and might again, but she speaks incessantly to keep from listening, and he shoots her in a mixture of confusion and anger.

Barker: 'The suicide pact is not an event of despair, caused by the futility of further mutuality, but a necessity given they have exhausted the potential for life itself ... they have lived and died in each other ... and the [joint] suicide would have been a triumph of the will and a repudiation of the Chekhovian'[35].

Cornforth:

> [Vanya's] desire takes him beyond ... definition. He does not shoot himself because it is a role that has been created for him, rather than a role he has created for himself ... He effectively has the choice to either shoot himself, which is ultimately to conform, or to transgress the boundaries of the play itself ... Ultimately, he is not inscribed within the terms of kitsch, for although Sonya and Maryia say 'He'll be back', his continued absence as the play closes denies the epitaph its validity.

The final line of the play foregrounds the fact of Vanya's disappearance rather than any assumption that he may return.[36]

Rabey: Vanya seems ready to accept the judgment and death sentence issued by the chorus of remaining characters, now a mixture of dead and living ghosts, but at the fateful moment he chooses not to yield, not even to his grief at the wasteful loss of Helena. He chooses to live further (like the prospective suicide in section four of Barker's poem *Gary Upright*), in instinctive imaginative defiance of the limitations which all the other characters have enviously sought to place upon him. Struggling with shock and grief, he nevertheless identifies a responsibility towards his own potential and towards something which Helena once represented and offered but no longer can: a necessity for continued existence. He pivots on the hinge of confusion as to his destination (**'Where am I going'**), and repudiation of his current situation (**'Where am I'**), to discover the *'effort of will'* which takes him out of the room.

Barker: 'Vanya's "Damn, missed..." is a failure, a cowardice, not matched to Helena's bravery. He has backslid. But the option of backsliding to chess or guitar-playing is to repudiate all he has achieved. Thus he departs into the night with absolutely no destination'[37].

Thus Barker views Helena as a figure comparable to Draper in *Ten Dilemmas*, the tragic heroine of a thoroughly tragic play, whereas Cornforth and I prefer to propose Vanya as a catastrophist hero of a thoroughly catastrophist play. Whichever perspective the reader or audience member prefers, *(Uncle) Vanya* offers them an experience in which the dramatic and theatrical conventions of Chekhovian naturalism are incisively exposed and vividly dramatised as a stylistic signifier for an imaginative (national British rather than Russian?) ideology of systematic self-abnegation and limitation, even the dignifying of depression, through a formal reinscription and neo-religious veneration of 'the melancholy of unlived life ... exquisitely redeemed' by the collective 'forgiveness' of internalised 'self-murder', 'self-betrayal' and 'self-disgust'. *(Uncle) Vanya* dramatizes and demonstrates the possibility of broaching the definitions inherent in Chekhovian form and, analogously, its attendant enshrinement of social and personal determinism in the theatres of both national culture and personal consciousness, instilling the sense that possibilities are intrinsically, inevitably, and perhaps fortunately, limited. It does so with a combination of ferocious humour, a fully articulated spectrum of passionate emotional reach and an ultimate extension of the promise of indefinition which is unique even in Barker's canon, making it one of his most thrilling, ingenious and moving plays.

The manifestation of delimitation in this play importantly extends to the space and setting of the play, the plasticity of which reflects the human upheaval in terms of spatial poetics, for example in the fragmentation of the house and the incursion of the sea, whilst replacing the stage

conventions of theatrical naturalism and realism (which are explicitly identified with conformity in objectified imaginative defeat) with a painterly
sense of composition for the stage which includes unpredictable effects of
rupture and transformation.

Our 1995 Aberystwyth production staged the seismic disintegrations of
the Chekhovian house through the use of a revolving stage; Chris House's
set design incorporated a back wall of gauze, with a centrally situated door
through which Vanya could exit to the garden, and from which the ghosts
could enter. With each assault upon the architectural structure, the set
shifted under the characters' feet and revolved several degrees, the incur-
sive sea being represented by a simple wash of green and blue light. By Act
Three, the set had revolved 180 degrees, so that Helena and Vanya's dialogue occurred behind the gauze (lit to be transparent for the audience,
opaque for the performers) with the former doorway standing in for the
mirror before which Helena poses and speaks. Vanya finally exited the set
through the "doorway" of the mirror and out through the central aisle of
the auditorium; the curtain call was not taken (the remaining characters
stayed in position on the set, though Telyeghin began a casual but discernible fascination with the abandoned gun as the audience vacated the
auditorium).

Barker's own 1996 production for The Wrestling School was differently,
startlingly, scenically inventive: the action was prefaced by the first of
Barker's distinctive theatrical *exordia* – an exordium literally meaning the
introductory part of a discourse or composition, here specifically a wordless
(but often sonically intensive) masque-like spectacle or performed activity
which creates an atmosphere, alludes to themes of the play and effects the
transition of the audience, from the conventions of daily life associated
with the foyer, into the world and space of the play, which they enter on
the company's terms. In Barker's production of *(Uncle) Vanya*, the servant
repeatedly scaled spiral staircases and dropped a clamorous succession of
tin trays and metal crockery to the floor, shocking the sleeping Astrov into
consciousness. The scenographic design, by Robin Don from a concept by
Barker, set the action on board a ship, the rusting metal hulk of which fractured open to reveal a seascape, suggesting the course of the play to be an
escape from a prison, disrupted by the elemental forces commanded
(Prospero-like?) by Chekhov when the onboard action strayed "off course".

This radically dynamic scenography, the incorporation of almost balletic
detailed stage movement and the use of disclosures of perspective and
space (perhaps initiated dramatically by sense of the imminence of aper-
tures to the world of the Brutopians) develops in *Minna* (written 1993, still
professionally unperformed), and further on in Barker's *Found in the Ground*
and *The Ecstatic Bible*. *Minna* is Barker's renegotiation of Gotthold Lessing's
1767 play, *Minna von Barnhelm*, promoted as 'the first major comedy in the
German language'[38]. Lessing's play, set at the end of the Seven Years War,

depicts a discharged, impoverished and ashamed officer, Tellheim, who refuses to marry his betrothed because she is an heiress, until she artfully tricks him into admitting his romantic feelings and letting them overwhelm his spasm of antiquated honour. From the outset of his play, Barker invests the post-war landscape with an unromantic Goyaesque detail (comparable to that of the Second World War landscape of his 1985 play *The Power of the Dog*): the war, peripheral to Lessing's action, is centralised in its visual consequences: '*An encampment in Saxony. A hanged man, a hanged woman, revolve in the breeze. A sprawling soldier tightens a drumskin. He hits it with a stick*'.

Lessing's play features a distant, fatalistic hero who feel himself irrevocably severed and transformed from his former honourable self (but haunted by the awareness that, after a war, the authorities find no one indispensable); his obstinate, wild, rough but faithful servant; and a heroine who almost succumbs to his tone of hostile melancholy, but who is determined to preserve 'reason' by dissolving the spectres of evil consequences, divisions of conventional social honour, and other such "misunderstandings" of humanist Enlightenment, through laughter. Barker's play takes as its starting point that possibility that Tellheim's crime or complicity might be more obscured, but more substantial, and that he hated neither the war nor its consequences, that he is not so much ashamed as beyond the reach of shame itself; that the servant, Just, might be one in a series of Barker's servants who embody 'unselfish love'; and that Minna might be a determined social investigator who is wilful but nevertheless (or therefore) susceptible to emotional complexity. The setting oscillates between 1767 and 1950s Germany, and shows Minna alternately confronting a man who declares himself unfit for her adoration – as in the original, but for different reasons – and collecting evidence for a postwar Commission on Atrocity, seeking the (Platonic, or Utopian) 'algebraic' resolution of 'crimes against humanity', punishment and justice through her prosecution of Tellheim, denounced by The Hanged as 'the Monster of Grosshayen who watched us die ... And laughed'. Minna is supported by a cluster of servants, 'the Franciscas', a comically self-contradicting chorus and 'barometer of banality' who anticipate the unison "flocking" effects of the peasants and novices in Barker's *Ursula*. She is also besieged by the alcoholic offers and attentions of the Landlord, a frank, self-styled 'slave to beauty' who continues Vanya's fetishistic devotion to underwear (which 'never disappoints'); and her odyssey is punctuated by the casual perambulations and offhand interventions of a dusty Cupid. Minna shuns whatever the Franciscas approve, but rejects, with Nietzschean eloquence, the accusation that this demonstrates will rather than heart: 'One must acknowledge that exceptional men rarely win the approval of common humanity, and how should they? It would demean the very quality of distinction...' (*CP3*, 201). She even finds herself pursuing moral accommodation of her intimations by rejecting the ultimate moral superiority of the chorus of victims, who maintain: 'We suffered, so we know'; Minna replies 'All right, you know, but what you know I might perhaps be forgiven for not knowing!' (*CP3*, 204).

Whereas Tellheim represents a discomfortingly absolute honesty, Minna seeks certainty through an algebraic resolution into harmony, 'The state of satisfaction resulting from the dissolution of two competing egos'. He recognizes, and she acknowledges, that this involves a destruction of individuality. Tellheim is mindful of the 'finite' nature of flesh in an 'accelerated' world, and the deterioration of all things. ('Love. Garments. Obligations'). Minna, like Helena, seeks reassurance of 'Self to the power of *n*' through her reflection in a mirror; but in the wake of Tellheim's admission that he 'supervised' punishment (which was 'undeserved' except by being), she proclaims him 'impossible to love' and 'the enemy of life'. She nevertheless insists she will follow him into and meet him on even this moral terrain, and publicly identifies herself, before her mirror, with the figure of Venus, a goddess whom she proclaims stricken with anxiety, pain and shame on discovery that the commanding powers of her beauty are not infinite (again, the comparison with Helena is relevant). However, '*euphoric party-goers …enter in black, as if in mourning, dragging with them a naked woman. It is the* FIRST FRANCISCA', who is instead identified by the revellers with Venus, and proclaimed as Tellheim's love. Minna is bewildered and outraged at this displacement: 'I am his lover. I am evil, Francisca, and therefore close to him. *(Pause)* You on the other hand, are good'. She points a gun, but then revokes her accusation ('**This isn't Tellheim**'), finally taking up position of association next to the dead servant, Just (an image of incurable pain rather than Lessing's facile reconciliation and dispelling of misapprehensions) with the recognition '**Evil requires goodness** / Oh, isn't that its poverty…?' (*CP3*, 240).

Minna's European and specifically German setting of its themes – the disturbing erotic attractions, collusions and mutual dependencies of the self-consciously "good" and the identifiably "evil", the will to denounce and the will to morally accommodate complicity in atrocity, the persistence of self-deception and its pretext for, and collision with, the surprisingly unfathomable – relocates figures from Lessing's German romanticism in a context of post-Auschwitz exposures and Adornian neo-romanticism, where they are obliged to contend with forces no systematic invocation of morality can dispose, and where understanding cannot be equated with reconciliation. *Minna* also constitutes one in a series of 'war plays' by Barker – *The Love of a Good Man, The Power of the Dog, Und* and *Found in the Ground* – which address attempts to seize the testament of war, in which the protagonists (a prosecuting lawyer, a grave-maker, a photographer of atrocities, an aristocratic Jewess and a Nuremburg judge) are separately overwhelmed by the scale of the material and experience they are trying to order; and thus the plays dramatize 'the death of the enlightenment in the ashes of Auschwitz'[39].

Moreover, *Minna* shows Barker provoking increasingly ambitious effects of stage composition, in which conventionally irreconcileable details are brought together in disjunctive non-rational yet haunting confluences of perspective, depth and rhythm: for example, the stage directions:

> *(Snow falls. TELLHEIM and JUST are seen in long coats, bearing shovels. They go to a spot and dig. JUST's rapid drum tap. The CUPID, on a wire, flies over the stage. The THIRD FRANCISCA, in a costume of 1769, enters laughing with a picnic hamper and a rug. She spreads it on the snow.)*
>
> *(He stops, seeing them kiss. He retreats silently up the stairs. A pipe plays. Water drips, as in a thaw. THE HANGED turn on their ropes. The kiss endures...) (CP3, 213, 220)*

The image and pose of Just, shot dead in girl's clothing, parodies Fragonard's eighteenth-century painting 'The Swing'; and the camps of soldiers and Seekers after Venus are visualized in locations recalling the long, ominous, prefiguratively shadowed gardens of Watteau's paintings[40].

Nevertheless, Barker's highest achievement to date in terms of re-evaluative text and production will be considered separately: the play, and The Wrestling School production of, *Gertrude – The Cry*, Barker's 2002 responsive meditation on *Hamlet*, will be analyzed in detail (alongside *Knowledge and a Girl*, Barker's retelling of the story of Snow White) in Chapter Eight.

5
Separation, Sacrifice and Sainthood

> What you describe as a rebuke to me, is no more or less than my desire
>
> *Hated Nightfall (HN* 19)

The triptych of concepts identified in this chapter title resonate throughout Barker's works, but are valuably characterized in his essay, 'Saintliness, Death and the Perfect Family', which appears only in the 1994 Calder Publications/Royal Court playscript of *Hated Nightfall / Wounds to the Face*:

> If the sign of the saint is sacrifice, it is a sign illuminated by the vehemence with which he repudiates the world. For the saint finds the world lacking, and his desire is focused on what can never be satisfied. His passion can discover no worthwhile object, and the more searching his gaze, the more contamination is revealed. For Dancer [the protagonist of *Hated Nightfall*], all is transparent, and this transparency is appalling pain, for we require to be deceived, it is the condition of social acceptance. Without deception there is perhaps no hope ... the saint is lured by human love only to discover its inadequacy ... the saint is first and foremost his own work of art, exhibited primarily to himself. But neither saints nor works of art are socially desired, for they are disruptive to the bourgeois and the collectivist alike.

The Eloquence of the Significant Exclusion: *A Hard Heart, Terrible Mouth, Hated Nightfall, Ego in Arcadia, The Brilliance of the Servant, The Gaoler's Ache, Twelve Encounters with a Prodigy*

By way of prelude to Barker's major statements on the theme of sainthood, *A Hard Heart* (written 1991, staged by Ian McDiarmid at the Almeida Theatre in 1992 and also broadcast that year by BBC Radio in a production by

Richard Wortley) depicts a strategist and architect, Riddler, who demands privacy, silence, manners and solitude even under siege, if she is to devise the means to save her city. She tells her queen and patron, Praxis, that these privileges of distance are culturally intrinsic and personally essential, 'or else defence is pointless, defence of what if not of civil virtues'; virtues without which the civilization under attack threatens to resemble the barbarians who resent it. Riddler and her location are classically Barkerian in their senses of tensile but fragile boundaries, and their potentially catastrophic (but possibly ecstatic) reversibility: war is merely an exteriorization of her natural emotional element, 'I've been under siege all my life ... At the edge of catastrophe you are so trembling and acute it is a sort of love it is an ecstasy of sorts' (*HH*, 9). In return for her strategic innovations, Riddler seeks dispensation for her son Attila to be spared military service: he is a petulant but dependent overgrown boy, who ascribes his own childish egoism to his mother's solicitudes, which he nevertheless voraciously considers to constitute her boundless obligation and duty. In her exultation, Riddler shares with him her profoundly seductive sense of fortification strategy, which is even more pre-emptive and aggressive in form than Krak's theories of fortification in Barker's *The Castle*. Riddler's constructions are based on the circular secrets, reversible challenges, apparently superficial abysses and *trompe l'oeil* horizons, combining appearances, provocation and the raising of stakes in an edifice which manifests oneupmanship and death, where the dividing line between victory of one side and defeat of the other threatens to become challengingly illegible. She claims 'Always, the enemy has an object', such as a wall, 'on which he concentrates his energy' and so 'the more passion he expends on it, the more magnetic it becomes' to the point where 'the object infatuates the army', the intelligence of which is disabled by a consuming passion. Riddler proposes to demoralise the barbarian enemy, by making its object '**worthless** ... You render the thing so wanted a perfect nullity. You make the love-object vile. All that he seeks becomes ash in his mouth, as a man seduced by so much effort can appear, in the cold light of morning, a comic rebuke to love itself' (*HH*, 11). Here, the seductive terms of war and desire overlap: the destruction of another's territory and existent forms of life involve the sacrifice and destruction of one's own.

This association is advanced by Seemore, a vagrant who proclaims himself '**roped to a love**', in his vertiginous desire for Riddler, whom he regards as his fellow 'genius'. His disclosure, 'I don't know who I am since I laid eyes on you', significantly awakens her sense of something to be scared of, the intimation of the impossible within: she admits 'I must know who I am. I need the city to know who I am. And the city must have walls or how can I – ' (*HH*, 16). Riddler is no less a Kristevan deject than Sleen: in her own 'divisible, foldable and catastrophic space', she is a deviser of works who never stops 'demarcating [her] universe whose fluid confines ... constantly question [her] solidity and impel [her] to start afresh'[1]; however, she lacks Sleen's abject invulnerability, demonstrating instead a vulnerably (fatally?) misplaced faith in her son and an intoxicated hubris in her own pandeterministic will to power: she proposes a series of sacrificial demolitions and detonations, including that of the city temple, as '**a matter of will**' in order to deceive the enemy. She deduces that the consequent inability of the crowd to 'separate their fate from their reason' constitutes an '**ecstasy**' which confirms her status as a god: 'It is only being more yourself than you dared to be' (*HH*, 34). When a woman objects that nothing justifies the death of her son, 'not even the greatest justice', Riddler can nevertheless placate her with the promise of a '**Revenge**' which is crucially separable from 'the greatest justice'. But Praxis is persistently anxious about the increasing fissure in definition: she tells Riddler her fear that what Riddler demands – the sacrificial and strategic appearance of the city falling to its enemies – means that '**It will have fallen to its enemies and the enemy is you**' (*HH*, 27). Praxis identifies Riddler as a subtler form of 'barbarian' who invokes essential terms of civilisation only in order to put them into play, where even the humiliation of the citizens is a tactical 'resource'. However, Riddler fails in her bid to be immaculately 'ready always for the impossible'; her son, whom she wished to exclude from military service, transpires to be a significant exclusion from her calculations in another respect: he has collaborated with the enemy, divulging her plans in exchange for his own self-preservation, a crucially unforeseen extension of her own project. Seemore catches Riddler in a double bluff, sacrificing 'thousands' to 'earn' their 'unity'. But even at the end, Riddler shows a 'reluctance to deliver [herself] into the hands of others' which may debar her from the survival which Seemore offers.

A Hard Heart is a sparklingly articulate demonstration of the dynamics of challenge and seductive strategy; indeed, the "confidence trick" of performance in order to deflect the opponent's energies away from their original objective has fascinated Barker since his very first play, *One Afternoon....* However, the play and characters of *A Hard Heart* are, by Barker's own standards, unusually *unseductive*: none of the characters has the emotional range of Riddler, and even she lacks the pain, wit, complexity and capacity to surprise the audience: the artful properties of outrageous but astonishingly engaging characters like Judith, Sleen, Vanya, Isonzo, even Holofernes and Judith's Servant. Riddler works to separate herself, but with the conscious ambition and vulnerable egoism of one who wishes to aggrandize herself to the power of a god, rather than having the charisma of a saint, whose sense of passionate personal sacrifice repudiates the world of social existence; this exclusivity of her focus proves to be her Achilles heel, and imaginative limitation. Despite the diamond-compression veneer of its exchanges, *A Hard Heart* occasionally resembles the dramatic demonstration of an analytic thesis, essentially consistent in its characters and form, and even potentially predictable in the *dénouement* of its reversal. However, it is important to note that the terms of its analysis seem even more prescient, in the twenty-first century, than on first appearance, in their exploration of how social systems and city edifices 'obsessed with their systemacity' are fascinating: 'they tune in death as an energy of fascination', both inviting, and demanding (in their flaws and exclusions), the dissipation of the 'terrorism of meaning' (Baudrillard)[2] through sacramental (and ubiquitous) destruction. However, the play and characters can be *reduced* to a set of rational propositions and counter-propositions, rather than provoking an *irreducible* series of *contradictory experiences* in the audience as Barker's work of this period increasingly aims to do. This vulnerability was compounded by Ian McDiarmid's Almeida production, in which Anna Massey's performance as Riddler, her mannerisms and even her hairstyle, inescapably suggested those of Margaret Thatcher. This made the play perilously reducible to a (not particularly apposite or convincing) contemporary political allegory, such as which Barker has increasingly set out to transcend, in favour of generating wider and less obvious resonances of surprising meaning.

1992 saw a further, more successful collaboration with McDiarmid and the Almeida in staging a BBC commission: the opera *Terrible Mouth*, with words by Barker set to music by Nigel Osborne. Barker's Preface to the published libretto identifies a starting point: 'Goya, for all the claims made for him as a social realist, was a contemporary of the Romantics, for whom pain, decline, tragedy, were subjects not for social propaganda, but for sublime feelings of melancholy, curiosity, even ecstasy'[3]. Barker's character Goya, unlike the historical one, witnesses some events in the Spanish Peninsular War at first hand, and is incarnated by two performers: one to perform the deaf artist in his worshipful and agonized self-abasements, needing to be 'rinsed' in 'extremity'; another to incarnate (and sing) his interior voice, hymning the surfaces and depths of his mistress and patroness the Duchess of Alba[4]. Racked and rapt by her self-ingratiation with another, and spared excuses by his deafness, he speculates 'I think I want to see my love in pain'; his impulse is challengingly realized when soldiers capture her, and she defies, then makes herself amenable to, their cruelty and 'infantile obsessions'. When she leaves with the enemy, Goya insists that '**To watch is love**', but the personified hooded figure of Atrocity maintains that 'There / is / no / love / like / yours / for / me'. He has maintained that hopelessness is 'Only the beginning' and also his artistic 'privilege'; in his desolation, he contemplates how to inscribe a canvas art with surgical instruments rather than brushes, grappling with the residual patronage and vanity of literally minded aristocrats whilst seeking to 'sing' the 'cuntlife' of his beloved and degraded mistress. Sidelined from full participation in events by disability, calling and temperament, Goya inhabits, and watches from, the margins, in this speculation about what may have occurred at or beyond the margins of his drawings – both the done and the undone – which nevertheless informed his flayed sensibility and work.

Terrible Mouth thus – as often in Barker's *oeuvre* – foreshadows and prepares for more substantial developments: the irreverent surgeon, whose detachment gives way to voyeuristic fascination, and the sliding panels in the walls, anticipate the protagonist and set of Barker's play *He Stumbled*; the periodic strategic throwing of babies over walls, in order to enlist the erotic pity and debilitating compassion of potential adoptive enemy parents, recurs as a central motif in *The Fence*. David Pountney's production of *Terrible Mouth*, designed by Nigel Lowery, was strikingly unEnglish in its boldness: The Chorus of the Maimed: writhing figures beneath a floorcloth forming a nightmarishly swirling, animated terrain. Ian McDiarmid's portrayal of the fixated, self-flagellating Goya was a memorable feat of physical, vocal, emotional and imaginative athleticism, justly lauded by Barker in the essay 'The Anatomy of a Sob' in the third edition of *Arguments for a Theatre*. McDiarmid infused and animated the role of Goya with an orchestration of warring but restlessly consecutive impulses, including worshipful desire and vindictive spite, self-aggrandisement and self-torture, resistance and self-abasement, scopic ecstasy and enraged Satanic exclusion, with each successive emotion bending his body and voice into the precisely required shape of its extremity, then reversing into the full contrary energy of its antithesis.

In an auspicious gathering of forces, McDiarmid joined The Wrestling School (for a first and last time) under Barker's own direction for the company's 1994 production, *Hated Nightfall*, for which McDiarmid developed his inspired approach and performance style to provide the central impetus of a full-length play. *Hated Nightfall* offers in many ways a classical example of Barker's work: a figure usually excluded from or submerged by history – the tutor of the Romanoff children, Dancer – pursues his own claim to significance, in an apotheosis of absolute play, murderous abandon and vengeful erotic dedication, with wry but lethal disregard for the agents of historical order and official ideologies who attempt to interrupt him. In a speculation about the last hours of the Russian Imperial family in 1918, the play imagines the brief incandescence of Dancer, whose sexual fixation on his mistress is permitted tortured expression amidst the social reversals; he offers himself as a self-consuming harbinger of death. As Barker observes in 'Saintliness, Death and the Perfect Family':

> Dancer is poised on the threshold of an age which the more desperately it announced its total novelty, the more sinisterly came the echoes of the past. For all its philosophy of materiality, the Bolshevik Revolution opened with a symbolic murder as antique as massacre itself, and shortly afterwards its unsentimental architect was embalmed with the hysterical palaver of an Egyptian god ... but Catastrophe is also opportunity, and Revolution's seductive fallacy has always been the promise of absolute licence ... in the exhilaration of the vortex the ordered life dissolves and what beckons through chaos is **permission**. Here, one who aspires to

sainthood might rinse his personality in the foaming hydrants of a collapsed authority, wring himself out in a new shape and discover performances of self which, however insubstantial their foundations, are incandescent with **invention** ... Dancer knows that while his immaculate temper might be **used**, it can never be legitimized. He is the **Transient Phenomenon**, use-value epitomized, function in the grinding mechanisms of History that substitutes new pain for old ... Why not, therefore, make of this transience a perfect epiphany? Since revolution is unforgivable except by reference to its alibi – the future – and since reaction is grotesque with sham dignities, it is the brief life of the present which is the refuge of imagination, an inebriating atmosphere where both saint and artist perform their seductive dance and encourage a similar motion in others.

Dancer seizes the opportunity for an orchestrated expression of his own precisely excessive eroticism to take centre stage ('My sexuality / Its lethal nature'); indeed, the very spatial dynamic of the play is a strong example of the literally "hell-raising" Barkerian trope, in which some force repressed or marginalized erupts into the centre of the stage action to wreak chaos (Dancer: 'The world abhors me. It writhes to know I walk upon its surfaces ... **I jump on it**'; 'Chaos. How it suits me. It's my medium', *HN* 4, 31). He is in turn catechized by the choric Eumenides representing the reclaiming and/or prosecuting moral-historical order (compare *(Uncle) Vanya* and *Golgo*), whose terms are eluded, questioned and/or reversed by his infinitely resourceful absolute play. Johan Engels's set for The Wrestling School production of *Hated Nightfall* featured plastic walls, apparently firm boundaries which could suddenly deliriously warp, as the clamouring faces of the Chorus pressed their deindividuated features through the fabric to pronounce on the central action before vanishing or being banished again by Dancer.

Dancer compulsively challenges platitudinous notions of primacy, as when he proclaims 'Perhaps I know more about love even than you, but love of a different kind' (*HN* 3). However, his 'performance', and its terms of reversibility, are not carnivalesque in sparking emulation, nor even seeking it; rather, as Barker's essay finally acknowledges, 'It is a final act of self-description by one whose passionate gesture earns him – as befits a saint – the half-reluctant pity of his erstwhile enemies...'. Dancer (like Sleen) defies the ostentatious moral incomprehension and terms of permission of his adversaries. When a Romanoff daughter suggests 'you are not really bad, citizen', he insists 'I am / I am that bad *(Pause)* / Believe me. / You merely cannot bear to contemplate it' (*HN*, 6). When the first of several "men of the future", emissaries from the revolutionary forces, appears, it is his bloodlessly bureaucratic assurance 'put your criticisms on paper and I will forward them to the appropriate committee' that finally provokes the bloody farce of Dancer stabbing him. When the Chorus protest 'We can forgive the revolutionary death but this' and denounce Dancer as '**insane**', he contemptuously replies 'You would say that! How necessary you should think I am insane! Think so if it relieves you!' (*HN*, 8). He refuses terms of legitimacy: 'I do not call myself History. The Agent of Destiny. Justice. The People's Will. I don't drape myself'; 'This is not History. This is the opposite of History'; he pointedly asks, 'The distinction between an honest and a dishonest passion being what precisely' (*HN*, 12, 33, 10). However, even Dancer's proclamations of determination to be a law unto himself (a principle of the Barker aesthetic) are bathetically under-cut by the 'rustic charm' of servant Jane ('You're Billy the tutor with the balding bonce as far as I'm concerned') whose action of swabbing the blood from the floor is an early example of a recurrent action/motif in Barker's plays, simultaneously pragmatic and poignant in its rhythm, sounds and resonances (the scrubbing or mopping and clanging bucket also feature in *Golgo* and *Ursula*). Dancer's command of the situation repeatedly lurches into savage farce as he is required to murder successive revolutionary agents, often by taking them in a terrible embrace, to the accompaniment of 'comic music'.

In the second half, the stakes escalate; the Romanoffs scrutinize the trap-pings of a wedding feast, including a cake and an ominous tureen. Dancer presides over his ceremonies like a besieged Prospero, master of a crucially limited time and space, to celebrate 'A transient phenomenon' in anti-official terms: '**The committee lacks imagination** ... I would go so far as to say it defines the thing ... it fails to grasp **The / Symbolism / Of / The / Sacrifice**' (*HN*, 26). Romanoff tries to bribe him, another daughter begs her to impregnate him, further officials of the revolution appear; one even contravenes the party line ('Pity? We are replacing that by organization') by offering the Romanoffs freedom, until Dancer kills him (because his 'performance' on the stage of revolution, 'however predictable, threatened to overshadow me'). The Empress Caroline strategically subordinates herself to Dancer's ritual, offering herself as a 'bride' to be given away by her husband, seducing Dancer with the promise of sex and escape; he reels in susceptibility ('Hated nightfall / As long as there was day / I sensed the power to distinguish / Will from appetite', *HN*, 40), swaying in pain when she proclaims herself unentered by an impotent husband; however, '*The spell is broken*' when their son empties his watering can over the cake. However, the Romanoffs turn the wrath of the latest party official against Dancer, stigmatizing him as a murderer serving up human remains (though Jane protests that the tureen contains only a soup for which she spent hours finding mushrooms). Romanoff reinscribes the hierarchy of their family relationships and forges a brief alliance with the revolutionaries to deny Dancer his apotheosis; this resurgence of moral concensus reconsigns Dancer to literal physical abjection for his crimes of '**Arrogant / And /**

Petulant / Disharmony', threatening a death which Dancer fears will be 'poorer than my imagination predicted' (*HN*, 47).

Barker's direction of The Wrestling School production of *Hated Nightfall* set new standards of unpredictable precision, scenographic composition and nimble dynamism, of which McDiarmid's performance was the virtuoso centrepiece. The arrangements of expressionistic lighting, sound, vocal delivery and movement sequences represented a new thoroughness and originality of aesthetic proposition which would increasingly become associated with the company's work. When *Hated Nightfall* was revived in 1995 for a European tour, in repertoire with Kenny Ireland's revival of *The Castle* and (the regular and strongly distinctive Wrestling School Lighting Designer) Ace McCarron's production of *Judith*, Barker's direction of *Hated Nightfall* was (unsurprisingly but demonstrably) the most authoritative work, countering the English theatre prejudice that dramatists lack the objectivity to be the best directors of their own work. On the contrary: Barker was developing a sense of pictorial stage composition, informed by his work as a painter, in which movement acquired a near-balletic choreographic precision and the soundscape became an additional "sonic character" alongside scrupulously delivered language of operatic reach and scale.

Barker had learned this confidence of overall vision partly through directing the premiere of his play *Ego in Arcadia* in Sienna, Italy, in 1992. The stage directions of this play indicate a new level of attention to orchestrated compositions of scenography, including sound. As Charles Lamb points out, *Ego in Arcadia* presents a Utopia that is 'diametrically opposed to those of More and Bradshaw [in *Victory*]'; 'Whereas their projections are functional and utilitarian, "Arcadia" is essentially aesthetic, the ideal of innocence and the pastoral life – a world of poetry and song'[5] from which Death is traditionally excluded. However, Death may also constitute a shockingly resurgent discovered presence, disturbing identity, equilibrium and order, as in the two separate paintings by seventeenth century French painter Nicolas Poussin. Barker's setting is a 'catastrophic Arcadia', which includes not only Death but many other things conventionally excluded: 'The sounds of social disintegration punctuate the action, which ... retains the essential elements of the classical tradition in that this is a utopia where the individual pursuit of love is placed above all other considerations – including social harmony'[6].

The opening action of *Ego* is in many ways an essential crystallization of several Barkerian thematic tropes: Dover, a *'resplendent, dishevelled'* Queen of a former monarchy, is a captive on a rope: she challenges her revolutionary captor, Sansom, to separate her physicality from her/his ideology: 'I'm History *(Pause)* / Flesh *(Pause)* / and History *(Pause)* / Separate them, why don't you? ... / The acts *(Pause)* / The body *(Pause)* / Part them, can't you?' (*CP3*, 269). Sansom (who shares his name with the senior guillotinist of the French Revolution) proclaims his intention of justifiably hanging his sworn enemy ('**it's your government spoiled this place**') but significantly over-emphasizes his own dispassionate instrumentality in carrying out orders of the revolution; Dover taunts him with the intense and costly labour involved in the intricacy of her underwear, perfume and make-up, and conjures her indulgence into a Cleopatran magnificence which successfully overwhelms him: he presses his face into her skirts in '*an ecstasy of surrender*', acknowledging the overturning of the weight of his former convictions. Then Mosca, a Minister of State, enters and kills Sansom, despite Dover's incredulous recollection of Mosca's murder by outraged citizens, which he does not deny. He demonstrates the principle of resurrection which holds sway in this afterlife, how characters are repeatedly driven to kill each other but only to trigger a Promethean regeneration: 'This is Arcadia...where nothing lives but love...and therefore...is a place of infinite suffering'(*CP3*, 271). Their reunion is interrupted by the incursion of Sleen, the character based on the writer Céline and reanimated from Barker's *The Early Hours of a Reviled Man*, here accompanied by his dancer mistress Lili and his narcissistic actor friend Le Vig[7]. Sleen exults in the confirmation of his criminal persistence ('**We flee the righteous / We evade just punishment**'). They encounter the figure of Poussin, who expresses his definite sense of purpose, 'The function of the artist is to make you hate your life', and his awareness that 'no one dies in Arcadia...That is the horror of the place ... You'll want to, obviously ... And Death is here ... but so discriminating, you will hate him for his impeccable disdain'(*CP3*, 274). Sansom revives, confirming this as 'the land of uncommitted suicides...and murders which bring no relief', a cosmic rebuke to all pride in conviction, a perpetual limbo like that in Sartre's *Huis Clos*, and the dystopian version of Nietzsche's ideal of Eternal Recurrence in Beckett's *Play*, but more imaginatively various and surprising than either earlier play because of the wider array of characters, details of setting and terms of situation[8].

Lili admits she worships Sleen: 'Any man who suffers as he does must be a god surely! I mean, let's not be fussy who we worship as long as we worship somebody!'(*CP3*, 274–5). As religions and ideologies go, Sleen claims this is relatively innocuous, compared to the carnage being visited on the surroundings in the name of one fundamental(ist) truth or another. Le Vig is as self-contained as he is self-preoccupied, finding himself 'both more attractive and more interesting than anyone else'. Poussin's model Verdun harbours a resentful fascination for the artist and, like Sonya with Astrov in *(Uncle) Vanya*, wants to confine him to the malleably, mortally finite: 'I'll kill him. Then I can mourn...and he'll be mine...am I not mourning him already?'. Thus the 'pleasures of intrigue' compulsively fill the 'preposterous minds' of the political refugees, this unlikely juxtaposition of human flotsam, isolated and (self-)excluded from their respective social orders, now denied the destiny of being '**common, undistinguished and the colour of earth**' even in the melancholy landscape of antiquity. These self-styled saints (or pariahs?) of imaginary or collapsed dominions occupy themselves with their impulses to love and its concomitant inflictions of misery on self and others, in a series of frenetic configurations briefly recalling the sexual rondo of the characters in *The Bite of the Night* and their exhaustive erotic/political alliances. Just as they fear themselves consigned to national History, they fear their irrevocable consignment to a past tense of sexual history, as Sleen observes when Mosca is rejected by Dover: 'Oh, she is consigning you to her biography...chapters two and three...not bad...some are only footnotes' (*CP3*, 284).

A new dynamic is added when Sleen confronts Tocsin, Arcadia's selective dispenser of death, who decrees a song contest, promising the relief of oblivion to the victor. This involves the contestants in a savagely humorous conflation of the Barkerian tactics of seduction (specifically, challenge and one-upmanship) and abjection, in which they "win" by most eloquently denouncing the terms of prolonged existence (notwithstanding Tocsin's infernally middlebrow tastes: 'What I like in a song is melody and words that I can follow...of few syllables'). Despite Sleen's conviction that he will 'romp home', he encounters fierce 'competition' in other contestants' self-excoriations and proclaimed deadness to all gratification. Poussin, the inventor of Arcadia, inadvertently wins by his observation that 'nothing follows' from its intolerable beauty but 'the most mundane and futile happiness'; however, even he is seduced by Dover, feeling himself despairingly 'In love...and terrified'. Poussin has exemplified the Barkerian will to imagination, 'If you hate the world...You must invent another' (*CP3*, 321), but is now potentially fatally drawn towards one of his own creations (it is as if Chekhov in *(Uncle) Vanya* fell in love with Helena and himself turned transgressive). Poussin fights Tocsin, and it is the latter's body who ends up hanging from a rope; but the hellish wranglings continue when Sansom suffocates Dover – apparently irrevocably. Poussin, *'with a supreme effort of will'*, recovers and begins drawing Dover's body *'with an artist's eye'*, as Verdun observes him pitifully.

With its developed degree of specified sound and landscape, its vivid and poignant array of competingly and variously eloquent characters, and its startling plot twists and reversals (culminating in the death of Death), *Ego in Arcadia* is a resoundingly melancholic and savagely comic dramatization of Barker's *aperçu* in his 1991 poem, 'Mates of Wrath': 'OH THIS MONOTONOUS HOSTILITY WHAT COULD BE WORSE ONLY MONOTONOUS LOVE' (*AMG*, 77). Its premise and observations provide memorable and boldly distinctive commentaries on classical themes in art and drama, and its compulsively alternating hymns to the personal revolutions of disastrous love are both inescapably searching and inescapably comic. Yet, at the time of writing, it awaits a professional production in Britain; as does *The Brilliance of the Servant* (written 1994).

Servant extends Barker's series of investigations into master and servant relationships: as first substantially broached in *Ten Dilemmas*, with Becker's relationship with her faithful retainer (of whom she remarks 'To this man nothing I do is ridiculous / **That alone is the justification of service**'). Contrary to modern resistance to ideas of hierarchical status (and mastery and elitism), Barker's dramatic master–servant relationships (of this period) depict an honourable equilibrium on both sides, in which the master commands, and the servant places himself (quite literally) at his master's disposal without judgment and opinion, with the effect of mutual substantiation of both selves, so that each party has their own particular authority. Moreover, *The Brilliance of the Servant* shows how a conventionally marginalized character can take centre stage, and imaginatively outstrip and relegate the merely conventionally powerful, by performing his role to the hilt and beyond. To quote the pithy summary on the back cover of Barker's *Collected Plays Volume Five* (2001, in which *Servant* first appeared), 'a wedding is interrupted by the arrival of enemies bearing a legendary and cruel machine, a mechanical manifestation of the will to sacrifice which underlies so many of Barker's painful transactions between classes'. In '*the war damaged room of a great house*', the play opens with arresting directness: Sunetra pleads with her mother, Camera, not to seduce her fiancé, Taxman – too late. Three female servants, who together comprise a choric flock like those in *Minna*, proclaim both loyalty and its increasing strain. Camera is a truly mature woman in her magnetic wilfulness, demonstrating 'the intolerable spectacle of unpunished pride', accepting and even enjoying the world's tribulations without concern for the 'consequences of things' (thus she is in some ways a preliminary sketch for the unapologetic female protagonists of *Gertrude* and *The Fence*), whereas her daughter and prospective son-in-law are burdened with conventional ideals of justice and their right to happiness. The male servant, Shoulder, observes how the bombing of the nearby hospital seems rather to confirm Camera's sense of the 'general malevolence of things', that life is 'pitiless, malicious and absurd', and yet (or therefore) pleasurable. Not that Camera is immune to potential danger, or pain; however she weaves this awareness into the embroidery of her allure, as when she tells Taxman 'Come to my room, often. I will be in such pain, so you must come often … This pain I possibly require' (*CP5*, 93). Even Taxman (whose very name suggests his strain and limitations) deduces war's beauty is 'permission', and steps outside Sunetra's conventional romantic and humanist hopes for their wedding, in which the enemy might literally be disarmed by witnessing the ceremony. Tellingly, they are uneasy around Shoulder, and his acceptance of his role and status, but Shoulder insists on the difference of his feelings; 'It is not that a servant does not feel pain or experience humiliation. It is the use to which the pain is put that distinguishes a servant from his master … I am talking about true servants, and true masters, obviously' (*CP5*, 97). Sunetra and Taxman are incapable of being true masters, and shrink from the role. In contrast, Shoulder is devoted to Camera, her combination of the beautiful with the tragic. Camera rightly foresees the killing of Shoulder as appropriate desecration; he unflinchingly agrees and maintains that there is no emotional contract between them: 'I am your servant and if they kill me, they merely spoil your property' (*CP5*, 98). The unseen enemy are, as in many Barker plays, notorious for the cool spirit of analytic investigation which they bring to murder, maiming and torture; they strike at the face, 'the seat of love', with a sense of curiosity as to the limits of the human will to life (compare Starhemberg's observations on the cool maimings carried out by the Turks in Barker's *The Europeans*, 'it is to do with fear of love'). Camera rightly notes that for Taxman to part with free will, and claim that he offers himself for (and even requires) torture and death, merely represents in his case a parting with his 'licensed indecision'. Shoulder has a more accurate intimation of the scale of sacrifice and ritual demanded; officiating in lieu of a priest at their wedding, he commands not only Sunetra and Taxman but also Camera to be naked. Sunetra again reads this initiative in conventional terms: 'this nakedness will either stimulate them to acts of further outrage, or, more likely, move them to pity' (*CP5*, 116). The ceremony however becomes a more complex event, in which the anxious Sunetra is succeeded in undressing by the shameless Camera, and '*The two women stare at one another for some time, then Sunetra turns her eyes away*'; noting Taxman's avoidance of Camera's nakedness, Sunetra further observes 'I interpret that as delicacy / Refinement / Sensibility / That is my interpretation but there is another which – ' she finds literally unspeakable: '*(Pause. She studies her mother, with an effort of will)* How different her body is to mine' (*CP5*, 119).

Taxman is drawn into marriage to Camera rather than Sunetra; Camera rightly observes the inappropriateness of her own potential self-sacrifice. Shoulder instead offers himself for consumption by the enemy's baroque contraption (redolent of Kafka's engine of punitive transfiguration in the story 'In the Penal Settlement') which disembowels its victims alive; and which Barker wisely leaves to the imagination rather than realise onstage. But Shoulder determines to transform the extravagance of his ruination into that of an apotheosis, telling Camera: 'It will be the perfection of your life and I am honoured to be the source, the object and the means of your transfiguration. How better could I demonstrate my love for you?' (*CP5*, 132). Taxman exclaims 'How vile', but Camera recognizes 'The saint is vile' and those absolved 'The instrument of his perverse and egoistic will'. Shoulder re-enters after the process, a dying yet still eloquent husk whose naked agony moved his enemy to tears, but also threw into relief their own will to servitude: 'My agony was quite simply, a necessity, a factor in the motion of the world' (*CP5*, 140). Shoulder expires, the enemy depart, the female servants rejoice with trite resilience; Camera, contradicting Shoulder's insistence on the formal and necessary lack of emotional contract between them, sobs with '*the heaving passion of a bereavement*'.

Thus, Shoulder expresses his 'perverse and egoistic' will to an inhuman perfection as servant which elevates him to the status of charismatic saint; he engages with the low-status role so thoroughly that he incarnates reversibility and burns through to its opposite, a high-status command of others' emotions and obedience. Though he is (offstage) turned physically inside out, his presiding over both the trial by nakedness and the anti-conventional marriage, and his insistence on his own subsequent form of ordeal, are all stages in a project which turns his immediate witnesses emotionally inside out. *The Brilliance of the Servant* is a remarkably unswerving proposition which demonstrates its central character's surprising consistency in pursuing his own appalling enhancement, and delineating its power of purchase on the imaginations of others. Its remorselessly narrow focus on a relatively small group of characters might show up those, other than the dominant figures of Camera and Shoulder, as being at least occasionally relatively bloodless, imaginatively speaking. The play moves with the anxious deliberation of a slow-motion slide into irrevocability, replete with the suddenly intensified detail such as accompanies the drift into a car crash or larger catastrophe: an effect which will recur in *Und* and *House of Correction* in particular. Its single, though drastic, process of a plot line make it more of an extended theatrical short story (novella?), an extrapolated proposition, or 'Possibility', played for full lethal deliberation, rather than a major play. However, its concentration of its resources and focus on the differing methods and results of the *déshabille* of the four principal characters is a further significant stage in Barker's experimentation with nakedness onstage: both his investigation of the complicating consequences (dramatically and aesthetically) of its *duration* beyond the conventionally brief and/or climactic; and of its differing forms of pictorial force, including anxiety, exposure, (significantly anti-conventional) erotic magnificence in disclosure, and sacrificial agony, in which the performers effectively become 'living paintings', iconic physical manifestations within an overall visual composition. The frequent anticipation of the ubiquitous ordeal involved in undressing heightens and compounds the spectator's implication: a mingling of anxiety for characters (and performers, and their own reactions), and an erotic fascination for, and will towards, the completion of the characters' (and performers') impulses to nakedness: a complex engagement in destabilising imaginative activity which dissolves moral polarities and demonstrates the dangerous promise of reversibility, well summarised by Shoulder's lines: 'it is a crime to look, and a crime to avert your eyes...a crime to protest and a crime to enjoy' (*CP5*, 125). Barker's theatrical uses of nakedness, the effects of suspense and anticipation, are discussed further in Chapter Seven ('Erotic Disclosure Through the Rules of the Game'), in relation to *The Twelfth Battle of Isonzo*.

One may identify two starting points for *The Gaoler's Ache for the Nearly Dead* (written 1994, published 1998): firstly, the poem of the same name in

Barker's 1988 collection, *Lullabies for the Impatient*; and secondly, Dancer's exclamation in *Hated Nightfall*: '**Let them eat cake!** She was not cruel ... She was struggling with incomprehension, Marie Antoinette' (*HN*, 43). Barker's Introduction identifies his themes: 'Those whom the state wishes to destroy must first be vilified. Because revolutions are sexually reactionary, the most efficacious crime is that of sexual delinquency'; the French Revolution's alliance of the police and a newly uncensored press insisted upon a regime of visibility, transparency and surveillance in which 'Neighbours inform; children are endowed with innocence to act as mirrors in which popular sin may see its hideous image reflected' (*CP4*, 186). In this respect, the play also reflects and refracts aspects of the British press furore and moral panic surrounding the 1987 child abuse investigations in Cleveland, which involved the enforced removal of a high number of children from their parental homes on the orders of professional social workers, until foster placements for children were exhausted; some parents rebelled against what they considered to be a slanderous and punitive imposition of middle-class values in an atmosphere likened to a medieval witch-hunt, and relationships between police, paediatricians and social workers stretched to breaking point. The British popular press swung between an outraged sentimentalization of children involved and at risk, and an outrage at possible misuse of state power by police and government officials, leading to the exacerbation in some instances of a climate of fear surrounding actions which might be construed as an unnatural interest in children (even one's own) and thereby draw allegations of child abuse.

Barker's Introduction also identifies the necessity for the Revolution to demonize and sacrifice Marie Antoinette, publicly *abject* her from itself, as an incestuous degenerate: 'If Revolution is the apotheosis of a myth, Marie Antoinette, slandered by the Revolution, must be the myth of its opposite – depravity against cleanliness, license against order. In incest – invented, imagined, confessed – the crime of crimes, the transgression over the collective, hope and horror become inextricably mixed...'.

The Exordium of *The Gaoler's Ache* establishes pictorially a child on a baroque throne, who performs a casual but exhausting execution; the clamour of an animated, quarrelsome crowd which surges across the stage (anticipating some of the large scale choric effects in Barker's 2000 Adelaide production of *The Ecstatic Bible* and his 2005 production of *The Fence*[9]); a panel opens high in a wall, permitting periodic surveillance, before it shuts with a clap (an effect which goes on to inform the Exordium in Barker's 2000 production of *He Stumbled*). Queen Caroline warns her son, Little Louis, not to be alarmed by the sights and sounds of her intended love-making, but does so with a vividness of terms likely to increase his sense of mystery and fascination. The demagogue Trepasser (whose very name suggests that he is an incomplete trespasser) contrastingly seeks to achieve an 'absolute compatability with History' through his strangling of his own urges (his judgmental psychology recalls that of Angelo in Shakespeare's *Measure for Measure*, with a more explicit and acute sense of 'the reverence which is only fleetingly concealed by desecration', expressed through urination on a throne). Little Louis wants his parents to elude the growing Republican vengefulness headed and conducted by Trepasser; he urges them to abscond. Little Louis is one of Barker's philosophically curious children, rarely conventionally childish, failing to perceive the attractions or satisfactions of a swing or a ball, conscious of the necessary but unspoken 'silliness' of much behaviour, and even existence. In one particularly humorous scene, he orders some servants to disclose their varying sexual testimonies, but their more childish delight in gratification and simple but various forms of excitement are, to him, an incomprehensible **'idiocy'** which bewilderingly defies description. Little Louis takes refuge in logic and egoism to insulate himself against the threat to his father, identifying the fallacies in his father's conventionally emotional pronouncements, but even so briefly dissolves in the fear of death and grief that his attitudes seek to keep at bay; Caroline is differently, self-defensively dispassionate in contemplating bereavement and execution, refusing to give in to feelings which could overwhelm her mind. The royal family are under regular surveillance by Trepasser, who tells Little Louis that this observation may be a form of love, 'The gaze is never without its ambiguities', as confirmed by his own secret dream, a nocturnal entreaty for the child to 'piss on' him. In the fervour of his self-repression, Trepasser casts himself as an austere Pilate against Caroline as a (simultaneously vilified and adored) Christ; however, Caroline disrupts his analogy, insists she is 'not a bit' like Christ. On the one hand, Trepasser vilifies Caroline as sub-human: 'As a doll is not authentically a child, however cleverly it resembles one, nor is a queen a woman, however convincingly she masquerades as one'; her very flesh, skin and hair are 'made of different stuff' (*CP4*, 206). On the other, Caroline desperately propagates the myth to the inquisitive Little Louis that her urine can be venerated in intimate adoration because she is a queen, hence superhuman, only for him to ask 'How can it...? Taste differently...?'.

In a pivotal scene of simultaneous action, Little Louis is gaoled with Caroline, and expresses his last-ditch curiosity about the illogical power of his mother's body. He demands she uncover herself, and kisses her adoringly. Meanwhile, Trepasser instructs the Gaoler 'we are engaged upon an exercise of intellectual hygiene', purging 'infectious elements of superstition which have sustained the irrational institution of monarchy for so long'; the Gaoler agrees 'I'll say...She / Is / Unnatural'(*CP4*, 228). Caroline discovers in her intimacy with Little Louis a defiant, unbreakable resolve which recalls that of Webster's imprisoned and persecuted Duchess of Malfi: '**A real queen does her will**' (*CP4*, 229). However, Trepasser raises the stakes in his own strategy, gambling successfully on admitting to the crowd his own wretched 'love' for Caroline: 'pity's powerful...with one blow of her hand she lays logic low'; '*In an apotheosis of self-disparaging*', he suggests 'Forgive so much humanity'(*CP4*, 232). His strategic, and sincere, confession is vindicated as an example of dangerous play (putting himself in play, with all to play for) when the crowd bear him out. The ecstasies represented by persecution and confession are the permitted and compulsory forms of Trepasser's state; he has effectively enthroned his personal interpretation and mediation of the public as a vengeful, all-seeing god: 'The Revolution has abolished secrets ... That is its first achievement, the elevation of transparency, the cleansing of the smoked glass known as privacy, privacy for what I always ask ... If it's fit to be done, do it in public, or the public will surely find you' (*CP4*, 235). Conversely, in the abject depths of Caroline's cell, Little Louis reads a sense of supreme purpose out of his birthright: 'We are not human, are we? We are the monarchy'; moreover, he insists to Caroline 'I am your God. There can be no other'. When the Gaoler proclaims 'The child's no child' because '**he fucks his mother**', even Trepasser reels at what he first interprets as an informer's fantastic excess of zeal. Little Louis expounds his theory that a God, and a monarch, will distinguish themselves by their disdain for shame and secrecy. He redesignates their cell a 'Heaven', where he can claim privilege to do the forbidden. Even the Gaoler seems moved, and assures them of his silence. Caroline seems inspired by her son; in court she provokes uproar, and Trepasser's horror, by exposing her breasts (thus physically recalling and subverting the image of Eugène Delacroix's pro-revolutionary 1830 painting 'Liberty Leading the People'), and claims her own privilege:

> How should I feel modesty for nakedness that is not nakedness at all? To be naked is to reveal that which was yours. And these breasts were never mine, for the simple reason that I don't exist
> **I'll strip to the waist...!**
> **Hips...!**
> **Arse...! [...]**
> **Do you think I am afraid to die? ... I require it nothing less.** *(Pause)*
> To be parted from a body which never for a moment did I possess...can only be...deliverance, surely? (*CP4*, 242–3)

Thus her performance identifies, and physically subverts, the way she is expected to serve only as an object, of abstract veneration and of creating heirs. Meanwhile in the cell, the Gaoler subverts the expectation that (like Leary in *Fair Slaughter*) he is a decent man, humanized by the vagaries of his trade, when he fucks Little Louis. However, he also subverts the associations of melodramatic villain when he refuses to testify against Caroline, humiliating Trepasser by claiming the crime of incest to be the prosecutor's invention. Nevertheless, Caroline has to prepare for execution, and responds to Trepasser's dismissal of the truth as 'all appearances' by seizing him in her arms, kissing him fixedly on the mouth, the feminine abomination broaching the boundaries of the ostentatiously clean and proper body: *'He staggers, she clings. Discordant notes. An absurd dance. The figures observe'.* Caroline's insistence on playing out the forbidden subtext of the state's repressed sexuality infuses the Judas Kiss with the fierce choreography of Dancer's murderous *pas de deux* in *Hated Nightfall*; Trepasser is disgusted by her saliva, but he is the more contaminated of the two. When *'with a surge of will power she launches herself out of the room'* to death, Trepasser seems contrastingly drained: he *'obsessively'* propels Little Louis, who plays childishly, for the first time, on a wheeled horse, and *'shouts in triumph'*.

As will be apparent, the speculation at the heart of *The Gaoler's Ache* overturns conventional morality and sentimentalities about children, morality and sexuality (perhaps the ultimate popular taboo of the late 1980s and 1990s), and paves the way for Barker's later sympathetic portrait of Oedipal incest in *The Fence*, in which a central couple who are sexually criminalized are similarly shown to be less injurious or invasive than the institutionalised State. *The Gaoler's Ache* also involves the performer of the role of Little Louis to perform actions and encounter complexities beyond those which child actors would be legally permitted to consider, and in performance the role would necessarily have to be played by a suitable adult actor.

The challenge of interrogating terms of permission for children, and the performance of them, seems to carry over from the role of Little Louis to that of Kisster, protagonist of Barker's *Twelve Encounters with a Prodigy* (written 1998, published 2001, staged 2008). Kisster, a twelve year old boy, reveals the scene of his birth, as catastrophic as the landscape of Barker's Arcadia: 'I was born at a roadside. Planes swooped overhead, firing cannon'; smoke from burning fields attracted 'looters, rapists and psychopaths'; his mother hid him in a drain, at the cost of her own terrible suffering. However, his monologue has an audience, and therefore the status of performance (compare Katrin's first speech in Barker's *The Europeans*, which is revealed not just to be an atrociously post-catastrophic self-consciousness but also a formal confession, perhaps partly intended to discomfort the attendant priest). Kisster is being schooled by his governess, Cologne, in the recitation of a fictitious self-exposition, 'For each time that it's told it must possess the quality of revelation and not be anecdotal', with the incentive of her favours: 'Then I. My black centre. I.' (*CP5*, 149). This is designed to give him at least an opening gambit and a subsequent self-mythologizing licence in engaging with Coercion, 'the first law of human behaviour', which wears the masks of 'Pity. Argument. And Shame', of which the greatest is the first. Thus, Kisster is practising to be a 'perfect liar' (to borrow the description of Placida from the character list of *Ursula*). His age might seem initially to exclude him from the arena of pains and pleasures which constitute adult life; his preparatory observations of adult dealings permits him both interrogation of, and curiosity about, their conventions; but his only partial observance, or strategic subversions, of these laws is partly what gives him his formidable, literally prodigious power in his transactions with others. He learns to taunt his blind, lascivious tutor, who suspects Kisster intends to make the entire world 'squirm on the rack' of his will (though Kisster later wonders 'Perhaps the world deserves the pain… that I shall give it?', *CP5*, 170); however, when the tutor suffers a fatal seizure, Kisster can affect a more convincing grief than his compliant friend. Cologne presides over Kisster's meeting with his mother, a woman 'delighted' in her life of vulgarity with a wealthy husband; she finds his *Bildung*-performance inescapably sexually alluring, then severs herself from him. Kisster encounters a wealthy landowner's daughter, Gabriele, and abjectly performs as one of her dogs, earning him a beating, not only from Gabriele, but from Cologne, who is outraged at his seductive complicity and consumed with sexual envy for Gabriele's attentions. Denounced as 'mad' ('because I have ceased to admire the world'), Kisster can prove charismatically commanding even to his denouncers because he nevertheless possesses a self-love which is a pillar of crucial support when 'I require things of the world which possibly the world can never give me' and perhaps 'does not possess' (*CP5*, 184). A philosophical writer, Marston[10], gives Kisster his book, a testament of cosmic pessimism, in the hope that his tome has found its proper reader, but this may be a further form of attempted coercion; as are the attentions of two angels who bid Kisster rejoice in his boyhood and innocence, in what Kisster interprets as a fearful bid to stifle his never-childish character. Gabriele's father fatally pinions her to her new lover. Marston reappears, offering another book, 'A Perfect Love', in which everything is 'Reversed' from his previous misanthropic aphorisms, and Kisster rejects him **'What use are you…?'**. Kisster uses the book to sop up the lovers' blood, wringing it out and working his way across the stage *'like a washer-woman'*, an image to which Barker will return in *Ursula* (similarly, Kisster's chess game with a vagrant foreshadows that of Photo with Youterus in *The Fence*, and the erotic atmosphere of the forest whipping scene is recast in the final playlet in *13 Objects*). *Twelve Encounters* is in effect a compendium play with a common protagonist in each episode, a precocious boy genius whose repudiation of the world reflects his temperamental preconditions to be a Barkerian saint: his desire is focused on what can never be satisfied, his passion unable to discover a worthwhile object, and the more searching his gaze, the more contamination is revealed. Perhaps even more distinctly than *Hated Nightfall*, Kisster's encounters demonstrate how, to return to Barker's essay 'Saintliness, Death and the Perfect Family': 'all is transparent, and this transparency is appalling pain, for we require to be deceived, it is the condition of social acceptance. Without deception there is perhaps no hope'; and how 'the saint is lured by human love only to discover its inadequacy … the saint is first and foremost his own work of art, exhibited primarily to himself. But neither saints nor works of art are socially desired'. However, the form of *Twelve Encounters* intrinsically militates against Kisster's development of a passionate project with a compelling arc and repercussions comparable to those of Dancer, Starhemberg, Vanya or Photo. *Twelve Encounters* thus frequently resembles a philosophical character sketch, speculating on the formative days of a Nietzschean nihilist who remains perhaps too much in control of surrounding events to gain as much purchase on the memory as the above protagonists.

At my time of writing, several of the 1990s plays under review in this chapter – *Ego in Arcadia, Twelve Encounters with a Prodigy, The Brilliance of the Servant , The Gaoler's Ache* – await professional productions (in Britain in the first two cases; the rest, anywhere). Given their traditional form of Arts Council funding, it was impossible for The Wrestling School to keep pace with Barker's writing output and give first productions to all of his works; but, allowing for this, it is remarkable that no other professional directors or companies have ventured into the direction of any the above published plays, for even the slighter works (*Servant, Prodigy*) have their profoundly problematic pleasures, and the best (*Ego, Gaoler*) stand as major Barker works, prescient of recent examples of social catastrophe and the internalisation of state surveillance, amongst other things.

Barker's fifth volume of poetry, *The Ascent of Monte Grappa* (1991), develops further some of the themes explored in *Lullabies for the Impatient*: notably the ironies of time, the decay of ideals, the dialectic of emotion and calculation, and the discovery of an implacable dignity in refusing assent to the comprehensive reductivities of the merely contemporary vantage point. His most important breakthrough of the 1990s in terms of refining his explorations of sainthood and dramatising sacrifice occurs in *Ursula*.

Bright River, Dark Sea: *Ursula*

> The darker streams of pity ... they are black with sex
> *Animals in Paradise*

The subtitle of *Ursula* – *Fear of the Estuary* – is particularly resonant. The play sites itself at the juncture where a river flows and is drawn to surrender its separable identity and merge with the sea. A delineation of the simultaneous compulsion and resistance regarding the loss and remaking of the terms of the self, in the solvents of desire, characterizes many Barker plays, but none more strongly than this one. *Ursula* as a play is both unusually easy and unusually difficult to summarise. On the one hand, the inexorable movement of its single self-contained narrative, based on the legend of the martyrdom of Saint Ursula, seems straightforward:

Placida, a mother superior (and a Barkerian invention rather than a character from any form of the legend), grieves for the loss of her favourite novice, Ursula, in promised marriage to a prince, Lucas; however, Ursula experiences what she considers a divine revelation, and meets but refuses to surrender her virginity to Lucas; when he delays the return home of Ursula and her accompanying sisterhood, Placida displaces Ursula in his sexual attentions; Ursula's denunciation of Placida drives Placida to ordain and participate in the slaughter of Ursula and all the novices. Compared to the multiple and non-linear narratives which Barker mobilizes in plays like *The Bite of the Night, Rome, Found in the Ground* and *The Ecstatic Bible*, the narrative of *Ursula* seems to have a crystalline transparency in that it apparently invites, yet remains invulnerable to, allegorical interpretation. The episodes of the narrative unfold entirely sequentially, but the effect is to make the play more strange and unnerving, in its mysterious inexorability, rather than less so. It parallels the way that Shakespeare's contemporary detractors criticized him for his non-observance of the traditional dramatic unities of time and place, only for him to retaliate with a wry joke in his play, *The Tempest*, which observes these unities, but which is, if anything, more resistantly outlandish in so doing. Similarly, *Ursula* has the unfathomably unsettling compression of events, and power of intimation, of a dream, which can move alternately with petrifying deliberation, and with bewildering and irrational speed. Its imagery of water and geological instability, of juncture of dissolution between contesting land, sea and river, serve to emphasise the comparison. Anne Barton has remarked how *The Tempest*, 'Like an iceberg ... conceals most of its bulk beneath the surface'[11]. This suggestion, with its associations of submerged facts about characters and situations as well as Stanislavski's theory of the subtext, may be a surprising claim to make for a Barker play, when the author himself observed in 1987:

> I depict the self as wholly articulated. Whereas most drama relies on the unexpressed moment – the loss of speech – my characters excoriate language for emotional exposition, it is a need and hunger in them. In a sense, this is itself a performance, within the performance of the actor, a self-consciousness and a self-making, for others as well as for self. So that the insistence on knowing, which characterizes so many of my inventions, is only subverted by the impossible pain of the effort. They rack themselves, falter, insist again, on the public nature of their experience.[12]

However, Barker's increasing development (in his 1990s and twenty-first century work) of what might justifiably be considered as an additional dramatic character (if not several) through the specifications of scenography, effects of sound, light and movement, serves to make the world around these continuingly and compulsively articulate characters less readily *know-*

able than in his earlier drama – which makes the experiences of the characters, and audiences, more disoriented and anxious. Their bids to establish an equilibrium of *definition* are more urgent, but less ultimately successful. *Ursula* represents a new level of confidence and achievement in Barker's paradoxical orchestration of an approach to the experience of *indefinition* through the precise construction of individually *definite* elements.

Again like *The Tempest*, *Ursula* is unusually demanding of imaginative activity on the part of the audience, through its insistence (in both form and events) on intensity, concentration and the apparent self-sufficient complexity which characterizes both myth and divine revelation, demanding interpretation and expansion whilst being ultimately irreducible. *Ursula*, being a Barker play, does not depend on the suppressed and the unspoken as much as conventional drama; but, like Shakespeare's play, it proceeds by a set of partial disclosures, so as to compound, rather than dispel, the aura of strangeness and inscrutability, by provoking more questions than they resolve; confusing entities and characters who are normally thought of, and initially appear to be, distinct (centrally and pivotally, the positions and stances of Placida and Ursula); 'undercutting the sense of superiority usually felt by the theatre audience'[13] when actions of apparent resolution remain ambiguous and 'there are almost as many opposed ways of seeing a given event or moment in time as there are characters involved'[14]. Though there are less (groups of) characters in Barker's play than in Shakespeare's, the characters in *Ursula* are (similarly) definitely separated – into Ursula, Placida, Leonora, Lucas and the novices – and, though they speak insistently, they communicate only imperfectly. Whilst all the separately identifiable episodic scenes in the narrative sequence of *Ursula* serve to advance the possibly familiar or predictable story of a legendary historical martyrdom, aspects of the scenes and characters insistently elude comprehensive predictability; instead they compound a sense of tensile foreboding and suspense.

This sense of tension and determinedly commanding but imperfect poise is a keynote from the outset: Placida is keenly aware, not only of her charismatic powers of leadership over the novices, but also of its limitations, as represented by both the imminent presence and prospective marriage of her favourite, Ursula. She audibly struggles to find the resolve to acceptance and appropriate interpretation, but cannot extirpate the contrary workings of her own will: '**I'm so I'm so … Delighted for her if that's what she …** It isn't however is it…? What she wants…?'. Leonora appears, apparently blind and self-consciously unlike Ursula, but determined to replace her through willed transformation: 'I shall be Ursula / So very Ursula you will exclaim how little the first Ursula became herself'; she then deliberately discards her performance as a flattering beggar, and (only then) Placida accepts her, in terms of conscious cruelty for which she blames the still-unseen Ursula. Placida shocks the class by insisting Leonora's blindness is further subterfuge. When one novice, Phyllis, suggests that marriage is 'a permanent reflection of the self' akin to a room of mirrors, Placida disrupts the proposition with her own anxious sensing of Ursula's proximity; then channels this anxiety into an overcompensatingly energetic taking up of Phyllis's terms, and a significantly intent deduction that marriage is sustained by vanity, 'the mirroring of self in self – and that forlorn hope that in another's flesh might be discovered that refuge from solitude which in reality pertains only to God'.

When Ursula makes her much-anticipated appearance, she seeks to undercut Placida's superiority with physical shoves and wilfully contradictory utterances; her relationship to Placida is thus more reminiscent of rebellious daughter or younger sister. Ursula's adoration of her fiancé's portrait does not conceal her own uneasy curiosity about the physical mechanics of sex: 'Perhaps it is authentically ugly ... because it's loathed by God'; nevertheless, and not for the last time, she takes up a rhetorical position of opposition to Placida: 'Tell me the beauty of virginity and I will tell you the beauty of a man'. Placida grimly intimates that the sighing of men around women in general (and her in particular) harbours murderous impulses. Ursula's performance gives way (to another?) as she flings herself down, sobs and pleads for Placida's blessing, which Placida withholds; Ursula, in response, performs her own superiority to it, in prospective maternal and romantic authority. Placida '*writhes*' at Ursula's exit, and calls out for '**The blind girl**' as if to displace and visit her frustrations violently upon the consciously abject, doglike Leonora. The conflict between Ursula and Placida indicates that there is more personally invested by, and at stake for, each of them, in their relationship with each other, than the surface terms of debate would suggest.

In the second scene, Lucas appears, supremely self-conscious and 'imperial', ostentatiously refining and demarcating his life by the challenges he sets himself: in this instance, by swimming across the river, thereby marking a passage of his life, to be associated with the foolhardy 'Lucas-before-Ursula'. The peasant women who flock around him (in a manner resembling the deindividuated chorus of servants in *Minna*) reflect back his sense of sexual intoxication, even as he dismisses them for lacking Ursula's nonpareil 'yellow hair'. The third scene occurs in (as yet, and literally) unshared reality, but '*upstream*' of the same river. Leonora dogs Ursula, and they taunt each other with their respective awarenesses of the vanity involved in Ursula and Lucas's resolved formal correspondence. It is important to note that, despite the folklore connotations of the story and courtship, Ursula is no conventionally meek and anodyne fairytale princess; and though the trepidation which the novices associate with Lucas and his domain may recall the sexual dystopias of Bluebeard and *La Belle et la Bête*, Lucas disarmingly combines his boyish egoism with a philosophical thoroughness in speculation, and will be capable of seductively offering willed mutual transformation into the animality of 'dog' and 'bitch', rather than demanding restriction to a single shape. *Ursula* evokes fairytale motifs (as will *Knowledge and a Girl*), but does so in order to subvert them and enfuse them with sexual complexity (in some ways recalling Angela Carter's 1979 collection of stories, *The Bloody Chamber*). In one of Barker's most inventive uses of the split focus on a single stage, a vision of Christ "speaks" to Ursula through the body of Lucas (literally present, but figuratively absent from her scene, on the stage), unseen by Placida and the other novices. This dramatises vividly and internally Ursula's sense of a revelatory confrontation, whilst also suggesting elements of displacement and fantasy[15]. Ursula deduces that 'the blazing banner of my sensuality', her yellow hair, constitutes the 'ordeal' which she shares with Placida, of compelling men to **'yearn for what belongs to Christ and to Christ only'**. When her revelation is interrupted, Ursula denounces the 'infinite tranquility' of the novices, who have retrieved the portrait of Lucas which she discarded into the river, as a suffocation. Placida interprets Ursula's experience as a complex blessing, a deliverance from the hope and false promises associated with sexual intimacy, and the 'solitude and melancholy' which follow from the inevitable consequent disappointment; rather, deliverance from the stagnancies of the mortal self are possible through Christ alone (this deduction is theatrically shaded with the piquancy of the sight of Leonora, sobbing with hope and relief, clasping Lucas, the subject of her infatuation who is nevertheless '*still as a monument*'). Placida determines to 'overcome' and 'pacify' Lucas by converting him to Christ.

The difficulties involved in this are suggested by the next scene, which reveals the extent of Lucas and Leonora's subterfuge, and her passion for him. Leonora is Lucas's former lover, who has infiltrated the convent under the guise of blindness to report back to Lucas (she remains desperately infatuated with him, despite his protests that he has no remaining feelings for her which she can provoke). Leonora detects an antithetical, elemental antagonism between the two principal women: Ursula's 'naturalness flows up against the Mother's falseness' and 'erodes it'. In imaginative self-delusion, Lucas tries to interpret the recovery of his portrait as a sign and proof of Ursula's love for him, "overruling" Christ; Leonora more accurately reports her intending refusal to be his bride; her switch back into compliance with her former lover's longings – 'Blind again' – is a trenchant image of her self-disabling accommodation in love, which remarks both on its own hopelessness and on its abject necessity.

Act Two opens with a comic interlude which demonstrates Barker's skill in deftly and humorously delineating status and rivalry within a group (such as the committee scenes in *No End of Blame* and *The Europeans*, and the bankers' scene in *Victory*). The novices consider their imminent voyage with mingled excitement and trepidation: it is apparent that Benedicta is particularly young and naïve in her enthusiasm, Anne is pragmatic,

Lamentina anxious, Cynthia an outsider, Perdita eager and obedient, and Phyllis is, amongst them, the most charismatic (and perhaps most beautiful), somewhat resentful pretender to Ursula's status as "star pupil". Leonora punctuates their self-reassurances with her discomforting laugh, and compounds the tension amongst the girls with her unsettling, if wryly comedic, performance which equates their voyage with that of the river, drawn to lose itself in the sea. She also goes on (whilst twice disingenuously avowing 'I won't go on') to ask whose self-performance is the more fatalistic and morbid: that of Ursula, who since her reported encounter with Christ has become so '**Deathly**', or that of Leonora herself, who mockingly draws attention to her own artful performativity, bathetically professing her own "liking" of Christ and noting her own ability to unnerve them, which even extends to acknowledging the ambiguous "flicker" in her own performance of blindness.

As if to confirm Leonora's dry ironies, Ursula prowls the deck at night, commenting on the ominously 'turbid' landscape. She tells Placida that the novices are 'convinced they will be murdered why don't you reassure them'; in fact, Placida demonstrates a charismatic fatalism, that of the determined seducer, prepared to hazard all; Ursula glimpses her persistent resolve despite, or because of, the intimation that they will '**be killed by Lucas**'. Ursula herself tries to wrench free of Placida's powers of compulsion ('**Turn the boat around**') but is nevertheless also mesmerized when Placida defuses an emissary of discontent from the novices; then Placida is '*seized by a sobbing which is also a laughter, a kind of repressed triumph*', and Ursula '*takes her in her arms*'. Thus, even the self-conscious limits of Placida's performances are charismatic, and it is this power to impose the fictions of her will on others, alongside her consciousness of their/her fictional quality, which makes her (literally) supremely artificial, a 'perfect liar' as the characters list describes her.

Meanwhile, Lucas develops his own rhetorical image of Christ as a charismatic paradigm to be emulated. He considers Christ's disappointment that the 'Humans' fell asleep when he requested them to watch with him before his arrest; he deduces that this constituted a failure of judgment and nerve; nevertheless, he ascribes to Christ, in his most successful moments, a ubiquitously successful stillness in '**Flirtation**' which prepares the way for command: 'Ursula / I'll have your cunt now I will say'. His pre-eminently strategic and tactical admiration and invocation of Christ is the delineation of a philosophical scaffolding for a subtle and certain campaign of self-performance[16]. He discards the fates (and lives) of the other virgins with significant disinterest.

As the virgins near Lucas's domain, Cynthia's repression erupts into proudly lurid boasts of her loss of virginity, and this break in the ranks makes the other girls turn on her, trussing and gagging her. Contrastingly, Placida seems to thrive on the proximity of danger, appearing in a hat which Leonora proclaims '**perfect**' in how it reveals her mouth; and Ursula senses her own authority displaced. Lucas meets and surveys them, in a charged atmosphere of barely repressed emotions, tragicomically expressed by how '*in the silence, the gagged and bound Cynthia emits strange and eerie sounds*'. Lucas's formal attitudes of chivalrous eligibility strike up against Ursula's assertion that she greets him, but '**Never as a husband**'. However, Placida retells the story of Ursula doing Christ's will by discarding Lucas's portrait, but also associates His omnipotence with her own disclosure of the recovered painting, prompting in Lucas a '*short laugh*' of recognition. When he leaves, Cynthia's further sounds of fear and anxiety are the most extreme indices of a slightly hysterical uncertainty which pervades the women. Ursula is appalled, and even Placida is nervously confused, by Placida's apparent assumption of the role of courtly recipient of Lucas's flowers and entreaties. Placida attempts to subdue the collective and individual disarray by instructing them to '*sing defiantly*', restating their religious resolve; though, in practice, the sight and sounds of Cynthia will prove significantly and irrevocably discordant, and irreducibly ominous.

Act Three – and usually, in performance, the second half – begins, strikingly if enigmatically, with another example of a psalm: in this instance sung as a solo by Lamentina, attempting to convey and instil an attitude of assured (resigned?) fortitude, even as the bound and blindfold Cynthia presents a contrasting image of strained repression. Lucas expounds a disquisition on how his 'lost face' (caused by Ursula's denial of him) can be replaced by another, strengthened by a sense of shame; but this is belatedly revealed to the audience as a rehearsal of the next strategy in his sexual siege of Ursula. His subsequent repetition of the speech to its proper recipient focuses audience attention on the physical and spatial nuances between the characters; then he discards rhetorical sophistication for intimate directness ('fuck with me be a bitch I'll be your dog') and other attitudes of power, both courteous and domineering, symbolised by the presence of the sword, intended to 'frighten and reassure' simultaneously. Ursula's offer of a limited friendship is for him rendered impossible by her sexual allure ('**But you have yellow hair**'), and, though they express a shared wish to be 'parted from' their selves and made anew, Ursula sees this as possible only through Christ, Lucas as a specifically sexual crash through heaven's door: '**Through you / Your belly**'. Ursula leaves, in defiant disobedience, referring Lucas to Placida.

When Leonora breaks in on the (now truly) discountenanced Lucas, he makes the bathetic admission that the 'monstrous' sword is borrowed from the local executioner and baker's assistant, 'I can barely lift it', even as he dismisses Leonora as a 'former intimacy' with no rights. Wrestling with her grief, Leonora suggests that Lucas address Placida, her central assurance and authority, raising the stakes and anticipation of their private encounter. Meanwhile, at least some of the virgins seem prepared to consign Ursula to defloration in order to save them, suggesting the 'hideous violation' and ruin of her 'private space' is nevertheless her personal 'ordeal' to be borne for the wider good, like Christ's crucifixion. However, Ursula recognizes and rejects the utilitarianism of this argument, claiming the primacy of her individuality. The appearance of Placida once more conjures the momentary relief of the 'melancholy flock', who '*laugh, encouraged, relieved by her own radiance*'. Ursula weeps convulsively, until Placida embraces her until the fit subsides, just as Ursula had embraced Placida on the boat after the nocturnal deputation: further evidence of how the two women, alternately and uniquely, provoke both turbulence and calm in each other. The deliberation in Lucas's announcement that he has an 'Appointment' clears the room but for him and Placida; when she partly lifts the sword to clear a space for the '*toppling cornucopia*' of fruit he carries, he seems enraged by this reminder of his own weakness and scatters the fruit (though she is unshaken, and may even seem mesmerized or fascinated by the phallic power represented by the sword). Lucas disarmingly claims the error is his, 'for having placed upon another the burden of my hopes'; however, Placida has already been settling herself into a carefully composed attitude from which to offer herself, claiming her age and virginity as enhancements ('I've known no ecstasy but nor have I known shame'), moving from composure to a near asthmatic seizure of urgency and trepidation. On the one hand, she maintains that virginity is the only means ('rope' or 'cable') by which one may ascend to Christ; on the other she provocatively urges him, 'Do with me what must fail', in terms of self-transcendence, setting up a resistance to be willingly challenged. Mistakenly but significantly offering her an apple, Lucas succumbs to her temptation, even as she gives way to being overpowered, both by his body and by the rising tide of fearful excitement within her.

The tension and release of coitus are succeeded by the appearance of Cynthia, still roped and blindfolded and hopping across the stage of scattered fruit, a comic image of the return of the repressed, exultantly naughty in her experimentation with uttering obscenities and sexual fantasies of Lucas. However, Lucas realizes her fantasy by rising from Placida, and bearing Cynthia away for further defloration. Placida is literally nauseous (at her own conduct? Or his? Or both?), and her frailty panics the novices, though only Ursula knows its cause. She repudiates Placida's sexual acquiescence in established terms ('Ugly because it's loathed by God ... The self comes back again ... Obviously it fails'); however, Placida's subsequent '*spasm of energy*' enables her to pick up the sword, as if in contradictory reply to Ursula's dismissal of the possibility of change in the self. Cynthia re-emerges, deflowered and desolate, her plaintive cries of 'Mother' going unacknowledged by Placida, even when they become shocked realizations of Placida's differently focussed presence; a difference which is further expressed by Placida refusing Lucas's bid to return the sword which she now holds: 'Do I not need protecting?'. This begs the question: *from whom* does she needs protecting? Not, it suggests, from Lucas.

Act Four Scene One formally echoes the first scene of the play, but with crucial differences: Placida announces that she (not Ursula) is marrying, and though the virgins enter as before, pulling their chairs across the floor, they are all blindfolded in a pointed avoidance of the sight of Placida, an act of defiance led by Ursula. Placida taunts them in their refusal to witness her 'ecstasy', recharacterising herself in terms of sexual liberation, proclaiming herself naked and with '**Nothing to disguise**'[17]. The stand-off is disrupted when Leonora proclaims in anguish that her pretended blindness has suddenly become real. Ursula interprets this as fortuitous, and incorporates this into her sense of the proper rejection of the sight and sound of Placida, as an image of (not liberation and discovery but) of 'The collapse of every love', opposed to the surety of Christ, '**who is forever**'. Ursula and Placida have literally changed places from the first half in their statements of belief: Placida proclaims her own incandescent self-reinvention in a collapse of religious authority and an embrace of worldly passion and marriage; Ursula regards and denounces Placida as an incarnation of contamination in an inadequate world which she repudiates, refocusing her own passion on an immaculate desire for the purity of sainthood, ultimately conferred by death. Lucas knows how to reverse the power of Ursula's abhorrence, to turn an intended rebuke into a badge of shameless desire, telling Placida 'Our marriage is a sin' and therefore 'perfect', a defiant transgression which her age and possible sterility only heighten. But Placida withdraws, suggesting that there is also something (or someone) from whom she is imperfectly severed and therefore finds recurrently offensive: 'Ursula however … She was my dear'. Just as Placida and her form of power are intolerable to Ursula, so is Ursula('s) to Placida; their mutual 'erosion' has been reconfigured, but exacerbated rather than lessened.

In a scene which atmospherically opens the play up from the claustrophobic interiors which have so characterized its second half, Leonora seeks to merge literally with the estuary, drowning her blind misery; an impulse with which Ursula sympathizes. Ursula describes her opposition to Leonora as purely formal, 'rather like chess pieces' who have no actual hatred or quarrel when 'at the end of the game they are thrust back into the same box'; Leonora recognizes in this a rather strenuously philosophical bid for reconciliation, redolent of the former Placida. Even as he rejoices in Placida's unfurled sexuality, Lucas can regret Leonora's blindness, free the sailors and unquarantine the boat for the novices' return; however, Ursula insists on 'the evil' of the marriage, and insists they also take Placida, 'If necessary tied with ropes': such is their horror at the unforgivably sexualized body of the woman they continue to call 'the mother'. The issues (and investments) of ownership implicit in invocations of love and sanity are foregrounded here: Lucas's newfound generosity to others, which he attributes to 'love', is mocked by

Leonora: 'They think you're mad. And jeer at you'. Their interpretations of love/madness are irreconcileable. Indeed, the end of the scene, in which the virgins assemble in attendance like a human cloud, and Ursula and Leonora give in to a tremor, which involves a laughter and sobbing which evolves into *'a radiance'*, suggests that Ursula's own frame of mind is far from conventionally balanced, whilst also demonstrating her charisma and power increasing in proportion to her self-severance from mere worldly rationality. In the context of the scene, the increasing pitch of her self-aggrandisements – 'We laugh as Christ might have...If Christ had ever laughed' – is profoundly ominous, as it attempts no less than to claim the terms of spirituality and abomination (and, of course, their mediation, which is sacrifice).

Even a pre-marital scene, which Placida and Lucas play out as an almost domestic attempt to confirm their own sexual idyll, carries the repercussions of this tremor. Placida cannot separate herself from the virgins and their claims, so she determines 'If they won't go then they must stay'. Lucas notes how (like the Lvovites in *The Last Supper*) their rejection of other worldly claims makes them adhere (fatally): 'They tie themselves together'; Placida is herself shocked by the cruelty of her own response to their inflammatory immobility: '**Then hoist them in a net**'. As Lucas observes, in owning their souls, Placida also owns their impurity, 'Their vomit and their blood'; though she professes a wish to '**wash [her] hands of them**', he knows '**You cannot wash your hands of love**', which 'sticks' and 'coagulates' like taboo bodily fluids. The virgins represent a destabilizing abomination to Placida, as she does to them. There is a growing pressure on both sides which suggests that only sacrifice can mediate metaphorically, but in necessarily and literally and lethally violent fashion, between their heterogeneous and incompatible terms.

In the remarkable climax of Act Five, the virgins are wheeled on by a workman, *'each strapped to a steel trolley'*; in *'the peculiar peace that follows an accident'*, Ursula leads them in a psalm, in attempted self-reassurance through their formal and technocratic subjection. Wearing a scarlet (or black[18]) bridal gown and successfully wielding the sword which Lucas cannot lift (because it is she who has 'the need'), Placida determines to play out the dark power associated with the sexual (what Kristeva terms the Célinian) mother, who marries 'beauty and death' and who, in 'giving life' and 'snatching life away', points to 'the ephemeral nature of sublimation and the unrelenting end of life'[19]; she also rejoices in the assurance that her status as 'mother' will be realized in secular and sexual terms, 'I am watered hourly / His river in my sea', claiming identification (not with Christ but) with the very landscape, '**We are the estuary**'. Even now, Ursula attempts to claim Placida through a forgiveness, which Placida repudiates, though stung by Ursula's insistence that her charismatic voice has degenerated into a 'rough grinding'. As Leonora exhorts the virgins to 'love' the abstract and transcendent 'mother', Placida draws the blade across each throat, finally confronting Ursula, for whose unrepentant cruelty Placida expresses a liking, even as she kills Ursula with *'an effort of will'*. Lucas appears shaken by what he witnesses, dropping a bottle and glasses at this last killing (and thus theatrically underscoring its shock to the audience). He attempts a mutual reassurance, claiming that he now owns 'the whole' of Placida; and Placida claims that the novices' martyrdom has saved them from disappointment by enshrining their perfection in virginity and faith: 'Their hope is safe because of me. What more can a mother do?'. But even now, the limits of control are emphasised, in terms of literal physical balance: the attempted closure of Lucas's exit is spoiled when he slips (not once but twice) on the blood, and Placida exits separately from him. As throughout the play, effects of sound – in this instance, the noise of a bucket kicked by a figure mopping up the blood, the curlew's call – contribute crucially to the sense of desolation[20].

Thus, at the end of the play, Ursula claims her own dramatic authority as tragic heroine and martyred saint, in terms redolent of Kristeva's: 'The killed object, from which I am separated through sacrifice, while it links me to God also sets itself up, in the very act of being destroyed, as desirable, fascinating and sacred. What has been killed subdues me and brings me into subjection to what has been sacrificed'[21]. This seems particularly true for Lucas, whose totalitarian 'ownership' of Placida seems flawed and qualified, as he is literally rendered off balance by what he has witnessed and colluded in. Indeed, Placida's words beg the question as to what extent, or on what level, she is always trying to give the virgins what they want: did she initially offer herself sexually to Lucas, and surrender to the act which 'must fail' and sever her own connection to Christ, as an intended self-sacrifice and attempted preservation of the virgins, or at least their 'hope'? Or does her alluring millinery and seductive voice and rhetoric suggest a barely repressed eroticism awaiting the right moment (and man) to flow? Or are premeditation and calculation drowned by the excess of experience? Her passionate imagery ('Dredge my darkness') suggests a thorough and eloquent abandonment to the erotic (like Livia's in Barker's *Women Beware Women*), and she claims her own dramatic authority as catastrophist heroine, who bursts through the constraints of her former faith to demand consciously excessive life. Placida partly recalls Shakespeare's Cleopatra in her regal respresentation of an erotic order – which is, in this case, not subordinated to material politics by some Caesar, but victorious on the worldly terrain she values; rather, the transcendent sphere is claimed by Ursula, and to this extent, *Ursula*, like *Antony and Cleopatra*, gives a victory – even a different ecstasy – to each side in the terms which they value and claim. However, Shakespeare's dramatic sympathies identifiably tend towards his eponymous lovers in affirmation of their festive tragedy; *Ursula* is more even-handed in its dramatization of the conflicting forces, and more desolate in the sense of pain and investment on both sides. Neither Ursula nor Placida is obviously, singly, identifiably "right" or "wrong" in this play, but each impels the other towards their respective course of action, or arguably renders it inevitable. Ursula is a compulsive irritant, who finally perfects the expression of her 'perverse and egoistic' will to sainthood; she embraces the tragic irony that decrees that, in preserving her womanhood for Christ, she consciously condemns herself to death. Placida may seem the more recognizably Barkerian heroine, lethally sexual adversary of the handmaiden of ideology (compare Dancer); though it is also true that Placida is (irresistibly?) driven by Ursula to become another form of ideologue, like anyone who is lethally committed to her/his own terms of salvation. But it is crucially important that Placida kills Ursula with a manifest *'effort of will'*, lest she seem a merely sadistic tyrant: the sacrifice requires, and costs her, her former 'dear', whom she 'likes' to the end. Ursula's image of the characters as chess pieces captures some of the formal strength and balance of the play's design, and its inexorable process of polarization, but does not account for the extent to which the characters' consequent hopes and despairs are fathomed (Lucas is not deeply emotionally imaginative as a principal character, but is nevertheless disarmingly eloquent, witty and sincere)[22]. In both Barker's (1998) production and my own (2004), the identifiable use of the Cynthia actress as assisting in the execution in the final scene helped with a practical problem of onstage (wo)manpower, but also compounds the sense of reversibility, in suggesting that Cynthia has gone over to Placida and Lucas's side, and feels herself able, and required, to preside over the deaths of her former "sisters".

Barker's Wrestling School production of *Ursula* used the full depth and darkness of the main stage in its premiere at Birmingham Repertory Theatre, creating a powerfully ominous sense that people, sounds, and things, emerged from an unpredictable dark. Objects (such as the portrait and sword) were flown in on wires, and unnatural sounds such as shrieks and discordant music (by György Ligeti) suddenly rent the air, and deep rumblings carried a sense of foreboding from the end of the first half into the interval. Such scenographic effects heightened the play's powerful sense of indefinite (and therefore limitless) danger, in disturbingly non-rational ways. This sense of suspense was reflected in the balletic precision of the performers' physical postures, which frequently exposed the effort involved in a character's poise, though physically manifested counter-tensions: the holding of a balance whilst exploring its limits, or the quivering maintenance of a "flickering" state of delicate off-balance. Victoria Wicks's portrayal of Placida was central and exemplary in its movement from one state of tensile poise (anxiety, effortful courtesy, command, abandon, elegant murderousness) to another. These physical and scenographic aesthetics would be developed in Barker's subsequent directing work, but they emerged and came together in *Ursula* to unprecedentedly powerful effect, providing a style of production which was as unique and as irreducible yet resonant as the play *Ursula* itself. This resonance might be likened to the trembling of an earthquake, involving a radical shifting of ground that is the effect of two heterogeneous elements or states colliding with each other, refusing synthesis: that which Lyotard calls the 'incommensurable' event[23]. The play and production constituted confident new benchmarks in Barker's theatrical achievements.

6
Facing the Wound

'No Moment of Unity is Ever True'[1]

I have noted earlier (Rabey 1989) in Barker's work a distinctive (and Shake-spearean) interest in the contrary tensions of the impulse to define (to confine in words or images) and establish meaning, and the experience of imagination which nevertheless demolishes meaning through a catastrophic explosion of its terms. For example, in Barker's 1985 play *The Castle* both the play and its eponymous edifice are 'by definition, not definitive'; the characters are (again, like those in *King Lear*, *Hamlet*, *Macbeth*, and many other Shakespeare plays) 'endlessly involved in attempting definition, for-mulating impulses to contain or commit, which fail, even as they provoke or challenge similar attempts on the part of the audience'[2]. Frequently in Barker's work the body is both focus and disruption of power. His very syntax compulsively raises the spectre of indefinition, in his use of the hinge word 'BUT' to overturn all preceding rationalisations, and his use of the incomplete sentence (*aposiopesis*) which provides the cue (and semantic scaffolding) for confronting the infinite: as in the line in *(Uncle) Vanya* **'freedom is a place somewhere between desire and'**; which, as Andy Cornforth observes, encapsulates the sense of desire as 'that which liberates the self yet fails to define it', and this cancellation of definition 'in a sense *is* its liberation, shorn of subject and object, liberated, in a sense, from lan-guage itself'[3].

This links with what Stephen Booth proposes as the special appeal of highly valuable works of art is that they 'are in one way or another nonsens-ical': a succesfully purposefully artificial literary construct can give us an 'understanding of something that remains something we do not under-stand'[4]. Such works of art characteristically 'combine wrongness in one dimension with rightness in another'[5], they 'are, and often seem to work hard at being, always on the point of one or other kind of incoherence – always on the point of disintegrating and/or of integrating the very par-ticulars they exclude'[6]. As Booth notes, this is a metaphysical act, in that certain poems and plays can 'free their readers from mental limitations comparable to the limitations of physics' (and from that which is associated with intrinsically reductive invocations of 'common sense') by surmounting (possibly without noticing) 'limitations inherent in our expectations about language and the human mind' so as 'to give (not *assert* but *give*) infinity to what is finite and to put limits on what remains boundless'[7].

This is, of course, a task and effect which has profoundly political as well as metaphysical resonances. Anthony Kubiak's book *Stages of Terror* proposes an approach to the relationship between terror, terrorism and theatre, suggesting that the theatre is a place where 'political coercion as performance' can be uniquely thrown into relief as something visible, in terms of its 'first impulse' which is 'a terror … basic to human life'; for example, 'the classical stage artic-ulates itself specifically *as* theatre within the omnipresence of catastrophe', and yet, in some ways, 'catastrophe is also suppressed in this theatre' in that its central violent acts are alluded to, but not depicted, and the theatre 'draws back in the face of what is literally unspeakable and so unthinkable'[8].

Thus, Kubiak suggests that theatre was/is primarily a medium for con-fronting and confining terror. The point here is not only that sociopolitical power is deeply theatrical, even 'impossible without some implied and already recognized structure of performance'; but also that theatre is the very locus in which the 'self-division' that is 'intrinsic to subordination' is produced[9]. However, theatre can also expose, dramatize and interrogate this self-division, as well as produce it. It can confront the terror without entirely confining it, or, more precisely, define some of aspects of terror on the stage (as part of an orchestrated collective effort of attention), in ways which release it to the unconfined infinities of the (individual spectator's) imagination. Thus, the tragic theatre promises the appearance of was previ-ously unarticulated, a (partial and provocative) glimpse of the shapeless (compare Nietzsche's sense of originality as seeing that which has not yet been named[10]).

Like the theatres of Shakespeare and Beckett, Barker's theatre deliberately incorporates and offers the experience of 'a radical uncertainty of per-ception'[11], through purposefully extreme examples of what Kubiak terms a distinctively modernist theatrical effect, dramatizing the recognized and conjured multiplicity of selves, and the shifting terror of uncertainty with regard to identity, particularly where terror is applied as a (concealed or explicit) socially and politically coercive force to the body.

Strategies of the Doomed: *Wounds to the Face, Und, He Stumbled, A House of Correction*

> The crisis will
> The crisis must
> The terrible cleanliness of the crisis
>
> *A House of Correction* (CP4, 330)

Barker's dramatic compendium *Wounds to the Face* is a logical development of his recurrent motif of disfigurement, as a trope of confrontation, inscription, transformation and cancellation which calls into question the boundaries of status (as in 'losing face') and even distinction between the human and the inhuman. Here I refer the reader to Elisabeth Angel-Perez, who writes brilliantly and extensively on examples of this at various points in Barker's work, as well as identifying the specifically theatrical power of the horror-building dismantling of the face, in which Barker 'shows the actor first gifted with, then deprived of, the extraordinary instrument of meaning which the human face constitutes', so that the character 'is therefore amputated of his/her signifying tools: eyesight, voice, facial mimicry'[12]. Rather than present separate events (like *The Possibilities*), the playlets within *Wounds to the Face* permit spillage and development from one scenario to another (like the interludes of *The Bite* and *Rome*), in reprises and deviations in the manner of music for the string quartet. A woman struggles at a mirror to register her 'disobedience' at being **'incarcerated'** behind a face she is determined to transform for (and into) her self, artificially and artfully; a disfigured soldier is confronted with the irretrievability of his attractiveness and sociability, which may even constitute his identity and character, and that of his dedicated mother. Facial anonymity is required in both the punitive policing of the potential erotic transaction ('Not Absent, Merely Hidden') and the impersonal and depersonalizing assassination ('The Holy Orders of a Terrorist'). A woman claims her lover and deposes his wife by scarring him permanently ('She Knew at a Glance'). Two political decoys, facial doubles of a deposed dictator, find each other intolerable. An Emperor visits upon a painter the cruelty which he sees ascribed to him (a contrasting action to that of the Doge in *Scenes from an Execution*, who strategically absorbs offence). A revolution executes plastic surgeons in attempted abolition of distinction ('The Abolition of Beauty in the Interests of Social Harmony'), but cannot account for how the soldier's mother miraculously takes his disfigurement on herself (in turn, making her intolerable to him). Even the humour of sexual teasing, provoking visible consternation in her lover, may protect a woman 'possibly from some terror [she] can't face'. Finally, the Doubles and the Woman at the Mirror are both consigned to a death-like (and fatal?) paralysis of expression, as memorably suggested in the stage direction '*She is a gargoyle*' (*HN*, 80).

The Wrestling School 1997 production of *Wounds to the Face* was – uniquely – directed by Stephen Wrentmore, and fell short of that company's usual compelling standards, notwithstanding the presences of company stalwart Séan O'Callaghan and a promising opening performance by Kristin Milward as the Woman at the Mirror. Wisely, the depiction of the soldier's disfigurement was stylised rather than approximated realistically, with the actor wearing a black mask which irregularly blotted out one side of his features. The production was not exemplary of the direction style established by Barker, and not strongly cast; one consequence was a tendency to compound an analytic tendency within the writing of the compendium, and permit the audience to view the characters with not only speculative philosophical detachment but critical irony. On this basis, *Wounds to the Face* does not surprise the audience with the unpredictable varieties of emotion and engagement of which the characters in the later compenium *13 Objects* prove capable; though a subsequent, more imaginative production might revise my impression.

Barker's own productions for The Wrestling School – *Ursula* in 1998 and *Und* in 1999 – re-established the primacy of his claim, and that of Leipzig's design, to interpret the writing in a choreographic and scenographic aesthetic which befitted and developed its surprising compulsions. *Und* was a one-person show written for the performer Melanie Jessop (Barker's first foray into the solo piece since *Don't Exaggerate* for Ian McDiarmid in 1984, *The Breath of the Crowd* for Maggie Steed in 1986 and *Gary the Thief/Gary Upright* in 1987, originally conceived for Gary Oldman after his triumph as Sordido in Barker's *Women Beware Women* at the Royal Court in the previous year, but unperformed by him). Jessop's half-Jewish background gave her a cultural authority, should it be demanded, in playing the role of a persecuted Jewess who determines to view her imminent collection for extermination as a romantic tryst with a capricious suitor. The protagonist Und is (like Placida, Gertrude and Algeria) another of Barker's 'perfect liars', determinedly elegant women who partake of the self-styled status of the *femme fatale*, a self-determining immaculate vision who embodies an ultimate secret of existence (her own and/or her lover's) associated with nocturnal overcoming, irresistible convulsion and the sacrificial tangling of love and death. However, Und is unlike the others in that she (more directly, perforce) strives and contrives to avoid acknowledging her own potential role as victim, and to rationalize the malice of others in order to preserve her tragically misplaced hopes. Her compulsive fabrications avoid the sentimental platitude and moral superiority of a Jew who might claim, or be nominated, to possess the sole authentic perspective on the Holocaust; as Melanie Jessop observed (in a 1999 post-show discussion at Derby Playhouse), Und's characteristic insistence on continuing (her name reflecting her insistence to persist by yoking one imaginative possibility to another, effectively embodying the linguistic trope *and, and, and*) reflects her awareness that she must never reach a full stop until she is made to do so. She engages in an inventive imaginative complication of her circumstances, opposing the imposition of a brutal simplification (in this case, the label of 'Jew' and the consequent sentence of death) – as we all do when we await the belated call of a lover, and seek to mitigate terror and dread with labyrinthine inventions of possibilities and excuses. Und even eroticizes her malevolently deterministic executioner (in this, she may have nothing to lose) and endows her experience of suspense with a defiant sensuality and mutuality of determination (as when she insists 'I do not wish to know how terrible you are / You rather / Need to know how terrible I am' *CP5*, 237).

Barker's programme notes to this production deserve reprinting here:

> The doomed are not without their strategies. Und's assertion of her cultural superiority feels like a bond with power, a shared contempt for the world which will spare her humiliation. But in admiring nihilism she exposes herself to its withering effects. Jew and aristocrat alike are consigned to extinction by a regime which makes a passion of its loathing of elites. Difference is identified, but only the better to be exterminated.

In performance, the swift incursion of a mirror into Und's space – a domestic interior, scrupulously prepared for tea, its 'patience, langor, idleness' which permits courtly conversation – both primes and taunts her narcissism. The premise of the play may initially recall the mechanical imposition and increasing malevolence experienced by the characters who are nevertheless both restricted and determined to 'kill time' in Beckett's *Acts Without Words* and *Happy Days* and Pinter's *The Dumb Waiter*; however, Barker characterizes the external malevolence historically and politically, rather than metaphysically. In a series of fraught but graceful imaginative pivotings, Und claims her own 'archaism' as a 'fragment of a dying class' to be 'the source of its fascination', which repudiates the world as demeaning; nevertheless, she admits of her anticipated visitor 'He gathers Jews'. Hence her choice of her most extravagant dress, to '**Overwhelm** *(Pause)* / If such a man can be / If such a man is ever' (*CP5*, 211). She acknowledges that many men 'do not care for the extremes of female fashion ... except for homosexuals', who may share an enthusiasm for an extravagance simultaneously vulgar and discerning; but then the thought of homosexuals, and their possible fate in her social climate, brings her to another brief pause; she recoils from this imaginative terminal *cul-de-sac* by constructing an image of an imperious military officer, striding resigned and objective through the (vague but ominous) toxins and mechanisms of his surroundings. Nevertheless, she acknowledges that her former thorough preparation has been unbalanced. She speculates that 'the Jews exhaust him probably' and distinguishes herself through her 'quality of aristocracy' ('whereas all other classes are inhibited by stifling conventions they choose to call morality we make laws of our appetites murder for example', *CP5*, 215). However, the persistent ringing of her doorbell and then the '*shattering of glass*' force her to the admission that 'I'm under siege'; she then determines 'The siege is desire / Desire / Expressed / As / Rage' (*CP5*, 216). Whilst she insists that 'every action is susceptible to *(shattering of glass)* / Interpretation', this is challenged by her being driven to confess 'I'm not an aristocrat I am a Jew' (*CP5*, 218); and so she may have a single, lethal interpretation visited upon her. She refuses to be reduced to panic, placing her faith in self-inventing performance ('the trick is **Not to be** as I appear to be **Not to be** rendered foolish by the antics of another', *CP5*, 220) and even a sense of romantic destiny ('**We share so many oh so many if not all**', *CP5*, 222), but the incursions into her room become more frequent and the objects more ominous and taunting. Deductive logic itself is now her enemy ('How I hate simplicity its little laws its two and two make four its black and white its tit for tat oh how loathsome', *CP5*, 228), and in the face of her persecution she takes refuge in the posture of an ostentatious sympathy ('**I'm weeping** / **Not for myself** / **For him**', *CP5*, 231). When drenched with '*sordid fluids*', she redirects her will towards admission: 'This man intends to murder me'. This is another turning point for Und, who now fortifies herself with a sense of her own triumphantly performative artificiality, which has rendered her 'Vastly / More / Aristocratic than the ... Aristocracy'; this in turn leads her to the contemplation of suicide ('The / Perfect / Truth / Of / Aristocracy / Is / Its / Contempt / For / Life', *CP5*, 236). A tray of freshly dug earth is her final visitation; and her determination '**Let him in**' presages her death at the apprehension of someone or something which remains indefinite, perhaps incommunicable: 'I died on seeing ... On / Seeing / Died' (*CP5*, 238).

Like *Minna*, *Und* undermines any reflex of historical superiority in the audience by presenting a poignant portrayal of the woman who suggests (like Minna) 'what you know I might perhaps be forgiven for not knowing', the persistence of the attraction and collusion of ostensible opposites, and the persistence of self-deception and its pretext for, and collision with, the surprisingly unfathomable. Whilst the scenario of *Und* might most readily suggest Nazi Germany in the Second World War, its resonances in performance extended into the persecutions and war crimes of Bosnia and Kosovo and beyond. In performance, the play presents an event and experience (and a character) which are gratingly beautiful and, quite literally, desperately erotic; both character and performer negotiate an imaginative tightrope with a poise which is both exhilarating and draining. Barker shrewdly locates Und's demonstration of what is identified in *Don't Exaggerate* as 'the lies you tell yourself' in an avowedly romantic context, establishing a dimension of imaginative torture and avoidance which is available to all; as Zimmermann observes, 'even when she weeps and suffers from the cold and the heat inflicted upon her in the course of these attacks, this in itself also implies a parody of the ecstasies of the Petrarchan lover'[13]. But Und is attempting to impose the structures of self-consciously (humiliatingly) conventional and familiar rituals and agonies of love onto a terror which she cannot face; it is as if she is striving to impose the mask/face of courtly love onto the shapeless and indefinable form of massive systematic death, an admirable effort of will directed to an end both ludicrous and poignant. The character is both infuriating in her strenuous self-deceptions and also profoundly pitiful in her pain and danger, which again undercuts audience tendencies to attempt ironic superiority. If her self-consciously and deliberately refined sensibilities prevent her from confronting the banality of massacre, they may also finally save her from the most humiliating subjection to its mechanical processes: the mere sight of the unthinkable finally kills her, a delicacy which permits a possibly fortunate, because more dignified, release. In Barker's production, Melanie Jessop gave a virtuoso performance, clad in a white PVC dress which matched Und's bone china; Leipzig's set surrounded her with a Newton's Cradle of invasive and swinging chrome trays, which in their mechanical rhythm and sound recalled the action of technically superior meat slicers; and finally Und slumped, as an ashy sand trickled down onto her outstretched hand, so that for one reviewer the acting space became reminiscent of the base of an uncontained hourglass: 'Like her hands intercepting the falling grains, her intellect can filter, but necessarily fails in the attempt to evade succumbing'[14].

After the (devastatingly powerful) chamber work of *Und*, The Wrestling School were poised for a second, broader initiative within the same year of 1999: a revival, directed for the first time by Barker himself, of his most internationally acclaimed and frequently performed play, *Scenes from an Execution* (originally broadcast on radio 1985, first directed for the stage by Ian McDiarmid at the Almeida Theatre in 1990). Barker's production was immediately distinguished by a Tomas Leipzig exordium contraption which confronted the audience on their entry into the auditorium: a mechanical pulley hoisted a great dripping white sail out of a deep iron well, establishing the (microphoned and amplified) sound of a rhythmic delving into water, which established the keynotes of human ingenuity and the unstable terrain of Venice, and its distinctive element of dissolution; simultanously, three maimed sailors (in wheelchairs and naval uniforms) looked on, their heads falling backwards as if on hinges. Kathryn Hunter played Galactia, a theatrical "star" casting apparently intended to attract a broader audience to a Wrestling School play, but which seemed sometimes at odds with the company ensemble and ethic. Hunter's *sympatico* performance was somewhat self-enclosed (even beyond that identifiable quality in the character) in that it seemed to elude some of Barker's characteristically strictly restrictive and concentrating direction: she was often reliant on playing her own distinctive voice rather than the specific demands of the language, and frequently (and hence predictably) opted for a comic head-twitch and extended arm. She was significantly better in the monologic moments, and moved with an unpredictable animality in an added mimed interlude in which she stripped and accepted a red dress and shoes from Carpeta in preparation for the final scene (the dress was drawn from the aforementioned iron well, and thus suggested a bloody sail). Lucy Weller continued her high standard of costume designs from *Ursula*, here clothing the characters mainly in cream and beige wools and silks. Victoria Wicks was a poised and manifestly sexual Rivera, and Ian Pepperill effectively brought a note of childish peevishness to the Doge, wearing a pair of glasses which further suggested a resemblance to the critic Michael Billington (who typified liberal humanist insistence on the utility of social realism, and a barely concealed irritation at alternatives). Barker's direction attempted to strengthen the status of the Catholic hierarchy rather than render them stupidly clerical. With some irony, the play had come into its own, in new historical terms: Blair's "New Labour" government was establishing a climate and discourse which would not speak of Art except in terms of justification with reference to municipal neo-utilitarianism, participation and access, an age which begat and privileged more Arts Council administrators (who found the very notion of Art unquantifiable and therefore embarrassing). In this way, the play sparked (even) more laughter of wry recognition (not an effect usually associated with The Wrestling School) and redefined the debate in fiercely intelligent terms. However, the theatrical experience remained that of a debate – rather than an experience supplanting polarities – and its acknowledgment of the power, sensibilities and intelligence of the Doges and Riveras might have been a prime factor in such people finding it an unusually digestible and readily laudable example of Barker's work. The production provided an impressive vantage point from which to consider a decade of Barker's work, but did not substantially broaden The Wrestling School's popular standing – as it seemed calculated to do, in response to the terms and demands of Arts Council funding.

This in turn suggested that the company had nothing to lose in terms of being more characteristically and resolutely artistically uncompromising and stylistically distinctive, as was exemplified in their next production, *He Stumbled* (published 1998, staged and directed by Barker for The Wrestling School in 2000). This darkly complex play realizes dramatically Janice McLane's proposition that to take phenomenology (the study of consciousness and the objects of direct experience) to its very limits, 'the philosopher must approach his or her life with the same degree of ferocious and pitiless scrutiny a scientist might bring to dissecting an animal brain', embodying 'philosophy as purposeful self-vivisection'[15]. In Barker's production, the text's exordium (a one-page-plus vista of mass abjection and thieved intimacy) was adapted to present the viewer with a series of enticing but enigmatic tableaux disclosed by steel shutters (invoking and developing the atmosphere of voyeurism and surveillance of the unstaged *Gaoler's Ache*): a female figure (later identifiable as Turner, played by Victoria Wicks) naked, in a variety of provocative poses. The protagonist Doja (whose name, Kiehl observes, may suggest a corruption of Don Juan[16]) immediately demonstrates his sexual readiness and remorselessness, by ravishing a nun. Doja, an eminent and infamous anatomist, has been summoned and employed to perform the last autopsy on a dead monarch; he and his assistants Suede and Pin settle into their new surroundings with a mixture of routine and aroused suspicion. Like some other Barker plays (notably *Downchild*, in which the protagonist similarly stages a climactic, personally fatal *coup de théâtre*), *He Stumbled* contains the discernible dynamic of the stage thriller or detective play, in that there is a central mystery to be penetrated or secret to be unlocked[17]. In *He Stumbled*, Doja and his assistants puzzle about the actuality and the ramifications of the neo-Jacobean edifice, surroundings which recall a more gothic and labyrinthine version of the world of *Hamlet*. The death of the king seems to have quickened rather than quelled the sexual appetite of the queen, Turner, who views Doja invitingly. The prince, Baldwin, is fixated both by the mysteries indicated by his father's cadavre and by Doja, the man who would replace his father in his mother's affections. Baldwin challenges Doja's scientific separation of body from self: 'The flesh is not the man? What is the flesh, then? Is the flesh not this man's and no others? If the flesh is not the man, why are you here?' (*CP4*, 257; questions also posed in the different contexts of Barker's *The Love of a Good Man* and *Victory*). Contrastingly, Turner claims no great talismanic attachment to her husband's viscera:

> It must be the same with character. We love what we see. But that is rather little of it. Really, life is solitude. We attach ourselves to surfaces. But what is intimacy, Mr Doja? A fiction, surely... (*CP4*, 260)

Turner's question and hypothesis constitute the central investigation of the play. Doja perceives the complication that, whilst mother and son dislike each other, both flatteringly appear to glimpse in him 'the alleviation of solitude'. His speculations are counterpointed by the disembodied interpolations of two dead priests, a posthumous mockingly detached chorus whose comically structured stichomythia nevertheless hint at the inscrutabilities of death (reflecting some of the mysteries impervious to the living, such as '**The sheer exaggeration of the world**' when '**Death is the end of freedom ... And the beginning of**', *CP4*, 282; speculations which Barker would pursue more seriously in *Death, The One and The Art of Theatre*). Doja's protective casualness, sarcasm and humour are overwhelmed by the spectacle of Turner, her body seductively dressed just for him, promising an oblivion for the doubts and questionings of his soul. Baldwin views the prospect of Doja's penetration of both of his parents' bodies as further evidence that 'everything conspires to make me intimate with you'. Pin and Suede suppress their anxieties through calm application to familiar method, but Doja is besieged, and his former omniscience confounded, by successive sexual assignations. Suede superstitiously denounces Pin as an enemy within, origin of a curse and danger from some vague infidelity to their '**compact with the dead**', and thus attempting to locate, separate and sacrificially eliminate the source of his nameless terror. A courtier identifies Doja's own attempt at poise: 'How you have loved to be beyond...keeping your intimacy for the dead' and 'just visiting' the living (*CP4*, 292); but now something, or someone, lies in wait, drawing him in. Turner literalises this sexually, agreeing even as Doja realizes 'All things here...even the confessions...are manifestly false ... And naked-ness...far from being...a testament to truth is...further ... Manipulation' (*CP4*, 293). Unsettlingly, Baldwin identifies a self-protective aloofness in Doja, for whom the knife is 'the alibi for ecstasy never quite attempted' (*CP4*, 296). Doja's professional detachment crumbles into an experience of disgust for the 'nightmare of anatomy'. In turn, he tries to crack the poise of the 'perfect liar' Turner, by forcing her to ingest the fluids of the corpse as an act of decisive desecration. But it is Doja who is sent further off balance by hearing the 'ecstasy' of Turner with a figure who turns out to be Tortmann, the king still living. Tortmann has embroiled Doja in a web of subterfuge, baited with his queen whom he acknowledges 'impossible to own', playing out his pursuit of a sexual 'obsession' which is a simul-taneous 'privilege'. 'No common husband', Tortmann's preservation of 'distance' and cultivation of pain has bestowed upon him an extraordinary 'ecstasy akin to God's'. Doja tells Tortmann and Turner 'It is not pleasure, is it...you two share ... But terror....?' (*CP4*, 315). In '*a spectacle of will, dexterity, endurance...of magic, therefore*', Doja anatomizes himself with an 'Infinite skill', mesmerizing his audience by making surgical cuts which nevertheless suspend his moment of death until a sudden climactic moment.

Thus, Doja's attempt at cool, distanced, rational analysis encounters a different distance, that of Turner and Tortmann (and even Baldwin), which – like the physical distance between actors onstage – conducts and ampli-fies a sexual tension and heat, demanding an ecstasy which proves fatal – but to those embroiled in their project/trap, rather than to themselves. As the title suggests, *He Stumbled* depicts the loss of a precarious balance, apparently Baldwin's but more profoundly Doja's, until Doja recovers some artistry by keeping the forces of his death poised in space like a juggler. Like the Castle in Barker's eponymous play, *He Stumbled* has a deceptive architecture, in which new depths are constantly and unsettlingly exposed, a trope which will be further realised in the enigmatically confounding design of the edifice in *A House of Correction*. The structure of *He Stumbled*, in which the protagonist strives to fathom and name a nameless 'some-thing rotten' associated with a dead king, and finally wilfully plays through an imposed scenario "to the hilt" of his own death, is of course also common to *Hamlet*, presaging and in some ways preparing for Barker's later more profound meditation on Shakespeare's play in *Gertrude – The Cry* (2002). The theme of the sexual conspiracy which extends beyond death, in which someone manipulates a sexual liaison (here, apparently) from beyond the grave, recurs in the subsequent *Dead Hands* (2004).

Barker's production of *He Stumbled* was appropriately meticulous in its orchestration of darkness and light[18], in a set suggesting subterranean crypts, hidden panels (such as those through which the priests erupted, as if in mid-air) and secret passages. William Chubb aptly performed Doja's mixture of distracted bewilderment and decisive strength; Victoria Wicks memorably incarnated Turner's languid promise, naked beneath a transparent dress of smoky grey (which she would wear again in *Gertrude – The Cry*); Ian Pepperill was strikingly mercurial and precise in capturing Baldwin's boyishness, his disarming curiosity and impatient demands; and Julia Tarnoky created for Suede an eccentric, compulsive precision which was, appropriately, simultaneously comic and unsettling.

Barker's *A House of Correction* was published in 1998, broadcast by BBC Radio in 1999 and staged by The Wrestling School in 2001. Barker's programme notes for his 2001 programme are characteristically provocative:

Crisis – the object of desire
The efforts of society are directed to the elimination of the crisis. The psyche of the individual craves the crisis. This paradox can never be resolved. It is the secret tension at the heart of all social order.
Will, fate, choice, inevitability
As the crisis unfolds it oscillates between causality and randomness...as the characters grasp at significance, meaning slips between their fingers. whatever purported to be evidence is chimera...this is the reverse of the detective story, which contains an immaculate truth, a solution...the longing for enlightenment is here seen to be a desperate and punishing misapprehension...everything takes the characters to a destiny, but whether this destiny is willed, a collusion, or an accident, defies analysis. All that is left to them is an emotional perfection, a sublime *arrival*...

Three women – Shardlo, Lindsay and Vistula – and an elderly bedbound poet Hebbel inhabit a castle, with their recalcitrant and berated servants, as war erupts. Lindsay acknowledges Hebbel's ominous utterings of 'Blood' with the wry foreshadowing, 'You will get all the blood you want, I promise you'. Hebbel (who characteristically and teasingly subverts his conventional claims to sagacity by suggesting he has wisdom, but refuses to dispense it) maintains there is 'Never / Enough / Blood'; which begs Lindsay's question 'To satisfy whom...?' (*CP4*, 322). This immediately establishes a sense of foreboding, a wry impatience to know the worst, and a presentiment that the worst can never be (fore)known, as well as the question of who wills the worst (the perpetrator? The victim?) and why. Shardlo is impatient with the 'intolerable' peace and, with a neo-sexual nervous exultancy, anticipates the deflowering eruption of the crisis which will challenge her to 'triumph' and 'discover the extent' of her 'magnificence'. Lindsay similarly expresses a febrile hunger to 'see your ... To share in your ... Can't say it'; Shardlo maintains the appeal of the indefinable, and its creation of new terms of life: 'Don't say it, then...wait...witness it...and describe it afterwards'. She anticipates 'We shall be overwhelmed' – ostensibly with casualties of war, but also perhaps personally – and hints at what may lie behind the evaporation of hope and the draining of energy: 'the things we shall see ... the sight of which might now cause us to sink to our knees beneath a canopy of horror we shall ... I don't know yet'(*CP4*, 323). This very outstripping of the terms of conventional life and expression holds a particular promise for Shardlo, who has attempted suicide five times, a compulsion 'embarrassing' both for her and for the servants who feel obliged to rescue her. The crisis promises a reversal of priorities (such as the maintenance of mutual tolerance) which confounds the very impulse to anticipation which it excites: 'to even contemplate the crisis is a contradiction, illogical, futile, since by definition it makes havoc of the very conditions under which it could be contemplated' (*CP4*, 324). Nevertheless: *A House of Correction* centrally dramatizes this very impulse to analytic reflection and consideration of possible consequences which its events defy, even as they provoke. This is imbued with an erotic urgency, a longing for new experiences and forms of freedom which may nevertheless be painful and unenviable. At last the first evidence of the crisis arrives, in the person of the courier Godansk. Liz Tomlin has noted how *A House of Correction* evokes and subverts the deadlocked expectations of Chekhov's *Three Sisters* (in which 'Chekhov's refusal to allow his protagonists to determine anything of their desires, coupled with the strong implication that it was the social environment which was to blame for their apathy, negates, in purist terms, the tragic dimension of his work') and Beckett's *Waiting for Godot* (in which waiting for an event which never arrives becomes 'the pinnacle of human endeavour')[19].

Thus, the arrival of Godansk might seem to give the women the opportunity to put a face and a name to their terror. However, Godansk himself is confounded by his 'magnetic' surroundings, and their atmosphere of characteristically deceptive echoes, losing his sense of professional momentum, military conviction and personal resolve in this place which ominously realises Kristeva's image of the 'fortified castle' where

> Separation exists, and so does language, even brilliantly at times, with apparently remarkable intellectual realizations, but no current flows – it is pure and simple splitting, an abyss without any possible means of conveyance between its two edges. No subject, no object: petrification on one side, falsehood on the other.[20]

Hebbel might here represent a particularly wry and irreverent figure of petrification, not only in his own decreasing immobility but in his occupancy of the room and an embodiment of its dominant (possibly sexual) associations for the women (Vistula descibes him as the man who 'both created and subsequently maimed' Shardlo's character). He combines the physically and morally abject: Shardlo goes so far as to describe him as **'the thing ... that makes me ache for death'** (*CP4*, 333); and he mocks Godansk's self-effacing faith in a historical dialectic that will make everything 'clear' in retrospect.

Shardlo continues to profess with a feverish energy and hope that the crisis 'will be nothing but perpetual interruption... interruption to the extent that we will cease to recognize it as such'; the 'evaporation ... in the heat of crisis' of order, 'which we now take as the basis of our life, the very thing that permits so much reflection', will bring a vital if unenviably raw sensitisation, 'we will exist not in our heads but in our fingertips' (*CP4*, 331). However, Shardlo's hope also rests upon a neo-religious promise of scarifying deliverance, though at least she recognizes (and anticipates) how this will dissolve former moralities and polarities in any post-castastrophic survival ('The crisis will almost at once obliterate any distinction between the enemy and those we call our own. That is surely an aspect of its magnificence?', *CP4*, 331).

Lindsay propositions Godansk ('Feel me beneath my skirt') in a way that bewilders him further; she seductively admits both her characteristic mendacity and its fragility: 'I feel certain that the crisis will cure me of lying but will it make me happy?' – a rhetorical question. Less naïvely than Godansk, and perhaps Shardlo, she knows it will not offer consolation. However, Shardlo remains pleasurably animated by the prospect of crisis, its 'peculiar effects'; Lindsay's romantic enthusiasm is more focused on Godansk, who imagines her 'sudden and unexpected' love for him to be another higher force (beyond orders, history and inscrutably fortunate error) to which he might subordinate himself and his will; whilst Hebbel tries to undercut and upstage their hopes even from his bed, playing the self-dramatizingly abject pariah ('I'm all the world detests'; 'The crisis will certainly dispose of me...and that I daresay is its purpose') with a paradoxical and even preposterous verve that recalls Sleen in *Early Hours* and *Ego in Arcadia*, luxuriating in the unforgivability of his insights: 'As long as I was loved, the truth was something I could happily forgo...is it not the enemy of love, in any case...but unloved I find it has the fascination of an outlawed faith' (*CP4*, 351).

Shardlo momentarily succumbs to a sense of lost possibility ('that bottomless exhaustion of the soul that's like some vast bay in which the sea has died') but recovers to defy Lindsay in their competing terms for self-realization (**'Am I to cease becoming me for your'**; like Barker's Vanya, she recognizes how **'They prefer to pity me than that I should step into my own character'**, *CP4*, 351). It is the sight of Hebbel continuing to exert over Shardlo a mesmeric power which permits him the liberty and profound possession of her body ('so intimate and yet so unresisted') which drives Godansk to insist on seizing a similar liberty. When the women do not prevent Godansk's physical ejection of Hebbel, the usurpation seems complete; however, Hebbel turns out to be, like all Kristevan dejects, something rejected, but something from which his nauseated ejectors cannot completely or finally part. He is flung into a well, but survives and even seems to revive, and thrive. This recalls the details of an episode in Barker's *The Europeans*, when Starhemberg insists that the Officer tell him a story: in response to this strange order, the Officer invents an almost Freudian narrative in which an old man blocks his drinking from a well, until he hits him with a brick. *A House of Correction* reprises and realises this confrontation – albeit "offstage", thus adding to the hallucinatory quality of the episode – with the added reversal that the old man refuses to be killed, making him even more a revenant figure of nightmare; the attempt at individuation is (worse than failed) horribly incomplete. In the hiatus, the question remains which of the women Godansk will choose. Lindsay is mortified to discover that, although she loves him, 'the bride' is not her. Vistula is pragmatically reductive: 'What is a bride in any case? ... Presumably it's ecstasy ... I've had ecstasy ... Never in a white dress, but I've had it' (*CP4*, 378). She claims the title and status of bride, submitting herself to it as an ordeal, no less decisively than Lindsay's drive to submit herself to her own frequently invoked ideal of love. Shardlo attempts to galvanize the flock of servants, who '*hang in a group, moving like a weed in a current*', into throwing missiles down onto Hebbel and sealing the well; however, elsewhere the momentum is running down, as Godansk succumbs to an apprently fatal entropy. In a wry parody of the Chekhovian departure, Lindsay and Vistula leave the building, only to be killed by the enemy's bombs (they have dropped only leaflets of Hebbel's poems up to that point). Shardlo, submerged by the servants' clamour, survives, and is drawn down onto the bed by Godansk.

In the published text of the play, on which the stage production was based, Godansk eventually releases Shardlo and leaves the castle, to the sound and illumination of gunfire (which would seem to support Shardlo's earlier interpretation of him, '**You are the crisis**'). In the substantially altered radio production version of the play, Godansk says of the women and servants, 'their loyalty was wholly conditional on the theoretical nature of the crisis; once the crisis ceased to be an aspect of your character and began to manifest its authenticity, they naturally took to their heels'. He tells Shardlo 'Do not move from what is now your bed', and though she pleads 'It horrifies me to be horizontal here – Please...' her tone suggests capitulation.

In the radio version, Hebbel's wryly humorous interjections are further developed (as when Shardlo says she blames Hebbel, and he ripostes 'Doesn't everyone?'), and David Bradley gives a stark and masterfully sly reading of the role (which suggests his eminence as a future choice for the roles of Sleen and Isonzo). Nicholas Le Prevost brings to Godansk some of the alternately dissipated and commanding performances of gnawing precision which similarly distinguished his Lear in *Seven Lears* and Charles in *Victory*, and Victoria Wicks memorably conveys Lindsay's romantic ardour. But the most remarkably detailed performance is by Juliet Stevenson as Shardlo, who achieves swift transitions from a vehemence in despair to a desperate, feverish appetite for crisis, as well as an impressive erotic abandon in her surrenders to Hebbel (suggested by the additional lines of self-contradiction: 'Don't touch. Touch if you want to. Travel. Travel over me') and Godansk (in additional lines which also make manifest a sense of reversal: 'You had no choice. No, wait. Aoh, aoh…'). Barker's stage production of 2001 was, along with Kenny Ireland's production of *The Castle*, the only British Wrestling School production I was unable to see, and Barker records his dissatisfaction with it in *A Style and its Origins*. Apparently a principal strength was the Tomas Leipzig set, in which a massive fan (redolent of aircraft propellor blades) dispersed the falling leaflets over the stage, and created a sense of danger to the actors similar to that threatened by the flying steel trays in *Und*. Hebbel's bed hung from cables (like the resting place of the corpse in Leipzig's subsequent design for *Dead Hands*) and swayed as if he were at sea (amplifying the ambiguities of location).

In the radio documentary *Departures from a Position* (broadcast by BBC Radio 3, 14 February 1999, to coincide with the broadcast of *House*), Barker further identifies the distinctive reflexes and propositions of the play:

> In *A House of Correction* the characters assess, agonizingly, almost to the point of their own destruction: whether what occurs to them has been invited by their own collaboration or whether it's imposed on them; whether they will their own destruction, collaborate in it, or whether it arrives from outside. This struggle to make sense of the arbitrary preoccupies them, leads to their quarrels and makes them, perhaps heroically, try to fathom the nature of their existence. The movement of the catastrophe is from the external to the internal: the catastrophism is not so much imposed on individuals who then have to improvise on the grounds of the catastrophe, but rather the catastrophe existed in them in the first place, and therefore they use the catastrophe, need it. The identification of it outside ourselves is of course a political agenda, but in reality, psychologically speaking, it's a necessity for our own spiritual wealth, and possibly a world without catastrophe is a poor world.

After the relentless events and surprising developments which constitute the dramatic plethora of *He Stumbled*, and with which the characters and audience struggle to keep up, *House* investigates the opposing tempo, in which everything slows down, as in a dreadful accident, to the point of separation close to where 'no current flows'; the feverish and furious initiatives of the characters usually result in incompletion (such as the attempted killing of Hebbel), or the resumption of apparently terminated stalemates or characters (such as Godansk). However, even here a curiously *erotic* current flows, and may burgeon even in apparent inertia, as the characters experience and express arias of desire, that which 'rots all calculation' (to use an image from Barker's *Women Beware Women*), notwithstanding their imaginatively hyperactive compulsion to – fatally limited and defeated – calculation and speculation. The combination of frozen momentum, delirious insistence, dreamlike space and misplaced hopes makes *House* more uniquely haunting than readily engaging, its characters occasionally reminiscent of struggling flies trapped in amber, unavoidably doomed notwithstanding their incessant imaginative activity. Its distinctive developments of what might be termed a Beckettian tempo of 'arrested dynamism' will be furthered in simpler but starker form in Barker's 2006 radio play, *The Road, The House, The Road*. But *A House of Correction* and the other plays considered in this chapter constitute masterful studies of the human imagination's bids to face the wound of its own terror before the death of meaning. To borrow Stephen Booth's terms of appreciation of Ben Jonson's epitaphs for his children, their 'careful insufficiency' offers a 'means of making the poor, category-bound human mind superior to its own limitations – the limitations that language reflects and services – and sufficient to an impossible mental task that remains impossible to us even as we perform it'[21].

7
Infinite Reversibility

All He Fears, The Swing at Night, Albertina, Knowledge and a Girl

In 1996, Barker published his fifth volume of poetry, *The Tortmann Diaries*, in which a 'solitary and diabolical' protagonist pursues his senses of interrogation and wonder, reflecting the recognitions and active seizure of 'alternatives': even extending to the limitations of mortality, as identified in the major poem, 'Infinite Resentment', which imagines the mutiny of the dead, having glimpsed: 'The infinite reversiblity of things / The seduction of opposites'. The work in this chapter correspondingly reflects Barker's embodying of 'Alternatives', repudiations of the dominant terms of life – sometimes jointly undertaken by two parties, but frequently involving an ultimate tragic isolation; however, even this may have wider imaginative resonances for those (others) who may perceive and consider its terms.

I first want to consider a minor work, Barker's first play for marionettes *All He Fears*, written for and performed by The Moving Stage Puppet Theatre (Artistic director: Greg Middleton) in London. *All He Fears* (staged and published 1993) traces the discomfiting odyssey of the philosopher Botius[1]. *All He Fears* dramatizes the simple premise that Botius's characteristic impulse to imagine and speak the worst seems both to invite the realization of disaster and to outstrip his terms of prediction: his loss of his mistress to another, his blinding at the hands of hooligans, the stealing of his money by an initially kindly prostitute and his consignment to an abyss after provocation of a policeman, all stem from his compulsive anticipation and expression of what he considers at the time to be the worst of possibilities: bids for intellectual and emotional self-armouring which seem to incite catastrophic blowbacks (compare the observation in *Albertina*: 'Perhaps you, predicting [pain], invite it'). The moon shines into Botius's cell or *'grave'* and illuminates a rope which he climbs in the hope of escape, even whilst formulating the philosophical paradigm of Botius's Rope, which suggests that what might be 'construed as deliverance' might also be a 'means to further pain'.

All He Fears partly develops the speculation in Barker's *The Bite of the Night*, 'Do you think you lose your mind? You find others' (CP4, 72), though Helen in that play means these words as an expression of (possibly unenviable) persistence and resilience, whereas the events of *All He Fears* most startlingly demonstrate painful extensions of Botius's sense of what might be (im)possible, as when '*His head splits open*', festooning the stage with its confounded contents, until his '*hands arise from the mess, and push together the sides of his skull*'. Typically, this is a prelude to more events and perceptions rather than a conclusion, and the text's increasing preponderance of stage directions rather then spoken words reflects the increasing choreography of the mysterious, including the emergence of Botius's more sexually successful *doppelgänger*, a female figure pregnant with a void, and a horse who mocks human notions of kindness by eating a baby, but refuses Botius's offer of himself; Botius remains excluded yet painfully perceptive, 'Undigestible'.

It is the play's pursuit of Botius's extreme separation from others which makes it surprising, but not especially dramatic. *All He Fears* finally broaches a scenographic exposition of simultaneous 'nightmare and idyll' and the play's production may have further informed Barker's sense of what marionettes and their surroundings can and cannot do, providing a preparation for *The Swing at Night* (performed 2001, again by The Moving Stage on their London theatre barge). *The Swing* is a particularly haunting example of Barker's investigations of a transgressive relationship between mother and son (which may have its first intimations in his early play *Alpha Alpha*, and develops through *The Europeans*, *The Gaoler's Ache* and *The Fence*). The '*aged baby*' Otto, who is nurtured and advised only to grow up when the world is fit for him, observes how his beautiful and habitually naked mother Klatura kills men. The male compliance in seduction is represented by the labouring sound of a freight train, the murder by the shock of a human-sized pair of men's boots dropping to the ground beside the puppet Otto, which drives him to deduce 'L'amour / Has / Occurred'. His consternation is quietened by his mother's voice and a pair of hands which lovingly propel him on a swing (an ostensibly frivolous object which also appears to striking and poignant effect in *Minna* and *The Gaoler's Ache*). By day, Klatura cleans in a museum: an activity itself reflecting something of 'The futility of all things', but also her meticulous care and delighted dedication to artworks (such as Fragonard's 'The Swing'), until her contemplation is brutally fractured by the interpolated irruption of a contemporary pornographic image, which prompts her embarking on another artfully fatal assignation to avenge the combination of 'The reverence / And / Rage' which characterizes the male gaze upon the female body: as Otto observes, the 'tragedy' for men, as for dogs, is that in their optimism 'For them / A smile has one meaning / Only one'. As in *All He Fears*, The Moon is a significant presence (in ways which recall the poem 'The Moon's Advice' in Barker's 1988 collection *Lullabies for the Impatient*, 'Be what the lunatic would be / If she were allowed to love'), encouraging Otto's curiosity as to what prompts the men's horrified exclamations. Otto acknowledges 'Possibly / What she shows them / I want to be shown', the promise associated with being 'grown'; whereas he is not 'grown', yet 'old', and continues to be *'charmed'* by the *'contented'* moments of *'love and mutuality'* he enjoys with his mother on the swing, even as he glimpses the 'Comic / And / Sad' uselessness of the swing, 'metaphor for man's pitiful and hopeless struggle … To / Be / Free / Of / The Law'. The Moon claims 'I / Draw / It / High / And / The / Earth / Brings / It / Low', disputing the law of gravity even as 'It / Drags / All / Dreamers / To / The / Floor'; it even *'giggles'* at the balance of forces (*SN*, 28-29). Otto is enticed by the Moon into donning adult clothes, then finds himself *'frogmarched by his own garments'* into an assignation with his own mother, who initially pretends not to know his identity. Klatura observes how men hope to glimpse, through her translucent form, the promise of deliverance but encounter – bathetically, fatally – 'THE / SELF-SAME / WORLD / AGAIN'. Klatura and Otto laugh intimately, in a profound embrace. Otto delights to his mother, 'How / Terribly / Old / I / Shall / Be / Knowing / Women / Have / No / Mystery' (*SN*, 34).

However, this bald summary of encounters, propositions and counter-deductions does scant justice to the dreamlike imagery, atmosphere and vistas of *The Swing*, particularly in performance. The Moving Stage production featured, as specified, puppets of varying size to depict the same character in order to permit carefully orchestrated, rhythmically unfolding stage pictures which would be impossible in another medium; as well as a vocal soundtrack expertly voiced by Barker stalwarts Ian McDiarmid (Otto), Victoria Wicks (Klatura) and William Chubb (The Moon), with a full sense of the play's comedy, wonder, shock and tenderness. I recall with particular admiration the effects of perspective and luminous lighting which suddenly recharacterized the dark performance space as the *'slowly rising light'* on *'an empty, tranquil beach'* on which Otto and Klatura were discovered, and the subsequent oddly comic and poignant counterpoint of the skeletal form of a bemused dead dog floating by in accordance with the rhythm of a tide. The play resolves into effects of sound and light (feet on pebbles, a dog's bark) and visual perspective, producing an irreducible musical complexity of tempo and counterpoint, *trompe l'oeil* and advancing yet broken visual framings:

> *As the dead dog goes one way and the other, the white hands appear and release the swing. A photographic portrait of an old woman weeping slides in and is still. The hands push. The swing creaks. With a crack, the head of Otto pierces the photographs and looks, first one way, then the other. The dead dog drifts.* (SN, 35)

The Swing at Night is a brief but highly imaginative (re)characterisation of the marionette play as *poetic landscape theatre*, generating alternately delightful and disquieting images through a sense of overt symbolism which nevertheless cannot be entirely or finally reduced to a single meaning, like the classic European expressionist and poetic dramas of Strindberg, Büchner, Wedekind and Lorca which it resembles more than any British drama. The Puppet Barge's simultaneously bold and sympathetic production exemplified and developed Barker's ambitious sense of the *animated pictorial*, increasingly important to his sense of theatrical progression and *béance* (the exordium, the interlude, the offset counterpoints such as the Cupid in *Minna*) and to twenty-first century examples of his stage direction (in which, as Roger Owen observes, the characters can sometimes, if still unpredictably, seem to be 'part of the scenography, growing into the stage rather than being projected out to the audience', so that action is fluid, with effects of discontinuity, rather than tied to any one social or scenic milieu[2]). These effects (including the flocking birds and doomed dogs) will find further ambitious orchestration in the landscape play *Found in the Ground*.

Barker's radio play *Albertina* (broadcast by BBC Radio 3, 13 June 1999) is regrettably overlooked. The play's subtitle, 'The Twenty Duologues', suggests one aspect of its formal challenge. In Albertina, a tiny state in old Europe populated by faithless wives and resentful bishops, two lovers are in opposition: the actor Gradisca wants to 'bite' the world that has failed to satisfy him, and his self-consciously false mistress Laibach laughs sensually as she proclaims she 'likes the world as it is', not least because it breaks Gradisca's heart, and frustration renders him endurable, even preserves his beauty. Gradisca lacks religion but finds consolation in the 'bribe' of Laibach's body. Her husband, the intellectual Rocklaw, is meanwhile in conclave with Olmuts, the bishop of Albertina, who seeks to quell any dissidence in his regime of benignly framed totalitarianism, dressed in populist sentimentality: 'the less tumult, the less anxiety'. Rocklaw is nevertheless too intelligent to be immediately amenable, offering such observations as 'everybody is insane', to which Olmuts objects that this fails to assist his neo-utilitarian programme: 'ideas have to help or they are not ideas at all'. Gradisca observes to Laibach how their location reflects their compulsions (in terms recalling the imagery of *Ursula*):

> GRADISCA: The river pours into the sea, the river is possibly drawn into the sea, or just as likely dreads the sea, dreads an encounter which can only have one consequence – its abolition. What is that?
> LAIBACH: Fear.
> GRADISCA: Or ecstasy?

Gradisca sees in the river a sympathetic ebb and flow, an alternating surge and surrender, resembling human falsity; whereas Olmuts characteristically associates their landscape with his leaf-shaded parks, a triumph of social engineering in eliminating the unspeakably abject: 'The river carries away, like a drain…whatever is carried away by a drain'.

Rocklaw confronts Gradisca, lamenting how 'servitude in some is ecstasy' but in Laibach 'it kills her intellectual agility'; he dismisses Gradisca's claims to authority or power, 'Nothing in your life is true', even as his own emotional disarray testifies to the actor's effects of destabilization, which has driven Rocklaw to forsake his uncompromising objectivity. Gradisca claims Rocklaw should be grateful for the discovery of his own jealousy, a deliverance from his usual 'arctic icebound state of contemplation'. Rocklaw orders Gradisca, along with the other ostensible "undesirables" of Albertina, to be consigned to a barely seaworthy Ship of Fools. Gradisca is confident Laibach will join him, and Rocklaw attempts his own separation: 'If she is a fool, the ship is the place for her'.

Despite the fragility of the ship, some of its prisoners find it conducive, even preferable to Albertina's 'dullness and respectability' (though their resolve is strained by one vagrant's persistent bagpipe playing). Gradisca seizes command of the consciously expendable crew of adulterers, beggars, prostitutes and murderers, through charismatic performance which circumvent Olmuts's binary manipulations of suspicion and panic. Gradisca can 'perform both knowledge and insanity, look wise and shreik, clown, and hide sensible decisions in a froth of barminess'; significantly, he is inspired and driven by the desperate courage and inventiveness of passion, not fear ('I love my mistress and her flesh, I drown in her, but not in this [sea]'). Laibach joins him on the ship of fools, leaving a letter for Rocklaw which explains 'I prefer to suffer the ordeal of being with a man I love, to suffering the ordeal of being without him'. Rocklaw is distraught that it was he, not Olmuts, who added his wife's lover to the list of those to be consigned to the ship, and that she has joined – in Olmuts's reductively dehumanizing term – its 'cargo'.

On the ship, disorder is 'king'; the baggage of Albertina ('logic, cause and effect' and 'even a few bags of morality') has been discarded by its 'monsters and innocents'. Gradisca holds fast to his (literally hell-raising) conviction that 'No one's insane who wants to live' and determination to demonstrate the reversibility of the terms of order which would exclude him: 'I want to live, and make some heaven out of it, or a hell I don't care which'. He is unperturbed by the objections of fundamentally conventional observers ('I never saw a man so infatuated with his own existence, it's undignified'). Contrastingly, Albertina is plagued by an outbreak of female suicides who have had their happiness stolen by the cargo's inclusion of 'all those gifted in adultery' and the art of seduction. Rocklaw perceives how some element of transgression may even have been an ingredient necessary to the appearances of order in Albertina: perhaps 'the smiling mothers in the park smiled at other things than the capers of their infants'. In his own murderous assignation with a prostitute he claims to have glimpsed, in her 'wilful blindness', a 'profound dissatisfaction with the real'; 'we have outgrown God' in a 'repudiation of the world, of Albertina, even'.

Gradisca's apparent success – 'I rule the fools, which must be the apotheosis of the arbitrary' – is, however, not ultimately heroized either by Laibach or the play. Whilst he has developed a strategy of survival, whereby successive ports bribe the ship to set sail again, thus avoiding the prospect of an incursion of criminals and lunatics, his appeal for Laibach has foundered. As he recognizes, as long as he was powerless, he charmed her; the 'discipline' and 'savagery' of his current rule does not encompass her, as she acknowledges; 'how hard it is to love madly when all around you there is madness'. The element of deceptive transgression was a necessary ingredient in the sustenance of her passion, and she has lost 'the laughter of a faithless wife'. Gradisca concludes from her formal valediction, 'If that's a kiss, a dead leaf blown into your face by autumn winds is twice as succulent'; 'I'll stop demanding what you have lost the power to imitate … We have lost the river and now we have the sea'. Laibach plans to return her self to her husband, an admission of her 'inability to want'. But Rocklaw has despaired, and destroyed himself. Laibach accepts Olmuts's paternalistic care, and thanks him for the shade of his leafy parks.

Albertina exemplifies many of Barker's strengths: it combines philosophical and political enquiry with humorous vitality, performances of alluring eroticism, surprising shifts of sympathy and memorable characterizations with thrilling powers of expression. Its taut narrative is deceptively clear, both highly original and classically resonant: it surprisingly reanimates a familiar artistic motif, the Ship of Fools, as both a reflex of political abjection and a madly coherent anti-society, suggesting the mutual dependency of order and chaos, in their respective appeals. The stark depiction of Rocklaw's loss and reversal may recall Henrik Ibsen's *Pillars of Society*, how a respected subject comes to loathe his own policy when an unseaworthy vessel surprisingly endangers his more impetuous beloved; and Angelo in Shakespeare's *Measure for Measure*, when Rocklaw objects to Olmuts 'you have corrupted me' by 'lending me the power to indulge the basest instincts to which I, in common with all men, am subject'. Gradisca's malcontent-trickster energy is seductive and exhilarating, but congeals, on the ship, into another form of authoritarianism; in contrast and counterpoint, the former imperviousness of Rocklaw, his perception, logic and analysis, fractures as he embraces beauty and sin (his lack of repentance when called to account recalls that of the similarly insightful transgressor, Orphuls in *The Europeans*), then audibly and movingly shatters in two when the dialogic progression of the play involves him in an articulated mobilisation of his wilfully self-destructive and recalcitrant selves towards suicide. In Richard Wortley's excellently evocative radio production, Gradisca and Rocklaw are masterfully voiced by Robert Glenister and

Nicholas Le Prevost respectively; Juliet Stevenson imbues Laibach with a disarmingly appealing languid sensuality, and Ian McDiarmid artfully captures the verbal and political acuteness of Olmuts amongst his indignation and surprise. The play's civic and oceanic settings are vividly conjured by its language, and its central dynamic, of intent and intimate dialogues against a wider, fraught background, makes it eminently transferable to the stage for a strong ensemble of ten performers, for the wider appreciation which it deserves.

Knowledge and a Girl: The Snow White Case (written and broadcast 2002) develops the theme of how performance and theatricality are associated with manifestations of power. In Barker's subversive renegotiation of the fairy tale of Snow White, the assertive female sexuality of the Queen has chaotic consequences for various forms of patriarchal order. The Queen, *'naked in a forest'*, appropriates disparagement and wears it as an adornment: when her husband calls her 'Infantile', she ripostes 'He thinks the word humiliates me but I like the word / I *wear* the word / I *walk* in it'[3] (*GTC*, 97). She teases and threatens her Servant and lover, challenging him to try to assert his male procreative order upon her (to 'RINSE / FLOOD / AND WRECK [her] TAUT QUEEN'S BELLY') even as she acknowledges the power and importance of her own reflective accoutrements of performance ('I don't know what I am without my shoes') and goads the servant to new lengths of ecstatic desecration ('If any man can stretch me that man's you', *GTC*, 98). Unusually, Snow White is depicted as sexually precocious and jealous of the Queen, and of her recurrent disposition to be 'eaten' by visitors as well as by the King. Snow White's malicious envy towards what she perceives as a competing mature and transgressive female sexuality parallels that of Ragusa, the character who subverts the associations of the passive and terminally self-effacing Ophelia towards another transgressive Queen, in Barker's *Gertrude – The Cry* (2002), which is in several ways something of a companion piece to *Knowledge and a Girl* – a relationship emphasized by Victoria Wicks's powerful portrayals of the imperious regality and proud unrepentant sensuality of both Queens. The King acknowledges the Queen as 'utterly beautiful … And / A / Rage / To / Me'(*GTC*, 100); the Queen both arouses and embodies an unkindness in her provocatively artificial clothing which nevertheless makes her body, which Snow White tries to diminish as 'rather old', more alluring than the younger woman's loose thin gowns. The fairy tale motif of the talking mirror is also wittily transformed by Barker: various servants (with varying degrees of success) hold the mirror and compliment the Queen as she bids to 'Contemplate the shape of pain … not mine … His whoever beholds me'. The King notes the Queen's impression upon his visitor, Askew, the King of All the Irish, her infliction of a suffering which she characterizes as contempt, but which the King recognizes as something more complex; to his observation 'I never know with you', she replies 'How good you never know with me'(*GTC*, 104).

Thus they maintain a transgressive equilibrium in which her untamed sexuality is an ingredient, as is the display afforded by the mirror when the Queen exhorts the King: 'Turn me to the glass / Let me watch your agony *(Her clothing is violently disturbed)*/ OH SAY MY BEAUTY IS YOUR DEATH' (*GTC*, 109). This display is facilitated by a comic and self-consciously decrepit 'blind and legless' old woman who holds the mirror in attendance, and occasionally lets slip comments which suggest she might not be blind at all. The Queen arouses Askew but also offends him in her avoidance of the queenly 'SINGLE TASK' of birth; similarly Snow White wishes to leave the palace, whilst admitting she is fascinated by the Queen and could have 'so loved' her, but the Queen's power is partly refined by her refusal of the maternal, as she acknowledges: 'But to have been loved by you I should have / had to cease being myself'. When the King suspects the Queen's dalliances with the servant, Askew and his son (The Prince of All the Irish), the Queen explains:

<pre>
QUEEN: I love you but (Pause. She breathes irregularly.)
        I love you but
KING: It's all right
        To
        Love
        Me
        But
        Is
        All
        Right
        With
        Me   (GTC, 116)
</pre>

Thus – not without a frequently refined erotic anguish (comparable to Tortmann's in *He Stumbled*) – the King maintains a distance which permits the Queen's (barely) controlled play within his own strategic display and control of patriarchal leadership. However, the Queen falls pregnant with the child of Young Askew, who has been promised to Snow White after his bid to 'free [her] from the gaol of [her] own beauty' by "rescuing" her from her defiantly bohemian sexual idyll in which she lives naked with seven men. The King blinds these men, not for justice, but to rinse out at least some of his 'offence'; he further proclaims how Young Askew's reversal of the Queen's 'legendary' sterility has travestied all decorum between prospective mother- and son-in-law, 'KNOTTED HORRIBLY THE ROPES OF DYNASTY'. The King's imagination is inflamed by the idea of the Queen's sexual play, which he strives to regulate and control, definitively and fatally, in an ordered dis-play which will refer to both of her competing modes of assertion, her regal power and her female sexuality, staging an extreme form of female disorder in order to impose an ultimate masculine order as its binary opposition[4]. The 'compelling rhythm' of the Queen's heels has both bewitched Young Askew and frightened Snow White; and the King notes the importance of the shoes which have defined the Queen's identity for herself and for others: 'Shoes also are an act of faith / Shoes also are devotions' and accomplices to 'men's agony' (*GTC*, 125). He publicly announces and ordains 'two ecstasies': his own, 'obviously', but provoked by that of the Queen, who is to don red-hot shoes and perform a dance 'As if to show … HER MAD AND UNGOVERNABLE SELF' (*GTC*, 133). The Queen takes on the King's injunction – 'You are a legend and a legend must' – in an exertion of self-overcoming ('I CAN'T DO THIS') which makes the King (not her) cry out and drives Young Askew to seek refuge in recoil into speculative theory; Snow White ascribes her persistence to 'hatred'. More precisely, the Queen achieves a paradoxical triumph of performance through the reversibility offered by absolute play (which 'flaunts society's cherished orthodoxies, embraces what the culture finds loathsome or frightening, transforms the serious into the joke and unsettles the category of the joke by taking it seriously, courts self-destruction in the interest of anarchic discharge of its energy'[5]).

A striking parallel to the Queen in *Knowledge and a Girl* is provided by the titular protagonist in John Webster's Jacobean tragedy, *The Duchess of Malfi*, particularly in the terms of discussion applied by Kelly V. Jones: the transgressive heroine 'recognizes that she is a victim of display, or an imposition of play, an actor in a theatre playing the scripts of others, exhibited to show "her" tragedy' in a way that would make her 'an exhibition and a didactic example'[6]; however, she 'embraces her role as a tragic heroine' so that 'As she becomes a willing collaborator in her own death, she takes control over the situation', asserting 'her own stoic power over the situation imposed on her' to 'define herself through her self-willed action'[7]. Barker's Queen determines 'I'm unforgivable / Even when I am punished I do not weep'; 'MIRROR / REFLECT MY PRIDE TO ME' (*GTC*, 136). When Snow White crouches beside her, the Queen, even miscarrying and surrounded by '*hounds in a chase*', proclaims the terms of her own final aesthetic victory over the King: 'The beauty of our struggle and my death oh don't think death's a truce / See the bright blood of his cleverness / It shines on you Snow White who is less now less white than me'(*GTC*, 137). The evocation and knowing subversions of the self-conscious fairytale world in *Knowledge and a Girl* can also be considered in the terms of Jones's analysis of Webster's play: the performers have no need 'to play against the text in order to draw attention to the theatricality of the play-event and the fictionality of their dramatic roles' because such theatricality is itself written into the dramatic text through the each character's consciousness of artifice, 'the instability of definition concerning each character's role is an intrinsic part of their characterization'[8]. Thus, if Barker's play (like

Webster's) foregrounds the dramatist's (repossessive) control over the dramatic action, 'it also exposes the fallibility of any form of despotic control over play and theatricality'[9]. Moreover, it does so in ways which may alert the audience to the 'despotic machinations that lie behind both political and theatrical dis-play'[10], as opposed to the wilful individual play which permits transcendence of the role of sacrificial 'passive two-dimensional scapegoat'[11] and evasion of the control others would impose, as similarly demonstrated in Barker's *(Uncle) Vanya*.

Barker's own radio production of *Knowledge and a Girl* involved many performers[12] regularly associated with The Wrestling School, its style and methods, to strong effect. Barker's direction deftly discovered a comedy in reflexes of sentimentality, such as the Old Woman's voluble abjection and the King's penultimate collapse into her embrace, as well as an evocative indeterminacy, not just in effects of elements and landscape, but in the various 'ecstasies' of the Queen: her gasps of eroticism, horror and finally the measured breaths and steps with which she carries out her ordeal. Whilst it is possible to conceive of a (properly ambitious) stage production, *Knowledge and a Girl* is particularly artful in its uses of the medium of radio, not least in its regular demand that the audience *imagine what the characters see*: the sights reflected in the Queen's mirror, and its angle of suspension, being the two central images, in a play which centrally addresses issues of power in display, spectacle, provocation, enhancement, blinding, pretence, comparison and disclosure – and characteristically suggests the potential instability of all instances of power (the King's, the Queen's), its susceptibility to reversibility.

Interlude: Erotic Disclosure Through the Rules of the Game

Lingis has proposed that the artfully strategic manifestation of nakedness may trigger an outbreak of Catastrophic Time:

> No doubt the abrupt denuding of someone in our presence is a small hiatus in the time of our everyday work. Yet the dense chains of taboos forged over human nudity in our society, and in all known societies, bear witness to the catastrophic effect denuding has been felt to have, has been found to have on the world of work and reason and on the sane identity of functional citizens.[13]

When the performer undresses before our eyes, or the lover undresses for our touch, 'she or he takes off the uniform, the categories, the endurance'[14] in demonstration of chance and promise of dissolution (in the words of one character in Barker's *Ego in Arcadia*: 'Let me undress you...let me remove your body from the cladding of a routine life...', *CP3*, 280). For some, clothes may be the literally conventional indices of a social power or self-sufficiency; when they discard their clothes, they drop major components of their body identity, to emerge as if 'peeled or flayed'[15]; by so doing, they may discover not only a vulnerability but also an alternative form of power. Barker the dramatist-director (and painter) often deploys nakedness both tragically and erotically (nakedness involving a problematizing visceral force that can induce in the spectator a complexity of feelings that is not immediately manageable, as opposed to the contemplative and sublimated attitude of the nude); often, he does so in such a way as to problematize beauty (the assumed social norms of the aesthetically appealing), in a way which Elizabeth Grosz associates with a distinctively modern art (such as Francis Bacon's) which is 'exploratory of how to view bodies differently, how to see them from angles and perspectives unthought of, to bring out latencies that may not be obvious in everyday life', or through conventional forms of sexual depiction[16]. In appropriate application of her propositions to Barker's work, nakedness has the effect of emphasizing how 'bodies are not simply the loci of power but also of resistance, and particularly, resistance because of their excess'; bodies 'exceed whatever limits politics, management, or desire impose on them'[17].

As Angel-Perez observes, 'The moment of nudity on stage is a moment of utter suspension as well as a moment of revelation; it is a moment of loss as well as an epiphany'[18]. Barker's naked characters (such as Katrin, Tenna, Gertrude, Algeria and Klatura) are self-conscious 'explicit body performers' within their (fictional, but more widely resonant) social contexts; moreover, they are striving to make themselves into works of art (as do other characters, such as Lvov, Sleen, Dancer, Hudaconceva and Sleev in *I Saw Myself*, in their existential projects) in a response to what they perceive as a redundancy in in the prevalent terms of the human/body. As Grosz observes, to make oneself into a work of art 'is not simply to aestheticize life', 'It is to live one's life and one's body in excess of what is required, in excess of discipline, and even in excess of aesthetics'[19]. Lingis describes eroticism as the discarding of 'the carapace of character'[20]; but in Barker's theatrical manifestations of nakedness, one might more artfully identify a double discarding: of the dramatic character's social persona, and of the performer's fictional persona, in an invested incarnation of courage, extravagantly but precisely manifested. The relationships between these issues – eroticism, suspension, nakedness, disclosure, identity, endurance, power and resistance – are the subjects of many Barker works, but rarely with the ferociously sustained focus that is provided by the negotiations between 'the naked and the dead' that are provided by his mid-length (but profound) play, *The Twelfth Battle of Isonzo*.

Two Against Nature: *The Twelfth Battle of Isonzo*

Lamb has noted how Baudrillard's reflections on seduction offer 'a way of describing the world that focuses on elements which "rational" social discourses marginalise or suppress'; and established how his theories are pertinent to Barker's plays, in which self-consciously performative characters often play 'teacher and pupil' in a duel 'where the participants attempt to drive each other mad'[21], as part of an investigative, and frequently erotic, exploration beyond the realms of social normality. These ideas provide another apt foundation from which to approach *The Twelfth Battle of Isonzo*, a play written in 1998 which received its English-language premiere[22] directed by Barker in Dublin in 2001, followed by other Irish and Welsh performances in 2002. The play was staged in a co-production between my own Lurking Truth/Gwir sy'n Llechu Theatre Company, Íomhá Ildánach and the Irish Touring Company. This constituted Barker's first direction of the premiere of one of his works outside of his own company, The Wrestling School (it also featured set and costume designs by Leipzig and Kaiser). The Irish actress Antoinette Walsh played Tenna, and I played Isonzo.

The play depicts the first meeting of Isonzo, a very old man (avowedly one hundred years old), and his prospective twelfth bride, Tenna, a very young woman (avowedly seventeen), who both proclaim themselves completely blind. Their probing, challenging, enticing and dramatic self-fashioning in relation to each other must therefore be conducted, not through sight, but through words and other ritualised forms of sensed proximity. Like other Barker plays such as *Women Beware Women, The Europeans, The Last Supper, (Uncle) Vanya* and *Gertrude, The Twelfth Battle of Isonzo* depicts a man and a woman in the sexual heat of an erotic *duel*, which is also a *dual* or joint project to defy the conventionally socialized prospects of the world.

This play initially shows Tenna awaiting the first arrival of her betrothed, with a mixture of anticipation and trepidation. Isonzo initiates a series of tests and mutual challenges: he details his own history of sexual encounters; he physically deceives her as to his whereabouts, but also becomes distraught when he thinks she may have suffered a fatal asthma attack; he delineates his aesthetics of women's garments and the process of undressing, insisting that Tenna remove her underwear, and later discard her bridal dress for him to 'see' or sense her nakedness. She deceives him in return; their mutual challenge is a poeticising friction through which they keep correcting their inventions of alternative possibilities. She artfully pretends and delays her self-disclosure; but when she decides to comply, he tells her that he is 'not blind', but that his sense of eroticism depends on elaborated description and proximity in preference to conventional sexual consummation. She feels humiliated and outraged, but talks her way through to her own appreciation of his perspective, in ways that elicit his delight, but then his disappointment. Suddenly, Isonzo falls dead or dying; Tenna pursues a sense of her self-dramatisation as a 'widow' struggling to incorporate the legacy of her experiences with Isonzo, a dramatization which he (as a prone and posthumous presence) continues to interrupt, undercut, subvert and instruct, until she exits the room, drawn by the sound of church bells.

Isonzo is spiritually animated by a project: '*if the world is poor, you have to reinvent the world*'[23] by frustrating conventional dynamics; he knows this, and Tenna has an inkling of it. Isonzo, for all of his advanced age, is still searching: his anger can come out of the sheer delicacy of what he is trying to reach, as he inducts Tenna, his pupil, into the project. Tenna is deliberately pushing herself into an area she does not know, and is thus more daring than the norm: Isonzo is attracted by what she will later proclaim her 'infinite capacity for abstraction'. He is undertaking a desperate attempt to resist the limitations of conventional sexual "nature" through artificiality, sensing something beyond the purported end-stop of sexual "climax", namely the possibility of a sublime infinite prolongation. He wants to achieve the power of sexual domination without the banality of the act of intercourse (the limitation of sex to intercourse, or even of eroticism to sex, being a misapprehension which characterises sex as 'part of the happiness racket'[24]). Similarly in *The Europeans*, Katrin and Starhemberg undress and describe themselves and each other but do not touch; their contact is through gaze and words, refined and developed, poetically and politically, into caresses of imagination ('politically' in that this represents a joint dislocation and unlocking from, and resistance of, received socialized aesthetics and state ideals of gratification and order). Isonzo is also proposing an erotic ordeal based on looking and speaking, '*resisting the body to keep it in a state of tension, thereby defying and teasing Mother Nature by resisting her terms; shamelessness is the method, the end is spiritual*'[25]. The most enigmatic moment of rehearsals came when Barker asked me if I thought Isonzo was truly blind: I had presumed not (he says 'I'm not blind / I merely shut my eyes' when Tenna finally capitulates to his insistence that she undress, and he briefly removes his glasses). Barker replied that his ability to see was one possibility; but so was the scenario in which he was truly blind, and claiming sight in order to compound strategically her sense of exposure and confusion. This comment necessarily exploded my sense of my character's possibilities.

Like Lvov in *The Last Supper*, Isonzo is looking for the perfect handmaiden – to help him to die, although neither he nor Tenna are aware of this ultimate objective[26]. Isonzo is a 'collector' of women[27], a seducer in the familiar sense of a man compelled to besiege sexually a succession of women; for him, marriage has been the ritual execution of a form that consumes its subjects (taking on what Baudrillard calls 'the aesthetic form of a work of art and the ritual form of a crime'[28]). However, in this instance the exercise becomes mythical and demands sacrifice, not just of Tenna, but also of Isonzo himself. The play depicts a seduction in Baudrillard's sense of the process: a mutually enveloping and dislocating imaginative vertigo.

When Isonzo has tried to crack the resistance of his previous wives, they have all withdrawn from him at some stage of his project, and withdrawn into conventional objections, values and demands. However, the further Tenna responds to the punishment of his imaginative trials the way she does, the further Isonzo adores her. When her will does not falter, his does (and on some level wants to, though *'the pain of this is manifest'*[29]), in seeking closure; it is possible that she will continue their search further, with other men (*'probably in vain'*[30]). The play offers a stark yet highly erotical demonstrative dramatization of one of Baudrillard's principles of seduction:

> If sex has a *natural law*, a pleasure principle, then seduction consists in denying that principle and replacing it with a rule, the *arbitrary* rule of a game. In this sense, seduction is *perverse*. The immorality of perversion, like that of seduction, does not come from abandoning oneself to the joys of sex in opposition to all morality; it results from something more serious and subtle, the abandonment of sex itself as a referent and a morality, even its "joys".[31]

The process of the play is marked by a tense determination, a momentous commitment and desperation within the gamesmanship, with the characters self-consciously aware, and suggesting to the audience, that *'the process begins **now**; we don't leave until we've gone through it and beyond'*[32].

Our rehearsals began with a strategically concentrated series of readings of the script, with Barker monitoring and correcting the breathing and energy patterns in our intonations of the lines, and the physical and imaginative locations of crucially specific words. For example, the unpunctuated sections of the text were to be rendered, not literally without pause for breath, but rather as being animated by a single specific conceptual engine of excitement; the bold sections of the text being foregrounded, not by additional volume, but by a differing quality of intentness.

Our physical choreography developed through particularly precise attention to balance: its placing and loss. Tenna and Isonzo contact each other almost exclusively through words. Walsh's training in professional ballet assisted her achievement and invention of some arresting transitions into compelling attitudes of transfixion and strenuous yearning, which Barker composed or montaged meticulously. My own challenge was to develop an unnatural contortion of the body, suggesting great age and infirmity had rendered Isonzo's physicality to be so restricted as to be described by Tenna (appreciatively) as residual; nevertheless, I had to be able to respond to verbal and sartorial details with an unnerving speed and agility. Again, Barker challenged me to explore and extend my previous limits of physical balance, so the spatial relationships between us often aimed for a dance-like delicacy and synchronicity within the overall pictorial frame of a metal set. This construction visually evoked a harshly predatory entrapment and audibly resonated to our every movement, however minute: through sight, sound and touch, it amplified the cruelty of the play. Our rehearsals gave us a rigorously precise 'score' and structure for performance, in which the technical challenge was for us to play for maximum contact and maximum nuance, hitting the beats and raising the energy of the line endings with consistent intensity.

The production began with an exordium, in which Isonzo 'traversed the back of the stage, reaching for two white sticks which floated in the air just out of his grasp, to accompaniments of distinctly unnatural effects of lighting and sound'[33]. Barker directed Tenna to enter, audibly tapping her white stick to find her seat, and then to break her stick across her knee, discarding it with an amplified crash, signalling her irrevocable commitment to the here and now of her first meeting with Isonzo, divesting herself of pragmatic assistance to independent social orientation (as if in artful foreshadowing of the discarding of her clothes). She then embarks on a list of her own existential transgressions in order to challenge the audience's presumptions and status. In the posture of exposition and confession, this young, beautiful and potentially pitiful woman proudly proclaims her defiant commitment to something more challenging than "normal" sexual preferences, aesthetics and sensibility. She is dedicating herself to a man who is very old, humourless and painfully devoid of irony. She is insistent on her unregenerate infringement of social codes (she proclaims 'I am a winter' – not spring or summer – 'of anticipation'). However, she is not merely smug or petulant in her contempt for conventional habit, fiction and relief. For the audience, as for Isonzo, she is mesmeric in her visible effort to be unforgivably decided and particular in her embracing the initiatory role of 'bride'. Tenna determines to break through the routine into the extraordinary, evacuating the conventional religious authority associated with marriage to invoke and command more transgressive sources of authority and sacredness. Tenna's assumption of the role of bride is itself seductive; it is a choice of means whereby she might '*die as reality and reconstitute* [herself] *as illusion*'[34].

The gothically ominous sounds of Isonzo's approach may evoke associations of Bluebeard or *La Belle et la Bête*, which are subverted by his narcissistic proclamation of his own 'Beauty'. Isonzo redesignates age and blindness as not obstructions but aids to concentrated acuteness of perception and appreciation; he is delighted that 'We are so different' (rather than 'we have so much in common', as is the sentimental ideal). Then, whilst expressing awe at Tenna's proximity, he shatters the initiative towards contact by flinging down his sticks. Tenna rightly interprets this as a ruse, and a test; Isonzo joins in her laughter, pleased by her perceptiveness. He resists her offer to collect his sticks, recognizing the potential condescension in this gesture of pity, and turns the tables again, dismissing the sticks as 'an affectation', designed to lend him a 'Peculiar / Authority'. He begins to instruct her in the strategic reversibility of every thing, where notions of authenticity can be subverted and trumped by poetic artificiality, a superior achievement in which he implicates her: '**Our passion never was susceptible to truth**' (*CP5*, 247). Tenna alluringly proclaims her will to subordinate herself to his superior invention: 'Lie if you wish to'.

Isonzo then adds a further refinement of his status as unfathomable sexual connoisseur by claiming that appeals to odour, sight and touch have been surpassed by 'the rumour', the activation of the imaginative speculation. Isonzo keeps up the pressure with a further test, tricking her into thinking him 'dead or dying'; Tenna is distraught that he has not touched her, nor 'Consummated / In any form as yet'. Isonzo claims the right to surprise her repeatedly, both physically (as when he unexpectedly divulges his location by craning forward and resting his chin in her hand) and imaginatively (in his enticing but menacing assurance that 'There'll be a form'). His soliloquy recounting his experience of war, his physical survival which nevertheless led to his haunted imagination, is best performed as a genuinely felt trauma. Tenna mobilises a contrasting reverie, a kitsch image of her grandmother as a child studying the alphabet; challengingly, Isonzo links study of the alphabet to the consequence of his experience of trench warfare, suggesting that education leads to civilisation which leads to warfare. However, Tenna is neither impressed nor overwhelmed by his sophistry, holding on to her image of her grandmother, to which Isonzo defers (unconvincingly agreeing 'She was a little dear').

However, when Tenna adeptly challenges Isonzo on a detail of his wartime reverie, he insists on the inescapable, irredeemable horror of his martial experiences, whilst claiming the right to aesthetic fabrication, unravelling the baroque image of Truth as a malicious serpent: a harshly compelling performance by Isonzo which proclaims a fierce pride in refining an aesthetic of lying and artifice to keep the truth at bay. He unpredictably moves to another refinement, commanding 'Describe your pants'. Tenna grasps that he wants her to discover his relish of the jointly embroidered narrative: together they conjure an epiphany, located in a store's lingerie department, refuting the so-called 'deeper order' of sexuality, and substituting 'the charm and illusion of appearances'[35]. When Tenna breaks the mood by laughing, Isonzo angrily snaps back to his role and status as pedagogical instructor, claiming a delicacy which involves physical nausea at inappropriate details. When Tenna finds a ludicrousness in a blind man's valorization of colour, Isonzo ripostes with an eccentrically philosophical investigation of the situation which impressively insists on the importance of the 'frail poems' she draws over her hips, and persuades Tenna to remove them. Again, but this time inadvertantly, she breaks the mood, and is abject: Isonzo threatens her with his ability to take back his love from those unworthy of it; but then reconstructs the moment as a performance of his own 'agonized devotion'. She starts at his word 'gaze': hastily, he recharacterizes the word as a sensitivity not limited to the merely visual, but his skill in conjuring surprising metaphors is severely tested. Isonzo reiterates his frustration at Tenna's demanding and uninventive predecessors who lapsed into a literal and mean-minded dependence on 'truth': this begins another rhetorical trajectory to persuade her that her bra 'must also be / Unhooked'. Tenna voices her terrified veneration, only to be shocked and alarmed by the deftness with which he replaces her shoe.

Now she surprises him, by launching into an intense performance of a soliloquy which feverishly anticipates sexual consummation. The soliloquy alternates between locating the sex act in pictorial domesticity (lawn and summerhouse) and sordid urgency (gravel pit and railway yard), but demonstrates a rapaciousness for both sexual and aesthetic triumph which silences him briefly. But he insists that the sensation of yearning is precisely what should be cultivated and extrapolated:

> TENNA: ... **I ache to be**
> ISONZO: Ache
>   *(She lifts her hands in a gesture of pain)*
>   As for the bra
> TENNA: I am not wearing one *(Pause)*
> ISONZO: Feel free to cross your legs   (*CP*5, 262)

Challenging his deferral, Tenna indignantly unfurls her own (possibly fabricated) sexual history; he hungers for evidence of the spiritual poverty of his predecessors, and she delights him by acknowledging the fictional aspects of such itemisations. The following utterance may represent their most perfect equilibrium of mutual respect:

> ISONZO: Did
>   Ever
>   Anyone
>   Place more of their life's hope in this one act
>   Than
>   You? *(Pause)*
> TENNA: Only you *(Pause)*
> ISONZO: Only me *(Pause)*   (*CP*5, 264)

His admiration of her impels him into a frenzy of impatience, '**Undress I have to see you naked**', even as he acknowledges 'What's bride but the apotheosis of delay' and halts her compliance to specify the correct sequence ('Shoes last') of emergence from her 'cascading' finery. Like Tenna, Isonzo locates the erotic charisma of the bride beyond social custom and religious authority: the bride can represent language as perfectly embodied, combining the promise of wantonness of utterance with a body which is simultaneously a tissue of metaphors. The momentousness of the hiatus induces a cardiac spasm in Isonzo, which he struggles to contain through metaphor, before she proceeds. Her teasing subterfuges comically stretch even his sense of deliberation to breaking point; she demonstrates 'Exquisite trickery' when she draws down a zip, then *'shakes her clothing without however exposing herself further'*. This combines, for characters, performers and audiences, a sense of game, and also the tension of potential disclosure within the game. Isonzo out-manoeuvres Tenna by his fantastic proposition that he can detect her refusal of nakedness by ear ('Blood flows faster in the naked obviously'). He is right to identify nakedness as anything but a neutral state when it occurs in the company of another: it is full of cultural shocks, trepidation and tension. The proposition that, on this basis, he can discern her state is far-fetched, but his metaphors triumph strategically where logic would not, particularly when they are rapidly succeeded by the correct deduction of her trepidation and resentment of his accuracy. However, as a consequence of the tension of the situation, Tenna suffers an asthma attack, and Isonzo is pathetically scared that she has, like the other eleven wives, predeceased him. When she audibly recovers, Isonzo cuts short her explanation by a sudden insistence that she carry on undressing: his *performance* of lack of sympathetic concern is so heretical that it makes her laugh, momentarily freeing her from her own self-preoccupation. She accepts what he proposes as the inevitability of her nakedness and the 'relief' it might offer. Notwithstanding the *'tension'* which *'suffuses her'*, she determines to play her part through to the hilt, and undresses.

However, the promised 'relief' is (characteristically) neither given nor sought. Isonzo issues a *'terrible wail'* at how his blindness denies him the revelatory sight and the appreciative gaze. He affects a grief, a romanticization of her youth[36], and the posture of despairing *roué* whose jaded inspection she is spared. But then he makes his most surprising strategic reversal, removing his glasses and proclaiming 'I'm not blind' (*CP5*, 269) (in performance, this was the one occasion when I removed my dark glasses to reveal my eyes, but I kept them unfocused and slightly crossed, suggesting a possible blindness in contradiction of Isonzo's words; here and elsewhere in performance, I developed, with Barker's encouragement, a sending of mixed but inconclusive messages as to whether or not Isonzo might be truly blind[37]).

Triumphantly, Isonzo claims a charismatic transcendence of God, Paradise and Death in this worldly apotheosis of erotic revelation. However, Tenna is surprised when he does not respond to her invitation to him to kiss her arse. Breaking this posture of sacramental erotic entreaty, she fluctuates between fear of his desertion and a self-undermining sarcastic indifference (physicalized under Barker's direction by Tenna performing a stately but eccentric naked march behind the bars of the set); she expresses anxiety and indignation at his 'Stretching the pleasures of anticipation to their breaking point'. When she scrambles for her dress, Isonzo insists that she stay naked, 'Because I have to suffer you' (*CP5*, 270). Isonzo repudiates the banality and limitations of conventional climax, the finality of a sex that exhausts itself in orgasm, preferring to extrapolate and refine the agony of yearning in a precarious but graceful balance. He also catches himself out through his chosen image, an echo of an earlier erotic occasion which propels him into recollection of nostalgic wonder and then disgust at the reductive sarcasm of a woman who found his insistent deceleration 'more than she could bear'. Isonzo rhetorically monumentalises his own resistance to 'Mother Nature': the 'School Miss' whose authority he would defy and whose wrath he would compulsively provoke. When Tenna deduces that none of his marriages was consummated, she is horrified at the extent of his labyrinthine abstraction, even as he is possessed by vengeful disappointment. However, Isonzo has not stopped *trying* to find a woman capable of receiving, elaborating and extending his concept of linguistic friction and erotic heat; even as he waited for them to fail, he has retained the hope that one exists (and she will; she will extinguish him, and he will die of her).

With the viciousness of a wounded beast, Isonzo suggests that Tenna's proffered nakedness is only the '**Down payment on a deal**'. Tenna interprets his goal to be humiliation: she lashes back with the propositions that '**Rape is better** ... Appetite / Frankly / Is / Preferable' (*CP5*, 273). In our production, Isonzo cringed at her words, but in injured disappointment at her self-righteous misapprehension. His goal is **not** humiliation, but rather an unusual adoration: he may be bitterly if silently distraught when Tenna recoils into proclaiming a defiant kinship and solidarity with the 'broken women' she never knew and traditional, prescriptive notions of legitimacy, epitomised by her invocation 'Thank God': this is the ultimate rejection of his idiosyncratic project to distil a proud divinity with/in and through human sexuality. This point of the play represents the highest pitch of antagonism between Isonzo and Tenna.

A turning point occurs when Tenna realizes 'And now I sound like them': whether in 'Pleasure or complaint', she is expressing the same feelings and values as her eleven predecessors. Therefore, she says she failed '**The test I set myself**' (*CP5*, 274). She recalls her initial determination to fly in the face of conventional sexual appetite, practice and aesthetic sensibility. She has wished to distinguish herself from Isonzo's previous wives and demonstrate them to be '**Sodden with conventionality**' in contrast to her own 'Infinite capacity for ... Abstraction' (*CP5*, 275). Isonzo listens to this transition with a growing sense of wonder, and rewards her recovery of pride in difference with further revelations from his always-surprising encyclopaedic sexuality. He explains that he is not a virgin, indeed was a frequent visitor of brothels and connoisseur of sexual detail in many countries and contexts, and, moreover, that he is 'pole hard' in his erect longing for this new proudly transgressive Tenna. She is bewildered, but then appreciates that his project involves the extrapolation of the ache of sexual longing rather than the reductive summary of its satisfaction, and that their mutual 'contemplation' represents a joint triumph. In her exultation at vaulting '**the turbid pool of intimacy**', however, Tenna becomes indiscriminate and reductive about her own nakedness: 'It's being without ... Clothes'. For Isonzo, however, nakedness is of sublime importance, and he, disenchanted, checks her self-satisfaction with the statement: 'They all said that' (*CP5*, 277). Tenna is outraged. Isonzo tries to draw her from her retreat into petulance with a compliment, that the others said it 'In less exquisite words', and by presenting his precise identification and vision of 'The agony of age': that every door opened reveals something '**Precisely the same**'[38] (*CP5*, 278). Barker directed me to be exasperated but contemplative in my delivery of this speech; Isonzo should still include Tenna, rather than retreat into a soliloquy; he is still working to draw her in and educate her, not just **describing** but **teaching** his distinctive sense of melancholy. He further compliments her, that one of her sensibility is likely to have sensed already that the world is not rich enough. Tenna grasps his argument: this sense of graduation into a new imaginative level is developed by her initiative 'I'll kneel on you', suggesting that she will release him from life into death. 'Her triumph challenges him to go where he has to go, to die; she suggests *"I'm a suitable way to die – are you up to dying? Are you ready to go through that* door?"'[39].

Her acquisition and playing with the word 'prone', as a poetic substitute for 'dead', demonstrates the linguistic refinement which Isonzo appreciates, but even he struggles against acknowledging the full import of the word: when Tenna insists that he does so, he '*goes down like a dropped sack*', faints and, according to Barker, dies[40]. The rest of the play is a posthumous dialogue of Tenna and Isonzo's imaginative challenges to, and claims on, each other, played out in a different form of emotional reality. This transition was suggested in production by a sound cue of a choral note and the fall, from the wings, of a piece of black material (termed 'the widow's weeds' in production). Tenna used this cloth to adorn her naked body[41], as she narrates and performs '*a scene Isonzo would like to watch but can't*'[42], her apotheosis in another form of ritualized and paradoxical erotic allure: the widow in mourning at her husband's funeral. This is conventionally a view of the bride that all husbands are denied, and Tenna taunts him with the loss of this new phase of her eroticism, whereby she compulsively challenges men to console her sexually and beat back mortality in the defiant sexual quickening which often follows a bereavement. Even now, Isonzo refuses to be outdone; he undercuts her cruelly rhapsodic performance by claiming his own enactment of the same scenario as the male partner to a widow in Lisbon. This sexual encounter is a formative revelation for him: the spirit of frantic conspiracy led him to unhelpful impatience and fumbling, until the widow emphasized the pleasures and importance of exploration. He anticipates how Tenna will correspondingly attract male mourners in transgressive, but familiar, compulsive negotiations. Tenna finds herself unexpectedly emotionally overwhelmed, and strives to support his (and her) sense of his uniqueness. Isonzo insists on the incontrovertible 'sheerly / **Ordinary** / Nature of all things' and orders her to suffocate him in a final fatal kiss, performed 'With [Her] Arse' (*CP5*, 282). He wants her to overcome his inadvertent spasms of resistance, and finally characterises himself as 'An / Inquisitive / Child', who has demonstrated a strange innocence in his incessant explorations. '*Tenna seems poised to accede to Isonzo's request but stops suddenly*': despite the power of his intellectual argument, the '*cultural pull*' of the wedding bells, the sound of life, '*draws her magnetically*'[43]. Isonzo is frustrated in his bid for his "aesthetic orgasm" (however vehemently I performed his desperate exasperation in the cry '**I'm on the threshold of oblivion**', the line frequently provoked audience laughter, which was astonished but not unsympathetic). Tenna denies him this consummation ('*Why satisfy the weak?*'[44]), not only out of the insistent splendour of her own vanity ('**A bride must be observed**', *CP5*, 283) but also out of recognition that '*In asking for completion, he's breaking his own rules; in wanting perfect closure, he is betraying his own system of deferral and abstraction*'[45]. When she collides with the objects and boundaries of the room, Tenna is '*surprised but not horrified*'[46] by the bathos of her nosebleed; and whilst the printed script might suggest that she is too contaminated by his pessimism to escape the room, literally or metaphorically, in rehearsal Barker decided to direct the ending so that Tenna persisted, despite collisions with the chair and walls of Isonzo's room, and finally escaped from the room, albeit bruised and bleeding (Walsh devised some horrific looking and sounding collisions with the sharp and rusty metallic set which she nevertheless executed without major injury). Sound and lights faded on Isonzo, kicking his legs against the metal set, winding down in the darkness like a desperate but fading mechanical doll.

Thus, *The Twelfth Battle of Isonzo* is a *pas de deux* through the incessant explorations of sexual enquiry, to the sporting aspects of death, suggesting that '*The only proper way to arrive at death is through having exhausted all possibilities*'[47]. As in other Barker work, the potential starkness and gloominess of the proposition, wherein Isonzo instructs Tenna in the elegance of melancholy and the 'agony of age', is offset by the characters' tightly-wound nimbleness of initiative and invention. Barker emphasized that the characters should connect, even when rhapsodizing or being dismissive – their purpose and direction is always to have a seductive effect. They should also demonstrate a care for detail and its effect on the other, particularly in cruelty: Isonzo and Tenna have (like the audience) chosen to be in this space together, not been forced, and when under attack, they should be visibly emotionally moved, often with painful delicacy, as they move their rhetorical chess pieces to enjoy brief victories and suffer defeats, but also discover ways to push and expand jointly constructed imaginative initiatives.

A loss of the sense of ordeal for Isonzo might reduce aspects of the play to a melodramatic demonstration of Tenna's ordeal (if I lost contact and delicacy in an essentially self-preoccupied bid for intensity, my performance degenerated into mere callousness and stridency, which threatened to unbalance the play by making Tenna seem merely pitiful). The sense that Tenna is caught in Isonzo's web must be complicated by manifestations of his own sense of the pain and poignancy of what he instigates and provokes: such manifestations, when achieved, would draw in the audience, whilst the ubiquitous cruelty of the play's demands prevented any sentimentality in this engagement.

The play's dramatic geography remains constantly internally surprising, and blackly incandescent in its development of a Promethean erotic heat beyond the limits of conventional fulfilments. When approached with the requisite spirit of fierce determination, the central premise, which renders artificiality sacred, propels the performer into the realm of the dangerous: like Baudrillard's vision of seduction, it is an uninterrupted ritual exchange in which the players constantly raise the stakes in a game that never ends, and 'cannot end since the dividing line that defines the victory of the one and the defeat of the other, is illegible'[48].

The Politics of Melancholy: *Animals in Paradise*

The female character name Tenna recurs in another Barker play dating from the same period of composition, referring to a young female protagonist who is again wilful, provocative, frequently naked and compulsively drawn to an older man – but blinded only temporarily. Barker's *Animals in Paradise* was written in response to a commission from the Swedish and Danish governments which sought a celebration of the two nations' connection in 1999 by the newly-constructed Øresund bridge, a symbolic resolution of international animosity. The play was staged in Malmö, Sweden, in 2000, and then revived in Barker's own production, in French translation (by Jean-Michel Déprats and Marie-Lorna Vaconsin, *Animaux en Paradis*), in Rouen, France, for 2005, in a further international collaboration between Rouen's Le Théâtre de Deux Rives, involving actors from Guillaume Dujardin's Mala Noche company and designers and actors from The Wrestling School. However, the play's association with international co-productions only serves to emphasise its startling rejections of forms of (re)conciliation in ways which recall and rival Barker's 1987 play *The Europeans*.

The play begins with the counterpoint of two Danish women: Mrs Norris, literally earthy in her insistence on the ethic of progeniture (even as she digs graves to consign dead babies to the ground), claims fertility as task in support of the war effort; Tenna, a woman of catastrophic sexuality, exposes the futility of how the work of fecundity seeks to circumscribe and structure time, making sense through reference to a future of goals and results: Tenna ripostes to Norris, 'if I'm a whore you're an idiot you have buried five babies for what' (*OP2*, 181). Whereas Norris's body is marked and sodden by repeated childbirth, Tenna's body (derided by Norris as 'dry', 'brittle', 'jagged' and 'sharp') refuses collaboration and provokes disequilibrium, not least through the pursuit of a wild sensuality and personal exhilaration in her frequent choice of nakedness (recalling that of the similarly, but fatally, provocative Klatura in *The Swing*). Norris represents a version of what Bataille would (somewhat reductively) designate *human behaviour*, geared to specific ends in work, and so tending to reduce us to objects at the expense of our sexual exuberance; Tenna exemplifies what Bataille terms the *animal nature*, the sexual exuberance which prevents us from being reduced to mere things[49]. Perhaps 'Chekhovian nature' and 'Catastrophic nature' might be more appropriate titles for this distinction; but Bataille's terms of distinction may offer a clue as to the origins of the play's title.

Meanwhile, the Swedish prince Taxis attends the deathbed of his mother, punctuating his postures of dutiful reverence with a sense of appetite (for a sandwich) which recognises but undercuts the urgency of the dying words. The servant Practice interprets the Queen's hand gestures as decisive and vengeful in their instruction to 'Kill'. Taxis recoils, seeking the consolations of conventional sentiment; but the instruction is unmistakeably resolute.

The Danish philosopher Machinist is taunted by a herd of cyclists, who ring their bells in an inane clamour. Machinist's anti-populism and 'contempt for innocence' has made him a reviled man (akin to Sleen); his student Fourteen represents a recurrent figure in Barker's drama, the downcast disciple in what Lamb terms the 'magisterial' relationship[50], in which he is self-consciously overshadowed by his master. Norris's prescription is reductive – 'Stop snivelling find a woman tip her on her back' – and based on a functional repetition: 'THERE IS A WAR ON BABIES DIE GET BABIES THEREFORE' (*OP2*, 187).

At the Swedish Queen's funeral, her coffin is hoisted (not lowered into the ground) to take its place in a dark tower, designed by Taxis to fill the enemy onlookers with horror. When one courtier suggests that 'The war could die with your mother', he is killed by a labourer. Taxis offers no reward ('If loyalty were rewarded who could trust the loyal?', *OP2*, 190), and the labourer recognizes how little an armistice would change the priorities of the population at large. Machinist comes to a similar recognition from another perspective: the mockery of the Danish cyclists seems to sap his intellectual and sexual morale, they represent a philistine future and barren ground for his conceptual research. Tenna, his mistress, remains (provocatively, challengingly) passive, until she to decides to cross over to enemy Swedish territory. When Norris tries to restrain her, Fourteen intervenes, and he is plunged into disorder and amazement by her slap, then kiss.

Taxis resembles Leopold in *The Europeans* (with an added boyishness and architectural ambition reminiscent of Stucley in *The Castle*), in that his attempted orchestration of national identity narratives and myths would be more objectionable were it not so artfully performative. Leopold disarms his opponents through a combination of self-disclosure and whimsical clowning, disrupting criticism by discarding formality and displacing pain. Taxis visits the dying soldiers not to express sentimental clichés of brotherhood but to insist on their difference: he proclaims one casualty a 'Delirium / Of / Damage' and, through sheer and impressive effort of will, performs abjection by lapping the wound of one soldier. In response, the wounded '*weakly cheer*' even as he shamelessly disabuses them of their picturesque images of convalescent retreat: 'Thank you then for applauding my untruth / WHAT GREATER COMPLIMENT CAN MEN PAY THEIR MASTER?' (*OP2*, 197). Darling, a female torturer, presents him with Tenna, who has been apprehended and beaten around the eyes; however, Taxis discerns a beauty even in Tenna's bruising. Like Saturninus in *Titus Andronicus*, Taxis conceives a fascination and desire for an enemy prisoner despite, or because of, her incarnate offence to his countrymen loyally slain or maimed in battle; and shamelessly exults in his own absolute power to do so.

Fourteen contemplates suicide, but Machinist's apparently homicidal rage towards him shows this to be a posture, a 'philosophic exercise' – indeed, just as Fourteen tends to interpret all of Machinist's actions. In fact, Machinist is genuinely and profoundly dejected by Tenna's desertion of him, which renders all remaining life an exercise in 'TRIVIALITY'. Fourteen does not confess his own presence and desire at the scene of her crossing.

In the Swedish palace, Practice attends on Tenna, reviling her as she luxuriates in safety when his own son was tortured and killed in the war against her nation. Taxis acknowledges his own dislike of her; but also that 'unfortunately I love her'. Taxis sends his former mistress Darling as a painful 'gift' to the Danish king, Calltold. A sense of seductive strategy pervades the confrontation: Calltold admits he is not 'completely indifferent' to her nakedness, but not mastered by it; she professes an admiration for this 'magnificent achievement' of self-control.

Darling's account of a new mood of Swedish prescriptive idealism seems to have substance: Taxis conceives a child with Tenna and uses the infant as a symbolic property in the state pageantry of reconciliation which accompanies the declaration of peace. Significantly, the child's crying, as if in protest at her annexation, and in contradiction of all promises of "everlasting peace", is the only sound at the uncanny silence of the guns (a first suggestion that, in at least one respect, she will prove to be the reincarnation of her grandmother). Taxis's reconciliation is to be symbolized further by the demolition of the Queen's tower, and its materials being requisitioned for the construction of a bridge, which Taxis proclaims 'is love and like an imploring hand will cross the water to our neighbours'(*OP2*, 216): but, as another courtier warns, it may be viewed as a further insult to the pain of many, invested in the war.

As elsewhere in Barker's plays (Ball and Bradshaw in *Victory*, Ann and Krak in *The Castle*), opposition breeds attraction, and collusion – in ways which shatter more facile and conventional alliances. Against all odds – and perhaps precisely because of this – Tenna seduces her bitter enemy, the elderly and supremely dutiful servant Practice; and the very depth of his capitulation moves her to sobs. This is succeeded by another characteristic and iconic Barker scene, in which Taxis pulverizes the talismanic (talking) bones of his mother (invoking 'the discipline of love'). In counterpoint, Tenna drops her compulsively crying child in a gesture of remorselessness towards the future (recalling that in *Downchild*), only for the still-humanistic labourers to spring to its protection *'with pride and delight'*.

Fourteen claims that his failure to commit suicide has inculcated in Machinist 'A TERRIBLE AND ENDURING RESPONSIBILITY'; Machinist provocatively identifies this as the latest instance of 'cowardice'; and, whilst Fourteen may show some promise in political melancholy ('all the charms of peace and civil reconstruction have no appeal for you'), Machinist counters his naïvely idealistic utilitarianism ('PHILOSOPHY MUST MAKE US LOVE') with the supremely Barkerian riposte 'On the contrary it must make us more discriminating about where we love if we choose to love at all' (*OP2*, 227-8). Machinist's heretical proposition that 'THE WAR WAS NO WORSE THAN THE PEACE' is contested by new configurations: Darling proclaims herself an adoring follower of Machinist; Taxis struggles to accommodate Tenna's adoration of Practice by bending all contingency to his totalitarian imagery ('If love were effortless we would not value it'), but still baulks at Practice's offensive 'DECREPITUDE'. But Tenna (like her namesake in *Isonzo*) loves in precisely unconventional terms (but whereas Isonzo insisted on his own 'beauty' and masterfully refined aestheticism, it is Practice's sense of subordination that renders him adorable to her). Taxis discovers that 'The bridge is also ... THE FULCRUM OF DESERTION' (*OP2*, 237) when Tenna and her daughter abscond across it; it is also the site of infantile malice, when a group of jeering Danish cyclists hurl Practice into the sea.

Darling has proved to be the object of Machinist's murderous instincts (previously expressed in relation to Tenna); Fourteen describes Darling's murder as 'the stuff of philosophical enquiry' even as he offers to execute and bury the philosopher. Norris insists 'A killing's a killing', but the action splits consensus (one student objects 'She was not ordinary she was a tor-turer', yet admits 'I have a problem with this word ordinary I could torture people am I ordinary' (*OP2*, 241). However, Tenna finds Machinist's appalling consistency and integrity alluring and decides to stay with him in Denmark. Fourteen is bewildered and objects 'I'm entitled to some clarity'; in response Machinist slaps him, and Fourteen thanks him for this timely rebuke of simplistic conventionality. Fourteen strangles Norris, when she is impertinent to Tenna, in a precisely Machinist-like 'rage and control'. Taxis tries to take back Tenna and his child, and asks 'Let me boast we two are reconciled', then threatens to blind Tenna to gain Machinist's assent; Machinist does not dispute the rivalry except to utter one devastating line, 'PITIFUL BRIDGEMAKER YOU ARE THE BLIND ONE YOU' (*OP2*, 249). Machinist's pursuit and articulation of the wilful and remorseless imagined possibility makes him charismatic, to Tenna and others.

In a realization of the Third Student's earlier scenario of servitude, Practice tries to bring a succession of milkshakes to Tenna's daughter, even though he stumbles repeatedly in his decay. The child is unchildlike in refusing to play or talk in the obvious ways, at the usual times; the cyclists characteristically jeer at her nonconformist eccentricity. The Third Student offers complicity in her wilful self-definition, in profoundly Barkerian terms which entwine the initiatives of refusal, shamelessness, eroticism and warfare, telling her 'Don't talk ... Don't play ... And when your body says so show it to me ... Let me lay siege to you ... Let me be your enemy' (*OP2*, 250). When the jeering cyclists pass across the stage once more, the child is left alone: and at this point in Barker's own 2005 Rouen production, the child reprised the hand gesture of her grandmother in the direction of the cyclists, the gesture and command to kill.

As this summary suggests, whilst some Barker plays offer protagonist-driven tragedies of transgression (such as *(Uncle) Vanya, Hated Nightfall, Ten Dilemmas, The Gaoler's Ache* or *Gertrude – The Cry*), magisterial duels (*Judith, Isonzo*) or variations on themes (*Wounds to the Face, 13 Objects*), *Animals* is a particularly fine example of Barker's ability – notwithstanding his some-time avoidance of lineal narrative as 'reactionary'[51] – to mobilize a wide variety of vivid characters in surprising permutations through hauntingly irreducible scenes which nevertheless can contribute to a strong sense of highly original, urgent, unpredictable and reverberative narrative drive. *The Europeans* is a classic example of this model onstage, *Albertina* on radio, and *The Fence* and *I Saw Myself* subsequent refinements of this particular form. Barker's 2005 French production also constituted an extremely fine example of his direction, with a company which included the notable French performers Jean-Marc Talbot as Machinist, Christian Pageault as Practice and Judith Siboni as Darling, alongside Wrestling School core members such as Victoria Wicks as Tenna, Jane Bertish as Norris, Séan

O'Callaghan as Calltold and Chris Moran (who had recently played Hoik in *The Moving and the Still*, and Istvan in *Dead Hands*) as Fourteen, with lighting by Wrestling School Associate Ace McCarron complimenting Leipzig and Kaiser scenographic designs. Particularly memorable moments included the exordium, with Wicks as Tenna adjusting the hem of her skirt and kicking her legs seductively on a wall (like a young Gertrude), defiant of military searchlights; Norris compulsively toiling to shovel soil; the recurrent flock of giggling red-bereted cyclists; the funeral descent of the Queen's coffin from the sky, swaying in the wind; the *'whispering hall'* of the war hospital, in which a line of men in white masks appeared, then groaned and contorted physically into varying postures of woundings, as a series of uneven bed-heads descended from the theatre flies to demarcate their eventually prone postures like headstones, and commence the scene of Taxis's visit; Tenna's seduction of Practice across a dinner table where he was painstakingly laying cutlery (later the cover image of Barker/Houth's *A Style and its Origins*); a breathless chase across (the immense lowered structure of) the bridge, and Practice's endless slow-motion fall through water; the child's adult-undercutting curiosity and fascination for the dead body of Darling, and her final hissing, spitting defiance as she raised her gesture of irreconcileability.

The play is also one of Barker's most artful theatrical enquiries into philosophical questions of value and limits. Various characters are ab-jected or extruded into what Scott Lash calls 'the wild zones', disruptively crossing borders of land, class, morality and decorum, and running counter to a temporality of 'omniscience and surveillance' of 'a determinate past and future' (literally in the cases of the characters crossing the bridge); rather they 'create indeterminacy in the future and resist its colonization'[52], even when, as in the case of the Child, this involves the renewal of an ancient antagonism rather than a capitulation to an infantile conformity as represented by the cyclists. Lash's description of a politics of melancholy might be designed for Tenna's daughter:

> The melancholic looks back on tradition. She invents tradition ... a practitioner of a hermeneutics not of suspicion (as is the perplexed critical theorist) but a hermeneutics of retrieval. Indeed one of the archetypal memories that this politics of melancholy is straining to retrieve is the memory, the ghost, of difference itself. The politics of the melancholic reunites in another register dimensions of both time and value.[53]

The Appalling Vitality of the World: *The Ecstatic Bible*

Barker's most ambitiously scaled work, *The Ecstatic Bible*, was staged by The Wrestling School in collaboration with Brink Productions for the 2000

Adelaide Festival, Australia (in the text published by Oberon Books in 2004, Barker dates the play's composition as 1993–4). *The Ecstatic Bible* outstrips even *The Bite of the Night* and *Rome* in length, range of experience and feverish exhaustion of all apparently possible permutations of characters and subsequent imaginative discovery of new possibilities of contingency. Gritzner's identifications of manifestations of ecstasy in Barker's work (cited in Chapter One) provides pertinent keynotes: not just 'being beside oneself; being thrown into a frenzy or stupor, with anxiety, astonishment, fear or passion', but also 'the idea of losing one's centre ... physicalised in the characters' unbalanced postures and uneven walks'[54].

This is particularly relevant to many encounters and testimonies in *The Ecstatic Bible*, centrally that of the Priest, whose devotion to Gollancz drives him to lose the former centre constituted by his official calling; and in his quest he forfeits his ability to walk without crutches, as his intense emotionality is physicalised by his broken, crucified body. However, one might also argue that his devotion to Gollancz becomes his new centre, like a lodestone dragging him (perpetually off-balance) ever onwards (in the spirit of Céline's analysis of human motivation, 'Maybe that's what we look for all our lives, the worst possible grief, to make us truly ourselves before we die'[55]). Lamb estimates that *The Ecstatic Bible*, with its 29 scenes and cast of 84 characters, requires a playing time of over seven hours, and observes how the term 'ecstatic' is a significant choice in both its sexual and religious connotations, when: 'ecstasy comprises an experience of extreme rapture which, becoming the object of an all-consuming desire, is thereby capable of utterly transforming the subject'; 'the proverbial instance of Paul on the road to Damascus provides a paradigm where a moment of ecstatic revelation leads to a life dedicated to a cause.[56]

This cause is identified when the Priest exclaims that he thinks desire 'the most painful thing in the world', and Gollancz replies 'why shouldn't it be, when it is also the most beautiful?' (*EB*, 32–3). *The Ecstatic Bible* also constitutes the example *par excellence* of Evans's observation that Barker regularly challenges the conventions of realism and naturalism by allowing his characters an unusual degree of 'geographic as well as emotional mobility'[57]. Moreover, the play allows the characters a remarkable *temporal* mobility, a longevity of Biblical and epic proportions which repudiates the tenets of the hygienic state by being not always *enviable* in what it encompasses, or demands to be endured. In the surprising discoveries of agonized devotion, characters frequently testify to their own previously inconceivable abilities to build, or remake, their lives on the basis of a look, a gaze held with another which opens up ecstatic, questionable and terrible forms of new life, in a wartorn and/or totalitarian landscape.

The Priest views the self-possession and contemplative ease of Gollancz and Tread as a self-conscious offence to the '*vociferous rabble*' to whom he ministers; as he later acknowledges, Gollancz calls forth in him a contradictory passion (entwined with his moral condemnation) to minister in devotion to her as an individual instead of to his human 'flock'. Two Nurses interpret the residual existence of Gollancz's 'catastrophic' war-devastated husband as a moral rebuke to the 'spectacle' of her infidelities. In contrast, the vagrant Shaw couples with Gollancz and dispatches her husband into death, an original sin in this totalitarian context which, as Lamb observes, precipitates expulsion from a self-styled 'paradise' of socialized but vengeful moral hierarchy and a subsequent wandering in discovery – not just for Gollancz, but for the men, Tread and the Priest, who both require the 'ordeal' of seeking further glimpses of her inflammatory presence. The release of the husband from the nominal and attributed terms of his life also parallels the provoked suicide of Savage's father in *The Bite of the Night*, which releases Savage into an odyssey beyond the limits of shame.

The Priest boils and rages with the plaited fulminations of his resentful moral denunciations and his sexual longing, persistently rebuking the world and Gollancz in ways which are both comically and engagingly at odds with the promise of serene deliverance associated with his garb and office. Gollancz births her child at the dilapidated chateau of a recluse, Davis, who contests the platitudes of humankindness; despite, or because of, this, he falls in love with Gollancz, her lack of sentiment towards her child (prefiguring that of Barker's Gertrude) and appetite for further life and discovery. In a significant discourse on pity, Gollancz states she never asks for it, but the Priest notes 'some of us feel it, and they are cruelly exploited by the rest' (*EB*, 32); however, he is disingenuous in that he has also invoked pity (unsuccessfully) in order to coerce Davis, whom Gollancz leaves with the child, McChief.

McChief emerges in her fifteenth year, after the death of Davis, like an antithesis of Shakespeare's Miranda, unused to human beings but quickly confirmed that the world is as infinitely disappointing and absurd as Davis suggested. Tread fetches up at the chateau, seeking Gollancz through the methodical matrix of a grid; he recognizes McChief as Gollancz's daughter; she subdues his advances and sets him to dig a grave for Davis, with the promise of her bed in return for his obedience, her first step in discovery of coercion.

Meanwhile, Gollancz has completed a quest of her own, to find Shaw. Shaw realizes her seductive power of reversibility remains incontrovertible, capable of deploying the weight of other claims in order to overturn them, with a wrenching power which he physicalizes, in one of many (what might be called) 'hinge moments' which manifest the decisive alteration of the course and terms of a life:

> I shall miss my wife profoundly. She was the greatest companion of my
> life and a perfect lover. Everything will be downhill from now on.
> *(He goes to leave, with a bitter twist of his body)*   (*EB*, 46)

Gollancz (in what the Priest calls her 'spine-curdling' objectivity) is mindful that this in turn lays Shaw and herself open to the possibility of domestic life:

'Stillness', 'Even stagnation' (*EB*, 47). The Priest now attempts to admire all things (pre-eminently Gollancz) rather than be indignant, but once more manifests tension in his body and movements, particularly as he realizes that no one cares sufficiently to murder him. Shaw's wife Fashoda strives to kill the usurping Gollancz, but at this 'hinge of her pitiful existence', which she articulates in ways which the Priest finds extraordinary and moving, she kills herself rather than Gollancz or Shaw, who embrace and leave her body unattended.

Two soldiers are momentarily put off-balance by a poor farmer, Trackell, who thanks them for murdering his wife, which he interprets as the dispensation of a demanding new freedom, obliging decision. However, the soldiers see this unforeseen stance as his attempted diversion of their attentions from a dearer target, his son. McChief, who presides over this campaign of coercion and accepts the attendant title of 'devil', orders Trackell to kill the boy, or be killed; when he falters, the same offer is extended to the boy (a human parody of Abraham's trial by God). When one soldier punctures the tension with a coarse and stupid laugh, McChief's appreciation of the moment is spoiled, and he is also placed under threat of death for his 'spiritual barrenness'. McChief takes the boy as a lover ('For war's about love, also', *EB*, 62). Trackell recovers from the scene more slowly than the soldier, and this lack of resilience proves fatal.

The next scene provides a further study of strategy, double bluff and the crucial faltering of will. The Priest attempts to rob a 70-year-old woman, Henderson, who disarmingly proclaims the value of the contents of the many boxes in her house, notwithstanding their neglect by other thieves. The boxes contain love letters, divulging a narrative of such 'APPALLING PAIN' that the Priest is indeed mesmerized by Henderson, whose nakedness he insists upon (even as the letters proclaim it 'YET ANOTHER DISGUISE'). The Priest is apprehended by men, and when he discerns his role in instigating the retelling and furtherance of Henderson's personal narrative, he concludes 'I am your pain', for which he feels conventionally ashamed. Shaw takes Henderson to bed, deserting Gollancz, pregnant at 50. Gollancz accepts the defections of her allies with a shrug (which the Priest finds enraging, 'A VILE COMPACT WITH CONTINGENCY', *EB*, 71). Like other Barker characters, The Priest has what he wished for, in his more imaginatively limited moments, visited upon him in terms which make his imagination reel, as he and Shaw are crucified by McChief's forces.

Whilst guarding Gollancz, the Boy becomes transfixed by her body, which utters a 'history' whose profile and grandeur 'must remain the subject of tortuous, agonized speculation'; in return, Gollancz abandons herself to him, proclaims him 'exquisite' and herself deaf to the audible sufferings of Shaw and The Priest. On perceiving her lover's adoration of another, McChief gives them three hours together in the warmth of her caravan, whilst she waits '*in a suffusion of pain*' (which '*propels her as if from a catapult into a relentless walk'*, *EB*, 77), and orders the crucified to be taken down. McChief's atrocities are driven not from an insensibility but from an acute drive to imagine, which tortures her also; she imagines the Boy with Gollancz, though Gollancz insists 'It's never the same. All that you imagined did not occur, and what occurred you could not possibly imagine' (*EB*, 77). McChief requires to know the terms of what she construes as her 'failure' (that her Boy could desire another), and insists on knowing every corridor in his head and eyes; Gollancz maintains the need for such toxic security is mistaken and impossible, and deduces 'how cold it is to be the devil...I did not realise how hard she is on herself', even as she recognizes a magnificence in McChief's penitent sponging of the Priest's wounds. Meanwhile, Tread, whom McChief has had imprisoned for fifteen years after his impregnation of her, is freed by their self-consciously spoiled and wilful daughter, Winterhalter (first name Miranda), who taunts and releases the gaoled, as if Gay from *The Bite* were unleashed amongst the cells of *Seven Lears*. Winterhalter leads a counter-revolution against her mother, whom she has trussed naked in a cradle of tightly drawn ropes, before Shaw (now on crutches, and Winterhalter's consort) pronounces her death sentence.

The characters of *The Ecstatic Bible* make a surprising (not least to them) dignity from their obsessions: the crippled Priest insists 'I have lived a life of beauty humiliations possibly but beauty also hardly deviating from the thing I made my mission ridiculous or not still I' (*EB*, 89); similarly Tread recognizes literal or figurative incarceration to be a form of concentration, where 'lucidity or madness' prevails 'and you are not equipped to know the difference' (*EB*, 98). Tread orders a team of porters to seek for signs of Gollancz in a blizzard, even as he acknowledges the Sisyphean futility (and, arguably, grandeur) of the endeavour, until he is killed by Richardson: a further irony, as Richardson is a young male prostitute for whom Gollancz harbours an unrequited adoration. Gollancz revives the Fourth Porter with the warmth of her urine and the sight of her body (which she proclaims 'THE WORLD'); the Fourth Porter embraces her, then continues seeking her, with the sense of purpose that accrues from the project of a life's work. As The Priest observes, this faith and acceptance brings the Porter closer to Gollancz's mysteries than his own failed interventions.

In a war hospital, an orderly, Gex, presides over the forgotten, shattered and delirious casualties, until the empress Winterhalter stumbles in following a military defeat. Gex refuses to seek transport for her, preferring to prioritize his own survival and that of his poetic mission, to describe the consequences of the war it was her mission to begin. She is arbitrarily wounded, and crawls away. Richardson cradles and ransacks the corpses, in alternating attitudes of grieving adoration and laughing resilience, the preconditions for a genuine fixation which he experiences when he proclaims Gex 'the one' after failing to commit suicide with a blocked pistol.

Richardson uses the pistol and the threat of shooting himself to coerce Gex into embarking on an unconditional love. But Gex, stricken with adoration of Winterhalter (notwithstanding, or because, she is 'the antithesis of the lovable', 'the single cause of so much human pain the mind panics to contemplate', *EB*, 125), absconds. Richardson observes that even this unpromising (and, for many, terminal) extremity can provide survivors with the substance of 'Great Educations' (*EB*, 126).

For all the characters' feverish and far-flung travelling, Richardson is right to proclaim flight, as a means to escape or feel less, 'the very definition of illusion'; the appalling vitality of the world lies in the plethora of demonstrations of obsession, coercion and faith (however misplaced) which confronts them in every context, in different forms. Lamb correctly observes that 'Gollancz seems to embody a principle of perpetual if amoral and indiscriminate creative renewal'[58], but even she falters for a while, and becomes an anchoress, all but entirely walled up in an abbey: she claims 'God placed me here / Knowing my fatigue' as a way to transcend the human body, described by the adoring pilgrim Torboch as 'THE THING THAT ACHES' (*EB*, 132). But all things prove significantly reversible: Gollancz's attempted transcendence of the world makes her iconically magnetic for the abject, such as the now-crippled Winterhalter; and the adoration of a man such as Torboch, whom she realizes is 'weak' and limited, stirs in Gollancz the appetite for further astonishment and the overturning of the apparently impossible in her capacity for pleasure (sensations of 'LIFE' in a body that was 'ROCK'). In a comic trope on flight, Richardson and Gex appear at the abbey; Richardson's devotion drives Gex to question himself in bewilderment: 'He says my running away was nothing but a stimulus to further acts of intimacy', even that a 'peculiar reluctance to sleep with him is a fear of ecstasy, but if one were to concede that all one's aversions merely –' (*EB*, 136).

Shaw has taken up with a peasant woman, Claudia, and their child threatens the concentration necessary for his seductive skills on the violin. Claudia's sisters attempt to coerce them apart, with threats of a pistol, and with the enticements of their bodies exposed to him. The Fourth Porter appears, digging a trench; Shaw's question, 'And are you not seeking a woman', draws the Porter's reply 'Yes and no...', which Shaw recognizes as 'the common condition' (*EB*, 147). Before he dies, Shaw offers his child and 'the ecstasies of parenthood', but the Porter shrewdly asks 'What ecstasies are those?' (aware that 'everything impedes your mission if you have a mission', *EB*, 149). When the Porter strikes the distracting child a glancing blow, Richardson and Gex appear in a moral fervour, proposing to hang him ('Fetch a tree'); however, the drive of the Porter's project makes him non-co-operative in his own execution, and the moral intoxication of (the reformed pimp, scavenger and murderer) Richardson drives him into proclaiming them 'chosen' for adoption by a child who he maintains already resembles its new parents. Gex seethes, but picks up the child, travelling on to take his place in a tableau of radiance with their 'precious burden'.

Gollancz's pregnancy and birth, whilst 'miraculous' to some, causes her to be reviled as a libidinous fraud ('Obscene … to so clamour for continuation') by others, who seek to burn her at the stake. Again, a fugitive child appeals to the moral vanity of some, and to the susceptibility of a poor mother who knows its survival will demand the starvation of one of her own nine children.

The next scene shows both adopted infants grown: the first to Ravilious, a self-hating attention-seeking aspirant suicide who has become Gollancz's latest husband, the second to Elizabeth, a young woman aware that her very existence 'extinguished' that of one of her brothers, a gang who seethe with the violence of urban frustration. Barker's depictions of the characters' children, including Winterhalter, seems to provoke, but only to elude, appeals to a systematic determinism which might rationalize or account for subsequent cruelties (such as one finds in Edward Bond's plays *Lear* and *The Narrow Road to the Deep North*); rather (as in Barker's dramatic compendia such as *13 Objects* and *The Forty*) the emphasis is, at this juncture, on a peculiarly Barkerian sense of 'divine comedy': ironic reversals which mock any sense of natural justice or discernible theory, providing rather a series of variations which might more consistently be associated with the grotesque tragicomedy of endurance, and enduring (but constantly refocused) passion. One such example of the latter is provided by the plight of Meg, who is buried in the earth for a reason which her lover Window interprets as a concealment of her body from him 'in order to discover the true worth of my love'; in fact it proves an incarceration which provides opportunity for him to be overwhelmed by Elizabeth. However, Torboch immediately venerates Meg, perhaps not least for her resemblance to Gollancz in her days as an anchoress; he is driven to excavate her with ravenous fascination. When Gex leaves him, Richardson attempts self-dedication to a young woman, only to be killed by her protective friends. Gex is devastated at the discovery of Richardson's corpse, but rediscovers a sense of poetic purpose and conviction which permits him to vanquish and subordinate two thieves.

The inquisitor Poitier challenges the authority of Ravilious and Gollancz, who is pregnant once more. As the ever-thwarted Priest observes, Gollancz 'MANIFESTS HERSELF', an embodiment of plethora peopling the world with what he terms 'MRS GOLLANCZ AND YET MORE MRS GOLLANCZ' (*EB*, 195) – not entirely correctly, as none of her children replicate her indiscriminate allure. Elizabeth's loutish brothers fling The Priest into the river (recalling the fate of Practice in *Animals in Paradise*), an infantile vengeance they visit on all old men. In a scene of potentially mesmeric slowness (in a frequently bustling, relentlessly forward-driving play), The Priest floats amongst the rushes where even the fishermen languidly seek to sink him with stones. Gex's thieves kill the fishermen, but the widow of one blames The Priest, and hangs him on fish scales by a hook through his mouth, until his deliverance by Elizabeth, who is moved to adoring pity and orders her brothers to release and tend the man they previously abused.

Poitier has wrought his own cunning reversal by substituting live rounds for the blanks in pistol of the 'exhibitionist' Ravilious. However, Ravilious has blinded, rather than killed, himself, and this seems to permit Ravilious a greater intimacy of perception with Gollancz, when she invites him to touch her naked breast (as elsewhere in Barker's drama, blindness promises an eroticism concentrated in the sense of touch); this constitutes a further ordeal for Poitier, who has sought to supplant Ravilious but instead lent him a new authority and imaginative powers of sightless focus. In a bid to nourish and prompt the deliverance of Gex's poetic "genius", the Two Thieves cook and feed him Ravilious's son – that is, the child of Gex's own adopted son. The sight and questions of the blinded Ravilious prompt Gex's anguished remembrance of Richardson, whom he had succeeded in forgetting. The Thieves attempt service, but kill compulsively, one helpfully suggesting to Gex how the recognition of his wretchedness 'might be the very spur to ... forthcoming achievement'. Gex parts from the Thieves, and from Ravilious, who sits, still and *'occasionally lurid in the detonations'* of a new outbreak of war.

Conti, who inadvertantly crippled Winterhalter and subsequently became her devotee, now travels the war-torn landscape as an itinerant barber. A farmer requests his death, and his wife requests hers (a proposition further dramatised in *The Road, The House, The Road*); when the farmer tries to countermand his wife's order, Conti discerns that this is not from love, but in order to instil infinite resentment in his supplanter. Conti kills the farmer and prepares to take his place, in household and bed, when Elizabeth's brothers, now conscripted into the army, appear and menace the wife. Conti enlists the casual homicidal artistry of Gex's Two Thieves, who have turned from being attempted facilitators of a poet's muse to spouting their own unpromising improvised verses. The Thieves dispatch the brothers, almost effortlessly. As Conti settles into the promise of a simple rural and sexual idyll, he senses his own imminent death, as if it enters on his relaxation. The Priest watches as the wife and Elizabeth squabble over the mirror which reflected the final images of their dead loved ones. He reminds them how the dead have, intrinsically and characteristically, ceased the valuation and estimation of all human values, hence the futility of acts of maintenenance performed in their name, or to arrest their presences: 'Their drawn-back lips are sneers which scorn the things we cherish ... They cannot be like us, or death's nothing' (*EB*, 243).

In 'Cloaca Maxima' (a scene named after the sewage system which flowed from Rome to the River Tiber, considered to be presided over by a goddess), Gollancz nearly dies after soldiers effect a Caesarean delivery of her child conceived with Poitier. He bewails and protests her apparent departure, strikes an attitude of enraged bereavement to the Fourth Porter; then Gollancz revives, to Poitier's emotional disarray. The Fourth Porter continues his excavations in search of traces of Gollancz, notwithstanding her unrecognized proximity. Poitier tries to persuade Gex to take on the newborn child; despite his protests, Poitier insists 'You are destined to nurse the consequences of other men's desires'. Elizabeth appears with The Priest, in whom she has found what Poitier calls 'the necessary antidote to suicide' (*EB*, 255), and Poitier himself recognizes an obligation to his own child which may demand the conceding of his own 'longing for oblivion'. Elizabeth, who characteristically finds her own ostentatious outbursts of pity and compassion a gateway to eroticism, fixes herself on the frequently (if protestingly) self-effacing Gex, spurning his reflexes of apology for displacing The Priest. As before, the dedication of others inspires Gex to dilate himself to a point of genuine authority: here he recognizes 'it is not a matter of choice, but of – MASTERING THE CONDITIONS THAT ARE IMPOSED ON YOU' (*EB*, 259–60).

Poitier's son grows to be a philosophical youth named Lamb, who tends a recalcitrant machine whilst pursuing his studies. Nightfall, an epileptic woman in search of a husband, has an assignation with Torbach, who casually penetrates her during one of her fits of unconsciousness. Lamb observes without becoming embroiled, a self-conscious man of destiny (who knows precisely his 'resources' and 'the choices to which they must be put'). Notwithstanding her misfortune, Nightfall's imaginative poverty is disclosed by her reliance on the clichés of social realist speech ('the important thing is to get on and have a cuddle sometimes', 'that's another nail in my coffin that's the straw that breaks the camel's back'[59]), with the effect that she can command neither pity nor affection. Torbach seeks to conceals his identity, just as he has sought to shed past sins with former names. The Thieves menace Nightfall, but she forestalls their attack by fainting. The characters cut cards to determine who shall have care of Nightfall, and two engineers win; Lamb deduces, 'Randomness... Apparently the very converse of the idea of the argument...produces in us the unease associated with the arbitrary but...looked at from ... a fresh perspective...could surely be felt as affirmation' (*EB*, 275).

Poitier assists at the suicide of a besieged emperor, taking a professional control and prompting the fatal moment so as to forestall the emperor's desecration by vengeful drunken women; instead, Poitier becomes the focus of their Bacchanalian rage, and and Gollancz provides the cutlery with which they devour him. Paraffin, the soldier who delivered Lamb, emerges as a revolutionary leader, mindful that 'Unfortunately, as sometimes happens, the mob makes history...on the other hand, it never writes it' (*EB*, 284). The Empress ingratiates herself with Lamb over the remains of his father, offering to adopt for him the persona of whore (a woman 'who chooses to have no choice', *EB*, 287).

This presages a reversal of fortunes for Paraffin. Ousted from power by Lamb (ironically, the child whom he delivered), he seeks refuge in the country, where The Priest dwells with the young orphan Semele. The Priest intervenes in a marital dispute between two alcoholics, and is beaten by both, becoming the pretext for a raucous reconciliation. Semele, contrastingly, refuses to be subordinated by her encounters, intuiting, as the sound of a searching helicopter swamps the scene with the sound of its blades, 'possibly I am destined to be one of those who damages, and not one who repairs' (*EB*, 301). This is confirmed by her subsequent disengagement from a disastrous marriage, killing the groom at their reception ('it was him or me').

Elizabeth and a widow, Germania, visit 'A Museum', a state repository of the dead, seeking her husband. The Two Thieves have become the agents of a police state in which people are arbitrarily "disappeared"; the ranks of the dead include the Fourth Porter, the apparently 'immortal ditcher' whose indefatigable sense of divine purpose rendered him, in Gollancz's words, 'a bald and passionate god' whose rendezvous with her is irrevocably forestalled. Gex finally delivers his solitary poem, a hymn to disarming sympathy amongst enemies; even as, attended by the police-Thieves, he presides over the winding of Elizabeth into the ranks of the dead, which proves to be his own last act.

The fugitive Semele and The Priest arrive at a tower amidst the debris, where Germania besieges Gollancz with professions of her love ('you already love me the vehemence with which you struggle to repudiate me is the proof of it', *EB*, 321). Germania washes the besmirched Semele, and tries to kill The Priest, but realizes the impossibility of this; as Gollancz says to The Priest, 'We are immortal and God knows what we did to deserve it' (*EB*, 330). Instead, Germania couples with the prone Priest whilst he fixes his eyes on Gollancz, who finally undresses for The Priest; they join in a triadic ecstasy which delivers Germania's wished-for impregnation. Semele's resilience proves fatal even to the Two Thieves. Gollancz and The Priest lie staring at the sky; he likens her laugh to a crack in a gaol through which 'a little sky fell in...a chip of day like ice that skidded on the floor' (*EB*, 330). Even now, there is an eruption of contrasting recrimination, as a cuckolded man arrives in search of one of the laundresses who laugh and fold sheets in the tower. Thus, laughter, which testifies to release and relief, may also attract catastrophe and murder, a final demonstrative example of infinite reversibility. The Priest makes a characteristic effort (*'he lurches, half-falling, half-balanced'*) to intervene, but his almost superhuman will to do so (*'defying every regulation of a degenerate anatomy'*) finally subsides into exhaustion.

The durational and experiential scope of *The Ecstatic Bible* makes it a uniquely ambitious theatrical undertaking. The closest original modern dramatic analogues in terms of imaginative will might be: Bernard Shaw's

Back to Methuselah (1921), which tries to follow the human will to self-overcoming 'As Far As Thought Can Reach' and similarly features characters of disconcerting longevity and an indomitable and wilful feminized life force, but which is essentially a metabiological treatise which affords relatively little engagement with relatively limited, predictable characters who crucially cannot achieve transformation; and, in the 1930s, Eugene O'Neill's unrealised plan for a cycle of eleven plays to be performed on eleven consecutive nights, yielding only *A Touch of the Poet* and the unfinished *More Stately Mansions*, which proposed to follow generations of a given family, in a course of events which showed them assuming increasingly monstrous forms; Ariane Mnouchkine's epic collaborations with Hélène Cixous, and director-dramatist Robert Wilson's incomplete twelve-hour six-part cycle *the CIVIL warS* (1984), developing the extreme duration of preceding initiatives such as *KA MOUNTain and GUARDenia Terrace* (1972) and *The Life and Times of Joseph Stalin* (1973).

The Ecstatic Bible was first performed at the Adelaide Festival in 2000, in a co-production between The Wrestling School and the Australian company Brink Theatre (who had previously successfully performed *The Europeans* and *(Uncle) Vanya*). The leading players were English – Victoria Wicks as Gollancz and Gerrard McArthur as The Priest – and scenes were initially rehearsed (with understudies) in separate continents, then re-rehearsed (with actual casting) at the Festival site. The set permitted an elaborate exordium, of bombed civilians running in crowds with bags: a half-naked Robespierre was dragged across the stage as if at a festival of citizenship, but his voice provided by a cracked record, whilst Gollancz and Tread flirted inanely in their deckchairs. The tight discipline of The Wrestling School English contingent (which additionally comprised Ian Pepperill, James Clyde, Alan Perrin, Séan O'Callaghan, Sarah Belcher and Julia Tarnoky) provided an exemplary basis (particularly in rhythm and diction) for the playing of all scenes in this collaborative alliance[60].

Seismic Form: *Found in the Ground*

Even the range and ambition of *The Ecstatic Bible* might possibly be eclipsed by an even more original and innovative Barker text: *Found in the Ground* (published 2001, planned for production by The Wrestling School in 2009), which, as its title suggests, refocuses *The Ecstatic Bible*'s astonishing imaginative reach of scope from extensive duration into excavatory depth, the *divination* of a temporal landscape (in that word's sense of boring down into the earth, as well as its allied senses of a literally supernatural search for knowledge, and discovery through intuitive insight). *Found in the Ground* is a uniquely ambitious landscape play, a scenic poem set in a continuous present which creates tensions and strange surprising resolutions through loops and variations of words, meanings, sounds and gestures; however, its dramatic structure and theatrical and choreographic specificity, 'surreal yet disciplined as complex music'[61], avoid the fragmentary, ironic, provisional and partial qualities which characterise much contemporary performance[62]; its closest analogue might be the theatre of Robert Wilson in terms of scenographic orchestration of a moving landscape in extreme forms of scale and time. Barker notes how *Found in the Ground* extends the principles and choreographic features of his exordia, sustaining their 'musical nature' 'so that the play is like a string quartet by Bartók...the themes keep returning always developed and furthered, the torsion increased'[63]. It is a supreme example of what he terms 'digression and a multitude of spectacles which are related ... by tension, psychology or atmosphere'[64]. As Gritzner observes:

> The play contains sounds of infinite distance; the imagery of burning books with its suggestion of burning bodies; the metaphors of mud and soil to suggest the dead and the frailty of human life. The sublime effect of these very particular and in a theatrical sense highly determinate performance images and movements nevertheless lies in the ensuing incommensurability between what is physically presented and what is imaginatively suggested.[65]

The perambulation and litany of Macedonia, a naked headless woman who articulates the multiplicities of the exterminated, provides a keynote and counterpoint, as she announces herself as the incarnation of the exterminated ('I am all the Ann Franks', *CP5*, 310). She testifies to the irrevocable loss of human variety, the ditches and pits full of 'Composers / Violinists / Physicists', which also include figures more difficult to sentimentalise: '**A kidnapper / A prostitute / The child molesting wife of a sadist**' (*CP5*, 335). Conscious that the death sentence is 'inadequate', '**A mere symbol of our collective / Rage**' (*CP5*, 299), the dying judge Toonelhuis literally consumes and ingests the sodden ash remains of the war criminals he has condemned, as if to engage with them, evacuate them and exorcise them finally; the process sifts through persistent hauntings which leads Toonelhuis to dismiss 'the flavour of the underlings' in search of '**the ultimate dish**' which is Hitler himself. Hitler appears in surprising forms, as a wailing infant and subsequently as a confident unfolder of disarming philosophical reflection. Thus, as Angel-Perez observes, time is rendered indefinite: 'Historical landmarks abound but are systematically blurred'; this play generates a sense of 'chiasmatic circularity that confesses the impossibility of escape from history', anchoring the play and the spectators in 'an ever-present Auschwitz'[66], that is literally and historically nightmarish (in the sense of the imagination continually losing control over the images it produces). The eruptively barking bandaged dogs and the Hitchcockian flocks of birds indicate another facet of Barker's developments in conceiving and orchestrating staged possibilities: his work with The Moving Stage Puppet Theatre has clearly informed a more ambitious sense of physical orchestration extending to the non-human, which is here incorporated alongside human actors, such as the nurses (who move in "flocking" configurations) and Macedonia, whose regular urination recalls the innocent abandon of a girl whom Toonelhuis glimpsed in a forest epiphany.

Toonelhuis is sporadically and sparingly attended in a retirement home by his daughter, Burgteata (whose very name suggests a municipal sense of cultured distancing), whose proximity to her father is ritualistically performed and is counterpointed by her compulsive couplings with dying men (a 'vocation' and contrastingly ecstatic proximity to death which takes her beyond conventional boundaries of 'Politics / Marriages / Careers', *CP5*, 293; this fetish drives her at one juncture to present herself swathed in bandages taken from a dying man, from which she is to be unwrapped). Toonelhuis is also visited by Knox, 'The Spirit of a War Criminal' who nevertheless sidesteps the noxious associations of his character by bringing flowers and emphasising his own sense of compulsion which took him beyond '**The rational pleasures**'. Burgteata is simultaneously adored and denounced by the librarian Denmark; Burgteata suggests that he does not know if the smell of death which he associates with her 'arouses or disgusts' him, and seductively bids him to contemplate marriage to her, in order to delay consummation.

If Denmark's susceptibility to Burgteata's deliberate indiscriminacy recalls The Priest's to Gollancz, his intermittent bouts of puritanical resentment and disgust, combined with an imaginative paralysis and sense of doom, recall (like his name) aspects of Hamlet. Burgteata professes (at least) to find a charm in his studied unworldliness, but later rebukes him as '**Vastly too complicit**' (*CP5*, 319). Denmark has also been a precociously inquisitive, isolated youth, taunted when uncollected at the school gates that his mother is a 'naked' prostitute (which makes him resemble a grown-up version of Otto in *The Swing at Night*). He is aware that he wants 'too much' in his process of questioning 'Where is alive? Alive, where is it?', his sense of the meanness of the world driving him to the hope that 'It's in another's eyes' (*CP5*, 302; compare Helen's *aperçu* in *The Bite of the Night*, 'I deny the body exists except within the compass of another's arms', *CP4*, 115). Denmark literally cannot keep his footing on this landscape; he falls, beginning a sequence of horizontalizations: the nurses sprawl languidly, Hitler will stumble. In Denmark's case, his ungainliness is physicalized by an increasing curvature of the spine, in a degeneration particular to the librarian and scholar which also reflects his loathing of self and others, and 'The consequence of undone actions'.

Knox reappears, a discomforting revenant only partly concealed by an image of the pyramids[67], testifying to 'an appetite for murder' persisting 'now the ending of the war had removed the usual time-honoured excuse' (*CP5*, 304–5). Knox distinguishes between his 'adolescent killings' 'committed in the service of the state', then the '**political phase**' which 'masqueraded in the garments of an extraordinary love'(*CP5*, 307): the 'extraordinary individuality' and miraculous love of his victims making them '**fit**' for his attentions: 'Extinguishing such fineness ... Was the source of your ecstasy' (*CP5*, 308). However, Knox recalls that his third phase might be 'the least complicated after all' ('with mass murder / Strictly speaking / Nothing is personal'): his victims were 'no more than the material substance' of a dispute with God. Since the 'ecstasy of murder had been shifted to another place', namely 'the cosmic quarrel between man and God', the sheer volume and 'inert mass of slaughtered flesh' was to be extrapolated according to principles fundamentally '**Mathematical**' (*CP5*, 326, 328).

Toonelhuis's valet, Lobe, tends his dogs and estate. He expounds how he is, despite his advanced age, still 'not disenchanted' with the idea that intimacy is awesome ('If God's anywhere ... He's there', *CP5*, 353). The nurses enjoy his sense of stern appreciative relish, and compete for his attentions; but he exercises his will through resisting their overtures, imagining them as hapless murder victims. Whereas Burgteata discovers and pursues ecstasy in the imminence of death, Lobe uses death as a means to imaginative separation from the promise and associations of ecstasy.

Denmark, the impotent idealist, tries to coerce the book-burning workman with a pistol, even as he protests, with a misanthropy which underlies his postures of liberal cultivation, that it is 'Vastly better', if 'burning must take place', that humans burn instead of books. Toonelhuis undermines Denmark's resolve through provocation[68], and Denmark confines himself to scowling at the sunbathing nurses, deducing that 'Absolute meaninglessness ... Far from being a disaster is precisely the condition to which the vast majority aspire' (*CP5*, 333). Lobe embraces Burgteata '*kindly*', helping her to articulate her frustration, her need to be 'in hazard of [her] life', with the assurance 'You are / Surely' (*CP5*, 332).

Toonelhuis proclaims himself beyond the nets of habit, fiction or relief: 'Lying is now impossible for me / Just as the truth is impossible for everyone else' (*CP5*, 339). From his near-death position, he does not find a world without books 'Intolerable': he commands Knox to conjure Hitler, who initially appears as a baby presented by a Burgteata, dressed as a bride, amidst a cataclysm of cries, bells and dogs barking. Denmark recoils from her offer of marriage, and throws himself on the bonfire of books.

Burgteata is disclosed tending the infant Hitler in a suspended crib, between the two dying men, Toonelhuis and Denmark (now swathed in bandages): to vanquish the suffocations of sentimentality, she reflects on how 'In the great plagues in the great wars' the simultaneous death of separate generations 'was commonplace', and pursues the imagination of their perspective to point which has some contact with observations in Barker's

Death, The One and the Art of Theatre: whilst the senses of the dying may be sharpened, they also approach a point of **'supreme indifference'**, as they enter 'a new realm of consciousness? / Unlittered / Unburdened / Un / Everythinged?' (*CP5*, 348). This gives way to what Angel-Perez identifies as 'One of Barker's most revealing images of the paradoxical coexistence of the inhuman at the heart of humanity': 'the Piero della Francesca-like virgin-cum-child Burgteata simultaneously breastfeeding Hitler and Nuremberg judge Toonelhuis (in fact her own father), turning them into foster-brothers'[69]. This ushers in the entrance of the adult Hitler, who divulges his perception that the principle of observation produces only 'lifeless imitation', whereas the visionary must 'Stop looking / Stop listening / **Stop believing the world / Invent the thing**' (*CP5*, 352). Knox finally suggests the trial is necessary to the murderer, **'Pity the killer who is never known'** and hanged (*CP5*, 354): an apotheosis which Hitler was denied. The nurses make a further disclosure: how the adolescent Burgteata was deflowered by Lobe, with Toonelhuis's knowledge, in a corresponding 'ecstasy' of complicity (*CP5*, 356).

Macedonia appears, still headless and naked but for black gloves, leading the dogs, cloaked, on long leads. A screen descends and rests on her shoulders, onto which film of *'extermination and execution'* is projected; Hitler becomes fascinated by the images; Burgteata appears with the nurses, now *'chic, hatted and veiled'*, bearing the remains of Toonelhuis on a tray. Burgteata has rehearsed his death imaginatively, but the rehearsal cannot (and perhaps should not) entirely contain her grief. Lobe is solicitous, revealing his liaison with Burgteata's mother before her. Denmark, bandaged and in dark glasses, enters to wreak a vengeance on the remains of Toonelhuis, 'the burner of libraries'. Denmark takes encouragement from Hitler's words: 'we take pride in conquering this pain', of the elimination of old values, 'in rendering it the melancholy music which accompanies all'; Denmark is flattered by self-association with this melancholia, which Hitler extends to grace even the attempted iconoclasm of infantile vengeance: 'It is the temperament of artists / Rembrandt / And the nameless students who deface his works' (*CP5*, 361). Emboldened, Denmark thrusts the prong of a garden fork violently and repeatedly into Toonelhuis's remains, and repeats Hitler's words, whilst Hitler buckles at the knees and rolls in the mud, as Denmark had done formerly. With newfound conviction, Denmark commands Burgteata to say goodbye to Lobe, who *'smiles'* with the resilient assurance that the nurses, no longer focused on Toonelhuis, will continue to love him. Hitler works to weave even his own loss of balance into his own myth, again likening himself to artists, who court disaster 'if only because they are obsessively ambitious they are balanced finely between failure and success I can assure you it is failure that really intrigues them' (*CP5*, 364); even the besmirchment of his coat might be turned to activate the imaginations of others: 'Where has Hitler been they'll say ... Let them speculate'.

This recalls the moment in *Ursula* when Lucas's attempted closure of the massacre is flawed, he is rendered literally off balance by what he has witnessed and colluded in; and Doja's failure in *He Stumbled* to maintain his artist's control over the forces of death. But the most informative parallel is provided by Lascar's stumble in *Rome*, narrated by its Chorus: 'I fell. / And lay. / I overcame the shame in falling. / I was still. / I smothered the compulsion to scramble up again [...] / What was absurd last week ... / And what was revered ... / Change places' (CP2, 254).

Thus Hitler, like Leopold in Barker's *The Europeans*, seeks to annex even the forms of eccentricity and turn them to a strategic performative advantage which crucially slips the knots of shame and increases his power, even from what might seem initially a random indignity. As the Nurses reprise Macedonia's actions and Toonelhuis's words, Lobe slits the throat of each dog, in a *'kindly'* manner. As they subside unprotestingly, *'the industrial sound ceases'* for him to tear free the animals' leads, in a strange form of release into the freedom of death and *'The sound of infinite distance'*.

Found in the Ground aims to be as irreducible as the European history of which it projects itself as the inescapably haunting nightmare, and constitutes one of Barker's masterpieces, in both thematic reach and breakthrough into formal innovation. It presents a specifically detailed and orchestrated cumulative experience which (like the more plot-driven *Animals in Paradise*) questions and opposes all forms and initiatives of simplistic dissociative rupture and facile reconciliation, liberal moral superiority and insistence on comprehensibility from hindsight. It orchestrates a continent's trauma in terms of a rhythmic but constantly surprising form in which themes and torsion are developed and furthered, both painfully and beautifully. It is a unique and pre-eminent theatrical example of what Baudrillard calls 'seismic form': 'The form that lacks ground, in the form of fault and of failure, of dehiscence and of fractal objects, where immense plates, entire sections, slide under one another and produce surface tremors'[70]. To paraphrase Starhemberg in Barker's *The Europeans*: how do we escape History, or, (if that is impossible) at least come to terms with it? We replay its mayhem, and the perpetual oscillation of its infinite reversibility, as the complex music of our lives.

8
Wrestling with God

Perhaps the world deserves to be disdained?
Perhaps
Loving the world
As so many claim to do
Is merely
A Failure
Of
Imagination?
What is there
To compare it to?

The Swing at Night

Barker's theatrical philosophy and philosophical theatre incline, at the hinge of the Millennium, to the overt challenging of the terms of life, mortality and the world: politically, in that all the plays locate conflict in relation to social power, but also metaphysically, in that all the plays examine the ways that religious edicts are invoked and institutionalized in order either to support or to challenge social power: with speculations as to the degree to which the configurations of social power may, or may not, reflect a divine authority (as Barker observes the 'twin poles', the terms and forms, of 'realism' in his work are 'coercion and decay'; 'Between these absolutes, men and women struggle to find love and meaning'[1]).

This line of enquiry further expresses and develops the spirit of cosmic resistance already identified in the poem 'Infinite Resentment', which exhorts a sporting defiance of the received wisdoms associated with moral and cosmic order, much in the spirit of Nietzsche's combative reversal of terms, 'Let God be the sinner and man his redeemer'[2], and Baudrillard's assertions: 'It is by being challenged that the powers of the world, including the gods, are aroused; it is by challenging these powers that they are exorcised, seduced and captured; it is by challenge that the game and its rules are resurrected'[3]. This has points of contact with what Bataille proposes as '*Man's fundamental contradiction*', which seems very close to Barker's sense of the (contradictory) dynamics of the human:

> We define ... barriers at all events, positing the taboo, positing God and even degradation. Yet always we step outside them once they have been defined. Two things are inevitable; we cannot avoid dying nor can we avoid bursting through our barriers, and they are one and the same[4].

However, whilst one might agree with Bataille that the experience (and perhaps the contemplation) of death involves a confrontation with the innate aporia in human perceptions, one might argue that the dominant terms of social conformism and control systematically discourage any self-conscious, or resonantly expressive pursuit of the 'bursting of barriers'. Barker's work suggests that social control tends, rather, to reinforce this terminal hierarchy: either to diminish death (perhaps the human experience which is most intrinsic and challenging, of self and others) as a matter for silent or euphemistic municipal management; to recoil from it as an individualised failure of decorum and hygiene; or to levy it as an ultimate threat of social and mortal excommunication.

Barker's key formulations of the duelling dialectic and compulsive dynamic of what we might term 'Wrestling with God' might be approached through his major work *Gertrude – The Cry*. Gertrude and Claudius arouse and experience unprecedented ecstasies when he kills her husband; their cries merge with his in '*A music of extremes*'. But Gertrude's servant Cascan knows about ecstasy, 'A haunting mirage on the rim of life': 'All ecstasy makes ecstasy go running to a further place that is its penalty we know this how well we know this still we would not abolish ecstasy' (*GTC*, 10). Correspondingly he identifies Gertrude's cry as supremely beautiful, 'But unrepeatable surely' (perhaps a rhetorical question or speculation: Gertrude's immediate weeping may suggest a sudden awareness that the effort to repeat or transcend the cry will become an omnivorous compulsion, that the game and the rules will have to be resurrected). Claudius wonders at her sustained sexual will and strained but resolutely elegant and charismatic self-presentation, which effectively secularises, and physicalizes, traditionally metaphysical terms of devotion: 'Oh the religion of it / The religion' (*GTC*, 12). Gertrude discovers and locates her existential authenticity specifically in her sexuality, and her cry of ecstasy which is 'NEVER FALSE'; this stokes Claudius's addiction and resolve to possess, repeat or surpass her cry, 'I must drag that cry from you again if it weighs fifty bells or one thousand carcasses', because 'IT KILLS GOD', 'And that is our ambition surely to mutiny is mundane a mischief which contains a perverse flattery'. 'Killing God' is the term and objective of their pact, notwithstanding the knowledge that 'He rises again', 'ALWAYS THIS RISING AND HIS FACE MORE TERRIBLE EACH TIME' (*GTC*, 22).

Thus, the redefinition of the boundary does not, for the questing existentialist, diminish the challenge and appetite for transgression; rather it compounds them, urging more strenuous bids to 'kill God' even in the knowledge that he will resurrect vengefully (and enticingly) in another form. Barker's plays even suggest that this is the properly dis/respectful response to the game of life, to anticipate adversarial contingency (recalling Burgteata's line in *Found in the Ground*, 'I like to beat life at its own game') and demonstrate an unpredictable Promethean inventiveness in discovering what one might *get away with* however briefly, enlarging the definitions of what it might be to be human, whatever the punishment. Barker increasingly locates this potential inventiveness in the self-challenging and self-renewing arena (battleground?) of sexuality. As Barker agrees, the developing sense of 'the religious' in his work is neither institutional nor conventional, but a passionate rejection of conventional worldliness:

> Why do I sometimes speak of the religious aspect of sexuality? I think because religion shares its ecstatic potential, but more, because religion is the study of secrets, and the secret retreats before knowledge and takes up residence somewhere else...[5]

'With my body I thee worship...': *Defilo (Failed Greeks), All This Joseph, Five Names, N/A (Sad Kissing), Gertrude – The Cry*

> Adultery is the apotheosis of one sort of personality
> *Albertina*

Several of Barker's plays at this juncture are strong examples of his interrogations of classical themes and stories, and departures from the conventional forms, but also from the conventional moralities with which these familiar stories have come to be associated, most pre-eminently in *Gertrude – The Cry*. Moreover, these plays usually centre on conflicts between the claims of religious or traditional social duty and the demands of a contradictory transgressive sexuality, which disrupt the former, more conventional discourses of primacy, self-definition and self-transcendence. *Defilo (Failed Greeks)* (written 1996, as yet unperformed) takes as its starting point a speculation about a Greek king, Filo, his son (called simply Boy), mistress and followers who are shipwrecked on the way to war with Troy. Thus, like other Barkerian characters, they find themselves in an anti-historical position and location ('Not Troy'): physically and perhaps terminally marginalised, consigned to the peripheries of the conventional grand historical narrative, but discovering there how a scurrilous and unforgivable eloquence may be discovered, as they recreate and subvert the dominant terms of life, society, morality and language. However, Filo's contrived failure to fulfil the conventional masculine objective of reaching Troy and fighting in the war also problematically transforms the terms of his relationship with the priestess Durer: the 'ecstasy and privilege' which is their secret (albeit 'INSUFFICIENTLY DISCREET' and therefore provocative), is now threatened by an envious and resentful lack of privacy. To counter Durer's rage, Filo welcomes the challenge of how they might discuss and arrive at new 'provisional description' of both masculinity and femininity. He also seizes the opportunity to re-establish his eminence in terms of the artful performance of charismatic will, mesmerising the surviving mariners with his offer of himself as a disarmingly wilful and self-inventive leader. His son remains mindful and prudishly unforgiving of how lightly Filo's performance skates on a potentially fatal absurdity (which the adolescent condemns as 'infantile'). Filo maintains that they have escaped the 'ECSTASY' of 'silly death' in war, and proposes a regime of confronting 'the very worst', from which to construct a reply. However, his son is moralistic and puritanically disappointed at Filo's disobedience to History and evasion of the common lot (the Boy's terms anticipate the sexually nauseated repudiations of Hamlet in *Gertrude – The Cry*: 'I WISH YOU WERE NORMAL AND A FATHER LIKE THE REST'). Filo seeks to preserve their existence and his eminence through his burstings of decorum ('I think the worst thoughts / Someone must / He who has most to live for, obviously'), as when he proposes cannibalising the non-survivors (a literalisation of the castaways' ubiquitous but less literal practice of 'CONSUMING ONE ANOTHER'). However, the very landscape against which he pits his small community tends to circumscribe his power: the 'inextinguishable' sea responds to his provocations by repeatedly returning the corpse of the drowned helmsman as a baleful talking revenant (described in the cast of characters as 'A Curse', and recalling the taunting reappearances of the dead Bishop in *Seven Lears*). A further challenge is provided by the Boy encountering Aitchison, resident hermit and 'tenant of the rock'.

One of the strengths of *Defilo* is its pace: an atmosphere of enforced languor governs the castaways' simultaneously gradual and desperate exploration of their newly restricted circumstances, without secrecy or privacy; the apparent freedoms of a wilderness divulge limitations and enforced proximities (in this respect, *Defilo* develops the posthumous arrested dynamism which the characters encounter in *Ego in Arcadia*). The Boy's encounter with Aitchison, whom he takes to be the god of the island, may bring to mind the regressive resurrection and animation of an ominous cosmology which occurs in William Golding's novel *Lord of the Flies*, with the crucial difference that this misanthropic "god" is interpreted by a frightened juvenile as insisting, not upon the inescapability of (socially predetermined) evil, but on a vengeful embargo on sexual intimacy ('You must not do what you most want to do')[6]. Durer, who has been beaten by the hermit, is fearfully compliant and bids to transcend her desire for Filo, convincing herself that, with imaginative effort, 'every charm contains its antithesis'. Durer is vilified as 'HELEN' by one of the sailors, who grow mutinous ('Kings take the credit...So they must also take the blame') and at one point plot a *coup* (in the manner of the lowlife characters in *The Tempest*). Durer demurs from contact with Filo, claiming 'I've changed', which Filo challenges as a restrictive determination to have 'ceased to love', insisting that they must keep their ecstatic 'madness' living, as a vital, innate and inescapable compulsion; Durer claims this struggle is not love but war, 'And my body is the field of battle'. The Boy maintains that the island is 'AN ORDEAL AND A DISCIPLINE' which 'schools' Filo in the appeasement of the "god" through the smothering of desire; then, the Boy is swept away by the sea, confirming Durer in her sense of an injunction to 'Bow down to what must be bowed down to'; however, Filo's maintains that her insistence, that he 'become another', is tantamount to a demand that he surrender his self. When he persuades Durer to defiant intimacy, Aitchison blinds one of the sailors. Durer leaves the island, presaging a final confrontation between Aitchison, who insists that his adversary is all 'pretence', and Filo, who acknowledges his reliance on artful play and inde-terminacy: 'To adore is to seek, and to know is to cease adoring'. The corpse of the Helmsman taunts Filo further, claiming that the Boy has escaped the island and found distinction in war, and that – worse than her drowning – Durer has found a husband on foreign shores. Filo sees this as a bid to 'wreck' him, as surely as the persecution of Job, but through alternations of 'Plenty / Poverty / Ecstasy / Despair', which he strives to interpret as 'neces-sary': 'I was delivered from the supreme humiliation of senescent love'. At this point, Filo is reminiscent of Barker's Vanya, who, alone, has fought off the moribund diminutions of the "god" Chekhov but sacrificed his sexual muse in the process. The further painful twists arrive with the brief return of Durer, to make her valediction; and the return of the priggish Boy, who gives Filo his death blow. Filo is defiant in his embracing the extent of his abjection ('Oh, how I want to pray but God's too poor to pray to'), insisting Durer admit his disintegration, 'THE HORROR OF MY FALL'. As he pours water onto her upturned face, she tells him 'I love you', as *'The waves break. The Helmsman drifts one way, and the other'*.

Thus, in its depiction of *'Failed Greeks'*, *Defilo* explores something in many ways more resonant and memorable than conventional heroic and military success: the thorough battering inside out of an individual, from imaginatively resourceful man to totemic abjection (the counter-name 'Defilo' is never spoken in the text, but refers to the injunctions and the process whereby Filo effectively becomes confounded into a fallen state). The sense of sexual incarceration, shot through with the drive to break through this, which characterizes the relationship between Filo and Durer is comparable to that at the centre of *Ten Dilemmas*, but the restrictive yet surprising details of the setting create a more powerfully specific context than the earlier play; the inescapable heat of the sun and the unpredictable rhythms of the sea provide elements of pressure and mystery (the image of the floating Helmsman recalls that of the Dead Dog at the conclusion of *The Swing at Night*). *Defilo* might also transfer very effectively to radio production, with its repeated insistence on the interpretation of visual details. The protagonist of this latter-day Book of Job is opposed to a god superstitiously made in the image of man who derives his sole satisfaction from surveillance, restriction and the inculcation of a love-killing shame. (De)Filo is not victorious, but claims his own form of apotheosis in the very terms and extent of his ruin.

The (as yet unperformed) short play (written in 2000) *All This Joseph* reexamines the relationship between Potiphar and his wife in the aftermath of her failed seduction of Joseph, as recounted in Genesis 39 verses 7–9. As in most Barker plays, the characters are striving to reconstruct and negotiate ways to persist when conventional order has been shattered: Potiphar observes, 'it takes catastrophe it takes sickness it takes the death-bed possibly to instruct us how exquisite the ordinary day / *(Pause)* / Such as today / *(Pause)* / Was'. Overcoming his impulses to violence, Potiphar insists that his wife 'Continue' in relating the details of her fateful tryst with his favourite slave, however painful and inflammatory the knowledge derived, in an utterance which formulates the drive of many Barker characters to seek definition (however consciously imperfect) of that which is at (or possibly beyond) the very limits of definition, particularly the sexual: 'I call it curiosity it is more terrible than curiosity but curiosity the word if not the thing the word will do'. Indeed, her account of her kneeling before the naked Joseph brings Potiphar to the edge and failure of language, to a gesture which is eloquently suggestive but intrinsically indefinite: '*His hand describes his pain. The woman watches the agonized gesture*'. In return, she acknowledges that she dresses her account in conventionally discreet euphemisms: 'By nakedness I meant his cock stood out / By worship I meant *(She stops suddenly...)*'. They agree that she has dishonoured him, not by attempting to seduce his slave, but by failing to do so; and that Joseph, though 'crucified' by his desire, 'ceased to be a slave' through the 'incredible' power of his will in refusing her. Potiphar attempts a similar effort of self-transformative will, proclaiming 'I'm Joseph ... a later Joseph', but his gaze and effort falter; she maintains 'I am your wife' and 'Never will I call you Joseph'. With him ultimately denied this mobility, consigned to the definitions of his identity in relation to hers, they achieve a further agreement that they both wish Potiphar dead (and that in some existential respects he already is). She seeks to be granted 'the gift of widowhood', because 'Whereas a wife is impatient for her seduction even from her wedding day a widow ... Is no longer property ... Property aches for its theft'. Notwithstanding audible pain and effort involved, Potiphar commits suicide offstage; alone, she discloses Joseph's words which held such tension and vibration for her: 'I WILL / *(Pause)* / But only when your husband's dead'. Thus Joseph has circumvented her terms of 'permission'.

All This Joseph is a deftly turned short play. Without the reach of a major work but too long to fashion an element in one of Barker's compendia (such as *The Possibilities* or *13 Objects*), its explorations of the balance of sexual power and double-bluff in a formerly familiar narrative which is pivoted on sexuality demonstrates some affinities with *Judith*, elements which will be worked to greater effect in *N/A (Sad Kissing)* and *Gertrude – The Cry*. Another work which seems to be something of a preliminary exploration of ground for subsequent deeper achievements is the unusually brief compendium, *Five Names* (written 2002, unperformed), in which five characters struggle against the definitions of their names and the circumscriptions of their identities. In a desert, the maimed soldier Corporal Webb implores a hermit for water, then persists in his rocking of a column in a conscious parody of Samson, in 'an effort to sustain the necessary irritation' presumably ordained by God when Webb 'might otherwise be one of the silent dead'. Sarah, aged 90 and pregnant as wife to Abraham, 140, strives to remain unquestioningly accepting of the miraculous burdens of longevity, the demands of a faith through which 'The yes replaces the why'. Charles VII, imprisoned King of France in 1461, insists a servant tastes his food to guard against assassination by poison; when the servant predicts 'your apprehension will deliver you to a more melancholy death than any poison', Charles suspiciously rejects the argument as ingenious and irrefutable, but 'the very fact it is irrefutable makes me certain only an idiot would heed it'; their gamesmanship and double bluffs persist to a point of intimacy which might serve as a description of friendship. Penelope aged 60 ponders the paradoxes of 'Belonging' to an 80 year old Odysseus, her loom 'remarkable monument to fidelity site of ingenuity icon of marital faith' now seeming 'encrusted with stained by rotten from ... Cleverness', which she recognizes is not the same as wisdom: 'I should have lain in bed under a younger man and he above a younger woman'. However, her servant reveres the loom as too sacred a monument to be destroyed at her behest. Penelope concludes to Odysseus 'I know the thing which even if we are not still we have to be', and they hold hands weeping, accepting the irrevocable losses in their melancholy monumentalization[7]. The last and best dramatic episode depicts Mrs Williams, a wife and mother who briefly and uniquely jettisons domestic familiarity: having dramatised herself for the extraordinary, *'naked but for a veiled hat and shoes'*, she returns to the house after an hour's expedition into sexual abandon with three thieves, seeking to assure herself 'how good it is how this is the only and if it were not good still it would be the only the only'; significantly her very terms of definition of her own sexual aporia falter, whilst her husband continues in his monotonous assurances through repetition of the single word 'Darling'.

Whilst the characters in *Five Names* question their terms of (self-) definition, they cannot escape or transform them. The five short plays are dramatic portraits in which some breaking of boundaries is attempted, perceptibly and significantly, but – with the possible exception of 'Mrs Williams' – ultimately contained by a sense of cosmic, political or domestic stasis. As my summaries should indicate, none of Barker's writing lacks significance (itself a notable achievement), but *Five Names* as a compendium suffers dramatically from the very self-restriction which it identifies: the characters remain creatures of pathos, rather achieving tragic development.

Contrastingly, Barker discovers one of his fiercest performative duels for sexual dominance, and thereby the terms of life and death, in *N/A (Sad Kissing)* (written 2002, staged in Vienna 2004). The initials refer to the characters of Neoptolemus and Andromache[8]. After Troy fell to the Greeks (through the successful stratagem of the wooden horse), Neoptolemus, the son of Achilles, avenged his father's death by leading the slaughter of Troy's revered king Priam; he further took as his wife Andromache, the widow of the Trojan champion Hector (Priam's son, killed by Achilles), having killed Astyanax, her son by Hector. This story of overthrow, the lure of desecration, seizure and strategized recrimination provides fertile terrain for Barker's divinations, tracing the frozen time of catastrophic aftermath and the mutual scrutiny which seeks desperate chance. The sense of traumatic suspension and pained self-exploration invites parallels with the themes and forms of classical tragedy (such as *The Trojan Women* by Euripedes) and also French neo-classicism (such as Racine's *Andromaque*, 1667), and the lethally high stakes of mutual seduction between triumphant warrior and enemy widow recall Barker's own *Judith*, but these points of contact are paralleled by telling departures. Neoptolemus is not philosophical, like Holofernes; he proclaims himself a destructive 'lout' compared to the more reflective Astyanax, who struggles to distil his experiences into the atonement of poetic logic ('Grief is beautiful wars make grief war is beautiful therefore … Poetry is necessary grief makes poetry grief is necessary therefore') even as he harbours the secret that he was sexually initiated by his mother on the day of his father's death (associating Andromache with the consciously excessive maternal sexuality represented by Caroline in *The Gaoler's Ache* and Algeria in *The Fence*). Astyanax knows that, for all Neoptolemus's brag that he ruthlessly kills the weak, women are 'not weak at all'; nevertheless, he tries to defend Andromache by addressing her as if a servant, though Neoptolemus finds her regality unmistakeable. She is aware that the situation demands that she be 'Dignified in order I may be more thoroughly defiled when Neoptolemus inflicts his barbarity on me'. As the city burns audibly, she '*gropes for an argument*' to save her son. Neoptolemus scorns argument and poetry, vauntingly proclaiming himself rather 'a poet of the woman / I /Sing / Her / Secret', though ironically unaware of the secret that connects this mother and son. Andromache recalls blind Hector, his 'silent ploughing' of her in bed ('me his terrible field a clay field me I turned like clay'), though she proves anything but malleable after Neoptolemus kills Astyanax, defying him 'can you / Pursue me … I am faster than you'.

The dreadfully deliberate action of *N/A* is punctuated by an interlude in different time (apparently both outside and between mortal time and perception of events) which pursues the emotional repercussions and consequences of death, speculating that the presence and pain of the dead (both felt and inflicted) can be infinite if not forcibly terminated. Priam and Astyanax appear as posthumous presences, intimating to Andromache (if only in her imagination) their reprehension of her 'way with the rule': the rule which decrees that women are property in war, 'spoil' to be harvested in their sadness. Andromache determines to 'speak the kiss' of this sort of woman, not angry or yielding but sad. The dead 'seek the shelter of the living as birds in snow flock to the eaves of warm houses', presences which would reside and sing in restrictive judgment (redundant yet needy voices like the "resurrected" ghosts of Serebryakov and Astrov who stalk Barker's Vanya). Astyanax resentfully deduces he must 'die twice': once by Neoptolemus, once by Andromache's 'trampling' of his memory in further life with his killer.

However, Andromache is strategically combatant rather than pliant, presenting herself to Neoptolemus smeared in the blood of Astyanax: in an eloquently suggestive, supremely Barkerian stage direction, '*N/A contest the space that lies between them*'. Both find themselves cleft into 'Two of me', contradictory and warring impulses in relation to each other, with Andromache seizing the strategic advantage of initiating intimacy. Stranded in the suspension of their residual afterlife, Priam howls, whilst Astyanax determines to play the poet and 'make sense of the howling'. But Andromache has stripped the suddenly gauche Neoptolemus of his security in self as ravager. When Priam warns that men 'kill what they cannot possess', Andromache rejects the 'tolerance' which Priam claims that Hector extended towards her, and significantly provokes Priam's impotent rage ('I could smack and fuck you Andromache'). Neoptolemus is in fact riven by elements beyond his control: just as Albert in *Gertrude – The Cry* will insist on his own agonized ecstasy when he exhorts Gertrude to strip, Neoptolemus commands Andromache to be naked and speak her sexual history:

ANDROMACHE: Are you sure naked because the names the names of
    all who uttered in the context of my nakedness might cause you
NEOPTOLEMUS: That's why I want it
ANDROMACHE: To be so
NEOPTOLEMUS: That
    Is
    Why
    I

Want
It
(They stare at one another)
To be so
(Pause)

Even as he suffers a failure of nerve, pleading 'not naked', *'she lets fall the towel'* which conceals her and announces 'Hector was the first'.

Subsequently, Andromache, clothed, reflects on the boyish 'frailty' of Neoptolemus's self-confessed 'despair' at the numerousness of her lovers and the 'terrible' nature of the day and their love. She has the strategic advantage (reflected in the stage direction *'She penetrates him with her gaze'*) through her extraordinary sensuality, which Astyanax recognizes ('exhaust him with your history rope his imagination to the rack and lean on all the levers'). Having killed Andromache's boy, Neoptolemus feels 'the boy in him' has been killed: 'to be a man is to know how little is or even should be knowable'; 'I am become not only a man but old an old man possibly'. Astyanax complains that the spectacle of them together 'both nauseates and depresses' him, but this seems less a matter of injured familial honour than a priggish distaste (comparable to that of the Boy in *Defilo* and Hamlet in *Gertrude*) at his mother's sexual activity 'at 43'. Andromache has shown her distinctly feminine strength in her decisively strategic reversal of terms: she discloses the secret of her relationship with Astyanax, and Neoptolemus recoils ('I have been dragged not unlike Hector dragged over such awful ground', 'I kill the weak but oh the weak kill me'); in giving him the required 'sad kiss' and permitting 'your ploughing me between my thighs', Andromache has in fact consigned him to a form of servitude, toiling like a labourer in a 'hard and beautiful' marriage where he is not the woman's equal, as he admits: 'like working the highest fields I shall lose my strength in you' (compare Claudius's dawning sense of Gertrude's body as 'ground trodden ground to which I'm bound a dirt poor labourer who tills and spills and fights and fails in his possession', *GTC*, 44). Andromache can be poised in the knowledge of 'Your fear of me ... My fear of you' and that 'Fear becomes us'. She *'removes her gaze from Astyanax, an appalling farewell'* which releases and exorcises his increasingly bitter presence, and exhorts Neoptolemus 'The road now ... After the city the road'[9]. Though he slings her over her shoulder and runs with her, *'as a thief runs'*, *'The cry of Andromache echoes down the flaming streets'*. Andromache's leaving behind the dead form of her father-in-law (*'Priam is still'*) emphasises her indomitability; and presages Gertrude's final striding over the dead to accompany her most recent consort, towards a future in which he already seems out of his depth.

N/A (Sad Kissing) develops and surpasses the form and theme of Barker's *Judith*: the seductive duel throws a tightrope over more profoundly sounded emotional depths of grief and anger, which make it one of Barker's most moving and resonant plays of the period. Whilst confronting the yawning gulfs of emotional loss experienced by both the living and the dead, the innovative presence of the posthumous characters manages to express their howlingly entwined psychic proximity and physical separation, whilst maintaining and asserting Andromache's existential drive further and beyond, as confirmed by the dead Astyanax's intimation 'I'm dead but are the dead sincere not necessarily'. The use of dead presences to express internalised impulses of regression may remind some of Bond's *Lear*, the subversion of the (neo)classical form may remind some of Kane's *Phaedra's Love*; however, both comparisons serve to emphasise the deeper emotional investigation and engagement with all characters in Barker's exploration of love, grief and war, which may make the former seem a predetermined political thesis and the latter shallow and sensationalistic. A profound, mature and surprising work in its own right, *N/A* importantly prepares the ground for the cataclysmic imaginative leaps and imagery of *Gertrude – The Cry* (staged by The Wrestling School at Kronberg Castle, Elsinore, in August 2002, prior to an autumn tour of British theatres).

Like *N/A, Ursula, Knowledge and a Girl, The Ecstatic Bible, The Fence, I Saw Myself* and other plays of this era[10], *Gertrude* depicts the mature woman as sexual existential heroine, in the Nietzschean sense of possessing the will to overcome her surroundings and her self – a woman who says 'yes' to everything questionable and terrible, defiant of a resentful majority who prefer to be only half alive and who sneer at those who choose to live dangerously, conducting experiments in living beyond the usual obligations (of family, class, nation and gender). Barker's re-visioning of Shakespeare's character of Gertrude is one of several ways in which, as David Kilpatrick observes, Barker's speculative rediscovery of Elsinore does not so much 'conflict with Shakespeare's story as shed light on desires concealed in *Hamlet* by socio-cultural necessity'; unlike 'Stoppard's and Müller's Hamlet plays, *Gertrude – The Cry* is not in a subservient relationship to Shakespeare's text', but 'bear[s] out a new relation'[11] to the story which Shakespeare used as the plot of his play *Hamlet*: itself a simultaneous and self-conscious development of, and commentary on, the familiar theatrical form of revenge tragedy.

The importance of Shakespeare's *Hamlet* as a starting point for philosophical and moral speculation is exemplified by J. P. Stern's association of three facets of Nietzsche's literary moral experiments with three quotations from the play:

1. The problem of morality versus life (... 'for there is nothing either good or bad, but thinking makes it so')
2. The morality of the will to power (... 'This above all: to thine own self be true')

3. The morality of strenuousness (... 'This was sometime a paradox, but now the time gives it proof')[12]

Nietzsche also reverses the idealist ethic of knowledge, as taken over by our scientists: 'It is not [wo]man that is to be the servant of truth and knowledge, but truth and knowledge shall be the servants of [wo]man'[13] (my brackets). Accordingly, Barker repudiates the ideal of 'truth', like the fundamentally commercial ideal of 'youth', in favour of a triumphant artificiality typified by the elegant and sexually provocative mature ungovernable woman, who assents to 'more life' rather than accept the limitations of morality and fecundity. Society proposes these limitations to regulate sexual conduct and criteria of attractiveness (bound up in survival, duration, numbers, biological usefulness) by preaching the internalisation of tiredness and despair of so-called "civilized life" (compare the provocative assertion by Leantio in Barker's *Women Beware Women*, 'No woman under 40 is worth entering'). But Nietzsche insists that Life's chief constituent is 'the discontinuous and catastrophic will to power, whose moral concomitant is ... personal authenticity'[14]. Nietzsche also proposes that 'belief in God results in an impoverishment of mens' lives; that the compensatory belief in Heaven ... reduces the value and dignity of physical existence ... preaches a denigration of the life of the senses and thus leads to a fanatical contempt for "what is real in the world"'[15]. Nietzsche's imaginary prophet Zarathustra proposes that the cardinal vices of lust, will to power and egoism be transformed into positive values, so that the Super[wo]man will continue to say yes to the prospect of the infinite recurrence of life, in all its greatness and triviality, and thus be self-reliant and self-determining.

Moreoever, Nietzsche proposed: 'The spiritualization of sensuality is called *love*; it is a great triumph over Christianity ... One is *fruitful* only at the cost of being rich in contradictions; one remains *young* only on condition the soul does not relax, does not long for peace'[16]; '*What is the seal of freedom attained?* – No longer to be ashamed of oneself'[17]; and (contrary to allegations of misogyny based on some of Nietzsche's less considered asides) that 'Stupidity is in woman the *unwomanly*'[18]. The relevance and resonance of these assertions throughout Barker's work should be apparent, though he would prefer the terms 'desire' or 'ecstasy' for the mutual suspension associated with 'love'. The forms of eroticism in Barker's works (perhaps most directly in the exploration of sexual suspense, *Isonzo*) demonstrate the resistance of received social notions of gratification and order, by sustaining, developing and elaborating arousal (often involving the centrality of the gaze and the mirror, as in *Knowledge and a Girl* and *I Saw Myself*).

The sense of a transgressive sexuality is the wellspring of the relationship between Gertrude and Claudius in *Gertrude – The Cry*, which, like the ecstatic achievement of The Cry itself, demands constant rediscovery and renewal. The character of Gertrude in Barker's play is, along with his depiction of Algeria in *The Fence*, his pre-eminent dramatization of The Seductress, as characterized by Baudrillard: 'The sovereign power of the seductress stems from her ability to "eclipse" any will or context', stepping and surviving 'outside' of any restrictive external psychology, demonstrating an 'immortal' relentlessness which insists that 'the game must never stop, this even being one of its fundamental rules'[19]. Gertrude's characteristic struggle and her mesmeric power are both erotic, frequently manifested in the play by her nakedness, beauty and pain in/of searching, as physicalized through visual contact and the effect of the gaze – on her, and on those around her – as her energy transforms and unsettles the surrounding characters, particularly the men[20].

Barbara Freedman has proposed that, if theatre has traditionally and conventionally curtailed and controlled the female gaze, theatre also 'offers the opportunity to reframe that moment' of attempted control in different terms: for example, some forms of erotic performance may generate what Freedman describes as a moment of 'quintessential theatre' in which the stage becomes 'the battle place of one's look'[21]. On reflection: drama often provides cues for the (avowedly fictional but compellingly convincing) performance of the 'battle place of the look'; theatre often provides opportunities for spectators to observe and relish these battles, relatively unobserved. But I propose that Barker's Gertrude – the play and the character – are unusual, and 'quintessentially theatrical', in the self-consciousness with which they meet the gaze of the audience, reframe a hitherto familiar scenario, interrogate the habitual terms of viewing, and return a chaotically expressive yet ultimately self-possessed questioning gaze back out to the audience, challenging their conventional terms of control, both inside and outside the theatre. The character of Gertrude in Barker's play is uncontainably sexual. She compulsively re-presents what Lingis describes as, not the female, but 'the feminine', obedient only to 'aesthetic laws of her own making'[22].

Barker's departures from Shakespeare's chosen perspectives in *Hamlet* are (dis)continuously surprising, with characters demonstrating a comparable (or superior) but different performative vitality and expressive range and depth, within an unsettlingly accelerative timeframe which periodically slows for intense exploration of minute details: of speculative, enraptured, or infuriatingly tangled, mutuality. In Shakespeare's *Hamlet*, the characters are principally viewed externally, and the audience's viewpoint is associated with that of Hamlet[23]. In *Gertrude – The Cry*, even the relatively minor or supporting characters – Isola, Ragusa, Cascan, Albert – are given dramatic opportunities to express and demonstrate surprising depths to their characters, often centring on their sexual responses or history (or lack thereof), and philosophical and poetic extrapolations from experiences and feelings which are specific to them. In a particularly bold reversal by Barker, the traditionally questing figure of Hamlet becomes the proponent of a puritanical (indeed misogynistic) social control based on surveillance[24]. Barker's Hamlet develops the attempted self-dissociation from 'flesh' which Shakespeare's Hamlet also expresses; but Barker's Hamlet seeks the social institutionalization of his own Jacobean disgust and hatred of the "unruly" body, in a way which characterizes political authoritarianism in various forms. Within this climate, Gertrude, in her alliances and defiances, compulsively stages and explores the power of the gaze: its devastating impact, the heat of its receipt, its grip across a distance which conducts and heightens its energy, its unsettling powers of return, provocation, capitulation, exposure and defiance. Barker's elliptical characters all scrutinize each other feverishly for confirmation, which is rarely forthcoming: the audience, in darkened response, are forced back onto their own choices in reaction, and ideally an awareness of the spectrum of the different forces in the gazes which they themselves return to the characters/performers. In terms which are deliberately and consciously excessive to both theatrical and moral convention, and unusually powerful even within Barker's work, *Gertrude – The Cry* emphasizes, fully energizes and activates the potential *theatricality* of the space between its self-performative characters: and, by further implication, between these characters and the audience.

Gertrude's cry becomes the supreme form of expressive dis-order. Her performative, theatrical and self-dramatizing qualities call forth and challenge those of the other characters (as when she says to Claudius, 'HOW GOOD I SEE YOUR COCK ADMIRES MY PERFORMANCE'); and, of course, they deliberately amplify the theatre audience's sense of self-conscious implication in the 'battle place of the look', the sense of seeing what should not be seen, and/or is (and/or has been) partly veiled, and the sense of the reversibility of this power, between spectator and performer. This is foregrounded in the very first scene of the play, in which Claudius commands Gertrude to 'strip' to show her husband why he is dying, and show Claudius why he is killing, to show Old Hamlet what he had stolen and Claudius what 'now belongs to me'; however, Claudius's terms of gloating over the 'DYING DOG' of the poisoned husband are immediately reversed by the realization, 'If anyone's a dog … It's me' (Evans notes the recurrence of the image in Barker's work of the man as 'dog' at the behest of its master, woman[25]), and Gertrude takes command of the situation, directing Claudius and provoking the orchestrated *'music of extremes'* constituted by the different but simultaneous cries of all three characters.

Gertrude's servant Cascan enters, admiring her dangerous disregard for decorum, and overtly challenging the audience's sense of mortal and moral limits in his aside on the compulsive draw of ecstasy ('eventually it lures us over a cliff why not a cliff is a cliff worse than a bed a stinking bed inside a stinking hospital no give me the cliff', *GTC*, 11). Gertrude, literally but marvellously unbalanced by her actions, pays tribute to the 'love' for her husband with which her murderous drive co-existed; Claudius wonders at her 'religion', in the sense of her pursuit followed with great devotion; Cascan raises the question as to whether Claudius, like Old Hamlet, will fail to 'meet' Gertrude, in her changeability in 'travelling'. Significantly, Claudius insists that he *does*; not that he *will*.

Hamlet is differently provocative: nihilistic, claiming to see through all things; his pain and nausea at the world undercuts his own performance, but he persists; like other Barker characters, he strives to incorporate his refinements into his performance, and is perceptive, proceeding from the notion that 'No man dies of poison who does not welcome it into his veins', he correctly deduces that Gertrude 'Choked' his father 'With / A / View' (*GTC*, 17). He strikes a crucially different keynote: 'Now I am king the entire emphasis of government will be upon decorum' (*GTC*, 15).

There is also a mischievous wit in Gertrude's revolt, as when the trappings, postures and clothing of pretended mourning excite Claudius and Gertrude, and drive her to fellate him at her husband's funeral in the near-presence of her former mother-in-law, Isola (like Cascan, an entirely Barkerian creation not in *Hamlet*). Gertrude goes on to acknowledge the effects of mute conspiracy in the shared gaze, 'I met your eyes / I said I would not and then I met your eyes' (*GTC*, 17), effects which contribute to her arousal. Isola is baffled and mesmerized by the unrecognizable 'look' on the face of her dead son, and recalls Gertrude's prodigious shamelessness even as a child 'I could not tolerate your gaze ... AND I WAS THE QUEEN ... You / Who was already shameless / Made me ashamed' (*GTC*, 20). Gertrude recalls how her own sexual curiosity was kindled by Isola's assignations with a man, in which the pained but collusive cry of his blind and crippled wife seemed to be a necessary ingredient in the sustaining of the arrangement. However, Gertrude also intimates the envious surveillance which her liaison with Claudius will draw from the unseen forces of the conformist state.

Hamlet seeks 'a refuge from humiliating criticism' such as conventional parents usually provide, but which his mother fails to; this lack drives him into 'sordid acts of intimacy I inevitably regret' with Ragusa, a young woman who is initially charmed by the forthrightness she associates with Elsinore. Hamlet contrastingly values 'the maze of manners', as exemplified by Cascan's service and loyalty (which, Cascan will assert, constitute his own form of, not decorum, but 'ECSTASY', *GTC*, 50). Gertrude notes how nothing Hamlet says 'is wholly ridiculous'; 'some of his wildest accusations however random and malevolent if we are prepared to contemplate them seem less preposterous than at first appeared the shock of an accusation so often conceals its insidious attraction' (*GTC*, 26); at least in this respect, he emulates his mother, who focuses this observation to allude to the (neo-?)sexual possibilities of Cascan's adoration of her. Claudius distinguishes himself from 'those melancholy proofs of masculinity' which men usually want: 'Kingdoms / Countries / Forests / Walled estates with speckled deer / parks / fountains / lakes of trout'. He seeks the cry, of which 'The woman is the instrument' (*GTC*, 33), suggesting less of an interest in the woman than what effect he can achieve in or produce from the woman (compare Cascan's deduction and question 'We prefer the wounds of women to the women / Why is that?', *GTC*, 57).

Thus, Gertrude's cry, a force of indefinition, unlocks and refutes (male?) meaning: the impossibility of understanding or repeating it excites bids to do so, but within the play it occurs only at moments of betrayal and transgression. Claudius is haunted by it, addicted to it, seeks to provoke it and command it, though it can never be the same; Cascan listens out for it, Hamlet (who associates a 'howl of wordlessness' with the dying of love) seeks to police it, subdue it and extirpate it; but it proves uncontrollable[26]. Claudius associates it with some primal force which he has sought, religiously, since or before boyhood, but this may be a sentimental veneration of something which is opposed or even inimical to human definition and form: the very juncture and meeting point of ecstasy and death.

Isola denounces Hamlet as 'A PRIG / A PRUDE / AND A MORALIST', and professes to admire Gertrude's unashamed sexuality, but it is debatable whether she encourages Gertrude to indulge the infatuation of Albert, Duke of the neighbouring rival territory Mecklenberg (and Hamlet's only friend), out of political strategy, or in order to protect Claudius. Whatever Isola's strategy, Gertrude's response seems to ignite the mysterious power of any compelling performance, as identified by the Servant in Barker's *Judith*: 'sometimes the idea, though faked, can discover an appeal'. Gertrude claims that Isola's urging 'prostitution' upon her showed little sense of her 'instinct and inclination' for the 'profession' (the inflammatory garments of which were 'already in [her] wardrobe', *GTC*, 38–9); as in *Judith*, the enfusal of tropes of performance with a genuine essential abandon (demonstrated by Gertrude's discarding of her pants, which fly straight into Albert's hands) provokes mutually surprising disclosures.

Cascan senses an unspoken weight between Claudius and Gertrude and attempts to mediate. Claudius, ever attempting to be definite and 'forensic', resists and dismisses the possibility of an 'impenetrable solitude' such as Cascan posits. However, whilst warning him of the terrible consequences of this knowledge, Gertrude reveals to Claudius the intrinsically transgressive nature of the power which she unleashes, to go beyond all limits: 'The cry's betrayal Claudius ... And it comes from nowhere else' (*GTC*, 44). Claudius, stunned, vacates the scene, literally clearing the way for Gertrude's seduction of Albert, her dangerous interlacing of the sexual and the political which will undo the dominant and received terms of both.

In the scene which usually follows the interval, which can and should also denote the passage of several months in the action, the audience learns, along with Ragusa, that Albert has seduced her (and a succession of subsequent women) in order to try to quell his own hunger for Gertrude (it is important here that the director and performer of Ragusa infuse her pathetic monosyllabic utterances with desperate hope, and then pain and despair, to inform her future developments). Cascan listens for the sound of the cry when Albert proceeds to his latest assignation with Gertrude. Isola tries to bribe Cascan away from loyalty to Gertrude with a desperate effort which leaves her '*depleted*' (and perhaps literally prostrated) before him, but Cascan holds fast to his devotion. Isola despairs of saving Claudius from Gertrude, consigning him to the past tense (to death on the 'battlefield' that 'was in his head') within his very presence. Claudius is now politically complicit in Gertrude's sustained liaison with Albert, whilst seeking to reassure himself that no cry issues from their couplings. In a wryly comic interlude, Hamlet proclaims his initiative to marry the haplessly malleable Ragusa in defiance of the usual ideals of love, which he disdains. Gertrude refuses the shamefulness which Hamlet seeks to impute to her, and hardens her resistance to him, reminding Claudius (who is considering conciliation) that they have a greater joint loyalty to her unborn child than to her vengefully adolescent offspring with another: her words 'but not our son' are an invitation to kill Hamlet, her greatest betrayal. Cascan reports that Albert is on the train back to Mecklenberg, but Albert reappears, hardened in his resolve to pursue the 'exquisite tension' and 'TENDER ECSTASY' which Gertrude offers him. Her warning 'Oh idiot I am your death' is both an erotic intensification (which he readily accepts, 'BE MY DEATH GERTRUDE', *GTC*, 59) and a literal intimation: she has lured him to a point where Claudius can assassinate him. But Claudius cannot kill Albert (a faltering which parallels Hamlet's inability to kill Claudius in Shakespeare's play) because of a devastating imaginative sympathy with *what he sees*, which is Albert's gazing at Gertrude in self-abandon and verbalized devotion: 'WHEN I LOOKED AT HIM I SAW MYSELF' (*GTC*, 61): Gertrude insists on his repeated proclamation of her nakedness as 'God', but the consequent relief (in which '*They are enraptured*') turns from Gertrude's reassurance that her copulation with Albert is merely strategically technical ('his body climbed in mine so what he might have been a surgeon') to her injured impatience with his 'kindness' which has squandered the opportunity created by her sacrifice and is, moreover, inimical to The Cry, which is intrinsically 'not kind' (*GTC*, 62).

Cascan observes how Gertrude exemplifies his sense of regal greatness: 'Her life is such a seeking and so beautiful is her pain'; in comparison, Ragusa feels devastated and irrevocably inferior, 'And my life is poor' (*GTC*, 64). Hamlet is propelled back into 'ranting' with a sexual nausea, worthy of the most satiric Jacobean malcontent, when he hears Gertrude's cry in childbirth, and baits Claudius with references to the 'fetid dungeon' of his mother's womb. Cascan proposes the murder of Hamlet, as Gertrude gives birth to a daughter, christened Jane by Cascan. Gertrude is, however, mutinously unmaternal and persists in her spectacular eroticism: to Ragusa's horror, immediately after childbirth, Gertrude dons her iconic high heels, applies lipstick in a trembling condition of post-natal fragility, and invites Claudius to suckle her, and it is the *sight* of this action which drives Ragusa to lodge a complaint against Gertrude, as part of Hamlet's panopticon society of litigation based on surveillance and reductive orthodoxy ('that is the law the most humble and despised citizens can call their betters to account', *GTC*, 69 – note, 'betters', not 'rulers'), sentimentally invoking 'the welfare of an infant' as the basis of the indictment. Cascan sees the threat and exhorts urgency in a pre-emptive attack on Hamlet, but it is Cascan who is killed (in parallel with Shakespeare's Polonius) in the struggle, to Gertrude's distress. Ragusa takes the child into care, all things confirming her in her crusading ideology of moral disgust. As a further taunting, Hamlet appears to Gertrude in Cascan's clothes (a bid to bring her to despair which recalls *The Duchess of Malfi*, specifically Bosola's satirical presentations); he accuses her of the crime of wearing the 'UNMATERNAL CLUTTER' of high heeled shoes on her feet, in his new order in which enforced amputation is the right of the 'morally offended'. With a *'defiant ... gesture of self assertion'*, Gertrude discards her gown and confronts him with the power and beauty of her naked body, which he resents, does 'not understand', but enviously seeks to destroy in its profoundly destabilizing opposition to his prescriptive ideology. However, Hamlet is mesmerized into compliance by the sheer performative power of Gertrude in the service of the temptations of a conscious and deliberate irrationality – as she tells him, in a defiant effort underscored by her sustained nakedness, 'I don't know why', he should drink the poisoned wine (*GTC*, 77). The killing of Hamlet constitutes a betrayal which produces a cry in Gertrude *'as if she were terrified of her own disintegration'*, an overcoming of self and law which Claudius associates with 'KILLING GOD' (*GTC*, 78); yet this cataclysmic upheaval does not depend on her being, in Claudius's word, 'fucked'. Gertrude insists on the equivalent, even superior power of the gaze which her body commands, and draws strength from, in this, her most transgressive performance: 'Your eyes... I was eyed...EYED WAS ENOUGH' (*GTC*, 80).

Ragusa sees Hamlet dead, and threatens to interpolate herself within Claudius and Gertrude's bed; bewildered (and somewhat vainly), Claudius ascribes her disarray to thwarted desire for him. Gertrude tells Claudius that she senses '[his] DEATH IS NEXT' and seeks to banish him from her increasingly fatal presence. With Hamlet dead, their principal adversary and main threat would seem defeated; however, Albert returns on the pretext of Hamlet's funeral, but is now more Fortinbras than Horatio: he lays claim to Gertrude, with the ultimatum of war if she refuses to present her body to his gaze. Claudius now identifies the cry with an inhuman force, a principle of catastrophic life or vengeful god: 'Always I thought the cry was in you / But it's not / It's outside / It waits / It walks / Some long hound pacing the perimeter' (*GTC*, 87): a demonized thing which he can no longer call, or feels able to command, as she bids him. Isola smugly proclaims her double vindication: first in her self-consciously 'silly' grandmotherly possession of the child Jane, and secondly in her proposal of Gertrude's seduction of Albert, which now promises a marriage to make Gertrude Duchess of Mecklenberg and cement an alliance (however unequal) between Denmark and its neighbour territory. When Gertrude goes to her marriage to Albert, Isola patronizes Claudius fatally, with the sanctimonious platitude 'forget her yes it can be done'; like Orphuls in *The Europeans*, Claudius repudiates his mother's infantilisation of him in lethal terms; he strangles her, whilst still cloaking the action retrospectively in (consciously strenuous) terms of kindness in an address to wedding-gowned Gertrude. After Gertrude leaves, '*Something is flung in*', which Claudius struggles to confront; he even calls upon, and impersonates, the dead Cascan and his attempts to render the unspeakable philosophically manageable, but Ragusa's harrowing wails compete with his desperate and ineffective abstractions. Raging and (perhaps insanely) distraught, Ragusa has drowned Jane as a punitive act of envy and self-righteousness. Claudius reelingly considers the poverty of such crimes, 'the little mischiefs which they heap at the feet of God', whose smile 'DRIPS COMPLACENCY AS A DOG'S JAW STREAMS SALIVA ON THE RUMOUR OF ITS FOOD' (*GTC*, 91). Thus Claudius associates God with a domesticated and demanding version of the destructive 'hound', but the unspeakable detail which further extinguishes his will to further life is sonic: '*The cathedral bells announce the wedding of Gertrude and Albert*'. As he struggles to die, Gertrude appears, smart in a going-away outfit, speculating on a honeymoon in a way which decisively consigns Claudius to the past (partly to put him out his misery, a kind cruelty such as he performed for Isola), in the form of the 'scattered ashes of burned men' ('OUR SLIDING HEELS WILL TREAD YOU IN', *GTC*, 92). Her realization of his death provokes a seismic cry which seems to come '*not from herself, but from the land*'. Albert continues to find her adorable precisely in the '*ruin*' of her disarray, as he orders the bodies to be burned and scattered, returned to the land. Gertrude's assent to his beckoning involves her self-recovery – '*She walks smartly to the door*', past and/or over the strewn corpses of Hamlet, Isola, Jane and Claudius. Albert's lack of awareness as to what has demanded such catastrophe, and how, suggests the fragility and brevity of his – of any? – confidence in power. In my 2007 production of *Gertrude – The Cry* (of which, more later), we suggested this by a brief glance from Albert, back at the strewn bodies whose 'forced cause' and fatal passions he could not – as yet – fathom: then, he directed his final orders to both sides of the theatre audience: 'BURN THESE / BURN AND SCATTER THESE'. This insistence that the memory of a stage image should be banished from the mind is, of course, doomed to failure when the character leaves the stage before the lights fade on the scene, as the stage direction demands. This provided an apt final image in a play in which *that which is seen*, in conjunction and collision with *that which is said*, challenges conventional terms of definition and comprehension[27].

Gertrude – The Cry is Barker's most profound dramatic exploration of the themes of ecstasy and death to date. Gertrude's initial transgression literally pursues a connection in Bataille's observation: 'Stripping naked is seen in civilizations where the act has full significance if not as a simulacrum of the act of killing, at least as an equivalent shorn of gravity'[28]; and Claudius's subsequent bids to possess Gertrude and her cry bear out Bataille's ensuing proposals:

> Possession of the beloved object does not imply death, but the idea of death is linked with the urge to possess. If the lover cannot possess the beloved he will sometimes think of killing her; often he would rather kill her than lose her. Or else he may wish to die himself ... If the union of two lovers comes about through love, it involves the idea of death, murder or suicide. This aura of death is what denotes passion.[29]

Whereas in Shakespeare's *Hamlet*, the principal activity of the characters is attempted interpretation[30] (of each other and the world), in Barker's *Gertrude – The Cry* the principal activity is commentary (on each other and the world); the difference reflects the way that Barker's characters do not work to fathom meaning in terms of motivation, cause and effect, but rather observe, and render more intricately fascinating and/or offensive, each other's performative actions and appearances, both for themselves and others (including the theatre audience). One of the ways in which Barker's Gertrude supplants (Shakespeare's) Hamlet from the centre of the play is her similar insistence on the importance of her subjectivity, what she really feels; like him, she adopts the mourner's 'trappings and suits of woe', but whereas Shakespeare's Hamlet emphasizes 'that within which passes show', Barker's Gertrude uses her self-dramatizing garb in conjunction with her transgressive actions to challenge and arouse both Claudius and God: the chameleonically stylish costume changes of Barker's Gertrude artfully exteriorize her constant self-reinvention rather than (as for Shakespeare's Hamlet) fall short of it; here and elsewhere, Barker is more encouraging about the possibilities of the conscious artificiality of the 'actions which a [wo]man might play' than Shakespeare/Hamlet. Barker's Hamlet develops the attempted self-dissociation from of 'flesh' which Shakespeare's Hamlet also expresses; but Barker's Hamlet seeks the social institutionalization of his own Jacobean disgust and hatred of the "unruly" body, in a way which characterizes political authoritarianism in various forms. Gertrude's cry itself becomes the ultimate form of expressive (rather than communicative) dis-order: as Mangan observes, in Shakespeare's play 'The ability of language to give order to thoughts and emotions is challenged from below by a pressure of thought and emotion which is beyond its normal capacity', constituting an irrationality which ultimately proves destructive of the "normality" and pragmatism of the Danish court[31]. In Shakespeare's play, this supernatural disruptiveness is represented in the patriarchal and martial form of the Ghost of Hamlet's Father; in Barker's play, Gertrude's Cry tears apart what Mangan associates with 'the normal settled routine of everyday life', by representing 'forces, emotions and desires which defeat the protagonists' usual ways of making sense of the world'[32] – which, for Barker, are also Gods to be challenged, seduced and exorcised.

Barker's own production of *Gertrude – The Cry* was first staged by The Wrestling School at Kronberg Castle, Elsinore, in August 2002, prior to an autumn tour of British theatres. This production was one of the high points of his post-*Ursula* breakthrough to a new directorial distinctiveness, confidence and authority. In a set design attributed to "Caroline Shentang" (like "Tomas Leipzig", another Barker alias), the back wall consisted of many white shirts (and a solitary black one) provided a surprisingly evocative, potentially mobile surface from recognisably commonplace objects: during the exordium, white mechanical "birds" shot across the space abruptly, with an unnerving irregularity and sound. Victoria Wicks as Gertrude demonstrated from the very first moments a remarkable urgency and courage, in theatres such as Birmingham Rep Studio where the very proximity of her frequent complete or partial undress, combined with meticulously choreographed physical mobility, was itself startling to the audience. Séan O'Callaghan was a virile Claudius who movingly explored the (increasingly apparent) limits of his own perceptions; Jane Bertish as Isola gave her best performance in her succession of female supporting roles (including Norris and Istoria) to Barker's more reckless heroines. Company newcomers included Jason Morell as a judiciously precise and poised Cascan, Justin Avoth as a bluntly engaging and ultimately authoritative Albert, and Tom Burke, who successfully suggested that the bottled, seething, racing imagination of Hamlet (here very much an overgrown schoolboy) might be partly the consequence of an unenviably delicate sensibility, however viciously censorious its political manifestations. Costumes were particularly elegant examples of the characteristic Wrestling School references to 1940s/50s classicism, shot through with modern *haute couture* eroticism[33]. The corpse of the baby was not thrown across the floor (except in the performance at Elsinore Castle), but swung down over Claudius suspended on a pendulum of death (its fatal motion recalling that of the swinging trays in *Und*). However, it was Wicks as Gertrude who provided the memorably dominant power and impression.

For my own production in Aberystwyth in January 2007, I favoured, and my scenographers agreed upon, a long traverse staging running the full length of our Castle Theatre (appropriately, a converted church hall next to the ruins of a castle which looks out to sea)[34]; the audience were seated on either side of a raised playing space. This appropriately involved a refusal of an authoritative single perspective on the action; the obligation on individual performers to command the gaze of the audience through their unconventional vocal and energetic investment in a long space, which conducts and compounds the promise of language and gesture rendered unusually sexual and electric by careful location of breath and balance; and the audience's additional awareness, beyond the performers, of a parallel row of semi-anonymous observers (whose evident reactions to theatrical events may surprise, as often as they reflect, one's own, producing an appropriately *refracted* self-consciousness).

In fact, I subsequently discovered that this scenographic arrangement was parallelled in Barker's account of his own first staging of the play in Elsinore Castle:

> Barker had asked for a low platform running [the baroque hall's] entire length, simply uplit from the floor... the audience were seated on either side... the audience were now slightly *beneath* the actors and this reversed the normal status-relations between public and stage ... [delivering] that sense of incongruity and anxiety that Barker asserted was the prerequisite of tragic experience...[35]

I encouraged the performers to find (occasional, rather than predictably consistent) moments in which they could turn out to the audience to address them directly: moments where the spectators could be challengingly included. Even at moments preceding speech, I suggested that performers could, according to their instincts, unforeseeably engage specific audience members with a gaze which expressed the challenge and enticement, *'You won't believe what I'm about to do next'*. Cascan's digression on ecstasy is the first instance of a character's tendency and licence to speculate directly, which he assumes in terms comparable to those of the Shakespearean fool or jester, as when he consciously acknowledges to both of his audiences, onstage and offstage: 'Forgive me I really must articulate sometimes the more curious and paradoxical perceptions that occur to me the rest I promise you I keep strictly to myself'. We found further moments of unsettling intimacy, as when Gertrude tells Ragusa that she knew a specific moment of conception, 'You do you know', a shockingly forthright disclosure which could be turned out to the audience in direct address, before it was promptly characterized as a verbal 'slap'. This was one of several junctures at which the audience would be associated with – or else challenged to deny association with – less open perspectives and terms of scrutiny, what Claudius later terms 'COLD WATCHING'[36].

Our Hamlet examined the face of his dead father, let fall the cloth, and spoke directly to the audience 'I had expected to be more moved than this' as a bathetic aside. He thereafter located his contending and imaginary interlocuturs, the "bitch" and the horse, on different sides of the audience, drawing them into a relationship with him which was redolent of the stand-up comedian, intimate despite the shocking nature of his observations on the world, conspiratorial even in his sly provocations which could include and refer to audience members[37]. Gertrude's line 'My skirt says everything to those who can read skirts' was delivered with a scan of the audience, seeking a gaze which bespoke such frank distinction. When Gertrude predicted the envious scrutiny which she and Claudius would draw from the unseen forces of the conformist state, in performance, their faces were underlit, their gazes reached out to the spectators on both sides (partly lit by reflection), challenging the theatre audience to associate themselves (each other?) with moralistic surveillance, or else dissociate themselves from it ('They are watching us / Like starved birds on a gutter / Heads on one side / They sense the clay depths of our uttering and hate us for it', *GTC*, 22).

Otherworldly Yearnings: *The Seduction of Almighty God, The Moving and the Still, Two Skulls, Acts (Chapter One)*

A group of Barker's plays at this time return to the themes of more conventionally institutionialized terms and locations of religion, its discourses of primacy, self-definition and self-transcendence, and investigations into the impulses and power of prayer. Barker has frequently been drawn to religious imagery: political invocations and annexations of it, as in *Pity in History, The Castle, Rome*; and individual responses to its existential challenge of exemplary apotheosis, as in *The Last Supper, Golgo, The Ecstatic Bible* and pre-eminently *Ursula*[38].

In this respect, one might draw some parallels between other primarily secular modern British dramatists who have nevertheless been compelled to study the historical forms of institutionalized religion and idiosyncratic challenges to it: from Shaw's *Saint Joan* (1924) to Whiting's *The Devils* (1961), John Osborne's *Luther* (1961), Peter Barnes's *The Bewitched* (1974) and *Red Noses* (1985), Arnold Wesker's *Caritas* (1981), David Rudkin's *The Triumph of Death* (1981), and Howard Brenton's *Paul* (2005) and *In Extremis* (2006). These are, similarly, primarily existential parables concerned, not with the promulgation of a faith, but of social responses to its promises, hopes and wilful (even eccentric) manifestations, which may (unforgivably) indicate and suggest the insufficiency of materialism.

Barker's *The Seduction of Almighty God by the Boy Priest Loftus in the Abbey of Calcetto, 1539* (written 1997, staged 2006) resembles many of the above plays in that it depicts a protagonist who 'repudiates the politics of *numbers* by declaring the sufficiency of *himself*'[39]. Loftus is uncomfortably excessive, 'embarrassing' in his faith, even to his religious superiors and fellow priests, but knows 'The same was said of Christ'. He is denounced as 'sick', meaning 'not modern'; the abbot Ragman observes how 'Everything about [him] ... Speaks of another age', in a way which will constitute both his power and his subsequent status of anathema. Loftus is another Barkerian deject, in Kristeva's sense of one who, 'Necessarily dichotomous, somewhat Manichean', divides and excludes others with the effect that he calls his surroundings, and conventional terms of belonging and desiring, into question[40]. He exemplifies the anguished dislocation expressed by the philosopher E. M. Cioran, in *Tears and Saints*: 'Can I ever forgive this earth for counting me among its own, but only as an intruder?'[41].

Loftus, in one way, recalls the title character in Barker's 1984 poem 'Genshu Hanayagi' (in the collection *Don't Exaggerate*): the greatest dancer in Japan who has questioned the terms of power and so finds himself denounced as 'out of date'. When Ragman is on the verge of dying, Loftus perceives his proximity as a privilege: 'God endows me with the ecstasy of walking with you to the door of death' (*SAG*, 15). However, the touch of Loftus seems to drain life from a monk in order to revive Ragman, and this begins his prodigious eminence, in which he insists 'We must move on from this / This / SHARING AND / FORGIVING' (*SAG*, 45). Devoutly celibate in a monastery which has accommodated sexual facility, Loftus is both admired and reviled in his authority: one philandering monk observes both his magnetism ('Women / Oh they love the man who won't / The man who could if / Would if / Only if', *SAG*, 22) and the atmosphere of 'Scrutiny' which he enshrines, through hating the received terms of life. But when the abbey itself is threatened by the forces of state control and deterministic economic materialism, Loftus seems able to act as an exterminating angel on its agents, striking dead the 'paper men' who lay economic siege to its walls. He considers himself 'ELECTED' to bear and express the hate of a 'mischievous' God, and attracts the devotion of two widows of men he has apparently killed (who both undress for him, insisting on his gaze and their own ecstasy, attributed by one to the erotic repercussions of catastrophe, a 'painful illumination' following on from sudden death: 'My husband's death and my proximity to the location of it created a delirium in me', *SAG*, 41). However, Loftus's apparent powers cannot save him from being raped by disguised intruders (not once but twice[42]), leading him to the deduction that 'Fatigue' has driven God to devolve his powers to Loftus, but only selectively. When Loftus's power falters once more, a woman fails to be delivered of her bureaucrat husband (who has, amusingly and misguidedly, urged the perverse young saint to plan for his pension), resulting in a manifestation of '*profound melancholy*' which she, and his female acolytes, find unforgivable: they castrate and kill Loftus.

Thus, Loftus constitutes a portrait of the man out of time, consciously 'not modern' and therefore 'the next thing to being dead', which has both unsettling power and the status of 'illness', manifested as 'blindness' to the conventional terms of self-interest and a feverish impatience (his 'skin burns in the presence of liars', *SAG*, 57). Unfortunately, *The Seduction of Almighty God* is, despite the title, oddly unseductive, and, by Barker's remarkably high standards, one of his least distinguished plays. Loftus's fall can appear conventionally tragic, like that of an over-reaching protagonist in a Marlowe play which (at least nominally) restores order by punishing the would-be superhuman. The figure of the precocious religious prodigy-irritant is explored with more depth and contextual complexity in *Ursula*; the liminal figure who convulsively dispatches invasive agents of a prescriptive social order is more thrillingly and eloquently dramatized in *Hated Nightfall*. *Seduction* is, comparatively, frequently static and dependent on offstage action; perhaps it was partly this resemblance to French neo-classicism which attracted the French director Guillaume Dujardin to select the play for his 2006 production for The Wrestling School, a reciprocal arrangement following Barker's 2005 production of *Animaux en Paradis*, at the Théâtre de Deux Rives in Rouen, in collaboration with Dujardin's Mala Noche company. Dujardin's production featured effectively intent performances from the two French actors, Christian Pageault and Judith Siboni (who had played Practice and Darling in Barker's *Animaux*), and a characteristically striking Tomas Leipzig set design (involving pillars coated in latex, appearing both intestinal and prophylactic), but the other performances (all by company debutantes) did not manifest the electricity of choreographic precision inspired by language, and energization of interpersonal space, which characterize Barker's direction of his own writing.

Barker's radio play *The Moving and the Still* (written 2003, broadcast by BBC Radio 2004) returns us to Abbey of Calcetto, but some years earlier, in 1450, and may take some inspiration from the speculation in *The Swing at Night*, 'If / Beauty / Were / Truth / Surely / It / Would / Keep / Still?' (*SN*, 15). This is a recurrent theme, particularly in Barker's poetry: how the dominant powers in particular historical moments feel (with a significant insecurity) obliged to insist on their own unique and unprecedented progression towards the project of social enlightenment, by belittling the achievements of the past. *The Moving and the Still* suggests 'Beauty moves', in the sense that 'there is no standard so absolute that it cannot be repealed', begging the question, 'These movements of the ugly and the beautiful can you say what causes them?'.

Hoik is another teenage prodigy, who proclaims himself at seventeen 'the greatest calligrapher of this or any generation' ('I exaggerate I love exaggeration'), and who proves 'hard to like' to his immediate fellow monks; one, Slee, resentfully alleges 'your perfectionism is a sickness and an obstacle to the fulfilment of the quota' but this, if anything, confirms Hoik ('I am not likeable was Christ likeable far from it I suspect'). Consigned to degrading menial tasks for his outbursts, Hoik deduces 'Something in the human character ... Finds distinction genius or even the modest signs of an uncommon intensity / Offensive / And longs to humiliate it / Extinguish it / And tread it into mud'; 'It is as if man far from adoring the exceptional as a gift from God / Shrinks from embarrassment / And runs into the arms of banality to find relief'. He entertains a disdain for the world ('poverty and banality render all efforts necessary to sustain life fatuous and humiliating'), but nevertheless repudiates the cynicism (along with the sexual advances) of his monastic superiors by insisting that the six books he has painstakingly inscribed by hand constitute 'Six ways to God' in their potential effect. There are, of course, modern analogues in the way that Hoik's faith in the power of the handwritten word earns him from some (such as the High Master of the Educationalists) the status of 'a relic from a bygone age'. When confronted with the latest technology, a printed (rather than written and illustrated) book, Hoik rages at this new climate of philistine utilitarianism which equates beauty with a 'necessity' which he cannot claim (indeed, rejects) as justification of his life and art. Rather, Hoik prefers, to the 'flaccid' effects of their utopian 'garden of goodwill', the contrasting 'profound anxiety' that governs the existence of birds and animals, knowing 'IT'S ANXIETY' that 'shapes my lettering / It forms my style / It's my identity'. Slee decries Hoik's 'negativity' ('you think it brings you nearer to God') and looks forward to a levelling unanimity when: 'WE THE INFERIOR WILL NO LONGE NEED TO SUFFER OUR INFERIORITY / All the pages / All the books / Will be identical and I for one will be freed of that resentment which has / Admittedly made me sometimes less than kind to you / *(Pause)* / It is not good to feel inferior it makes one vicious'.

However, another historically pivotal, socially levelling invention makes a decisive appearance simultaneously with the printing press. A lord of neighbouring estates, Toonelhuis (who shares his name with the judge who presides over war criminals in *Found in the Ground*), introduces the noise of a gun into the landscape for the first time; he explains 'Killing was an art until the Signor [the designer of the gun] came along ... any fool can work it unfortunately ... Lords will fall to idiots from now on and men to boys'. Hoik enjoys encouragement from the apparent 'lout', Light, who insists that however 'out of date' Hoik may be to the majority, Light nevertheless 'prefers' him. Setting up his writing stall in a supremely dejected setting, a disused lavatory, Hoik persists, despite the Abbot's insistence on 'demand', not for the mysterious complexity of gospels and 'indulgence' of calligraphy, but for printed 'primers', 'books which make other books intelligible' to those who are 'free' and therefore to be spared effort. Like Barker's Vanya, Hoik rejects the rhetoric of specious neo-liberal idealism and 'necessity' in the name of a fundamental totalitarianism ('too bad too late too everything'), and identifies himself with the disruptive power of gunfire: 'there is a sound created by redundancy I like it it agrees with me'. Hoik borrows the new-fangled weapon to shoot Slee, who has turned Calcetto over to the machinery of the printing press. At his trial, Hoik is accused of recoiling from the future (particularly criminal in the young) and thus drain-ing others of their satisfactions: he attempts a defence of the importance of style, 'the beauty of the hand and not the usefulness of it': indeed, the inextri-cability of style and beauty: 'Not only the words are God how the words are that is also Him'. This leads the Abbot to charge Hoik with 'REBUKING GOD' by suggesting the world has failed him and is 'NOT GOOD ENOUGH'. Hoik is, rather, rebuking the dominant terms in which the world is constituted and presented to him at this historical juncture, a political dissociation and denu-ciation which is traduced into a religious heresy by the combined powers of church and state (as is often the case). In an effort of will, Hoik literally burns his despised, 'misunderstood' hand, before the flames around his stake consume the rest of him. Light and Hoik call to each other as the fire grows, expressing their joint preference for each other. If Hoik's life has exemplified Cioran's philosophical assertion that 'Love of form betrays a partiality for death' ('The sadder we are, the more things stand still'), his resignation and death recall that writer's proposal that 'Boredom is melancholic stillness, while despair is boredom burning at the stake'[43].

'Preference' is a term with its own resonances of religious or supernatural endorsement (as in *A Midsummer Night's Dream*, 'our play is preferred'), and Light's name and nature suggest a holy fool of unlikely but penetrative insight. The final image, of Hoik calling and signalling to Light through the flames, recalls not only the story (and play) of Saint Joan, but also Carl Theodor Dreyer's classic film *La Passion de Jeanne d'Arc* (1927), in which the confessor who witnesses Joan's apotheosis is played by Antonin Artaud, whose subsequent visionary manifestos for a radical and convulsive art of theatre refer to the imagery of this moment[44]. Notwithstanding the Artaudian resonance in this passionate insistence on faith in art and the irreducible significance of witnessing, *The Moving and the Still* is one of Barker's most Shavian plays, in that it presents a parable located at a historically decisive moment, in which the protagonist (who is here, as in *The Seduction*, the most intriguing character by some distance) encounters forces of social stagnation, embodying a discernible thesis. When Shaw pits his higher evolutionary men and women against persecution by social convention, he presents them as anticipatory harbingers of a new age, in a conscious and conspicuous opti-mism of the will. By contrast, here and elsewhere Barker notes how the forces of restriction repeatedly dress themselves and their actions in a spurious novelty and ideological progressiveness, with particular and characteristic attention to how standards of beauty in art are moved to reflect social moral-ity: a proposition that is discernibly and significantly *reversible*, to propose, through art, that, in morality, there is correspondingly 'no standard so absolute that it cannot be repealed'. Both dramatists invite us to see beyond the wilful myopia and moral impositions of any given age; they resist ethical determinism and suggest that 'The golden rule is that there are no golden rules'[45].

Two Skulls (written 2001, broadcast by Danish radio 2005) is an artful return to themes broached earlier, particularly in *Ursula*, of the relationship between hope and deception, and the lure and the limitations of the reflected self. Hudaconceva, a Christian nun of the order of Lazarus, is a woman whose human passions discovered no reciprocity, as she realizes in retrospect, speaking from her unmarked grave: 'I was unloved by mortal man but not unlovable I fell between two shelves of time the one who might have loved me is not yet born or died'; whilst she maintains 'all my questions have been answered', immediately afterwards *'her cry rises'* in testimony to something, perhaps beyond (her) words. She recalls how she was entrusted with a holy relic, the jawbone of Lazarus, which has been encased in silver, with the explanation that, whilst this possibly made it more liable to theft, it forestalled tampering or the substitution of 'a common jawbone'. This reference to the possibility of deceit in conjunction with hope will prove prophetic, but in unexpected terms.

Hudaconceva is, like Ursula, Loftus and Hoik, distinguished by the vehemence with which she repudiates the world, which inspires in her peers a 'certain reverence' for her devotion but also 'a profound resentment' for the scrutiny and judgment of them which they suspect informs this. Therefore she is ordered to engage with the tribulations of the world beyond the abbey; her superior insists that she expose her cloistered virtue to trial, by confronting the danger of desecration. Secretly and unusually, Hudaconceva is acutely (self-)conscious of the way that prayer is an anxious demand for what the world will not give, and how, for all its need and passionate expression, it resounds in oblivion. Literally nauseated by the human social world, she hears a voice emanating from the jawbone, urging and informing her, and proposing that despite all appearances 'accident is only the dark side of design'; the voice explains that Hudaconceva is worthy of his advice because, whereas her fellow nuns and believers pray out of 'clamour for enlightenment', 'You know you are deserted'. She undergoes various misadventures, molestings and humiliations (occasionally reminiscent of the Marquis de Sade's *Justine*) at the hands of those she encounters; the voice explains 'They long to defile you / I would do myself / So the cleaner you are the more likely it is'. That central shocking admission prompts another, that the voice is 'not Lazarus' – and therefore Hudaconceva tells him 'from now on I must regard you as a threat to my sanity and not an inspiration'. Whilst the voice suggests that her vehement unworldliness is itself a provocation ('you wear faith … It's a lipstick to you'), Hudaconceva trains herself not to pray for deliverance, but to perceive each ordeal as 'only the precurser of another and still worse'. As she continues to be sickened by the smell of death which characterizes her pilgrimage ('DEATH AND THE ROAD ALL ROADS ARE DEATH'), the voice decides to tell her: he is 'Christ not Lazarus', proclaiming His love for Hudaconceva because she is 'sustained by prayer to an Almighty God' whose silence causes her no anxiety (although the voice, increasingly subjective and bathetic now it is personalized, adds 'IT CAUSED ANXIETY TO ME'). At her destination, Hudaconceva is even propositioned by the Bishop to whom she is bringing the relic; he entices her with the promise of desire itself, 'let us pursue an object we both know is unobtainable'; but he desists, conscious of 'the losses' when one's heart is led through the disappointments of a 'humilating pilgrimage'. Whilst Hudaconceva thinks he 'shrinks the world', she sees a beauty in his melancholy, which provides an analogue for her more faithful travails. On her return, she tells the voice that she accepts his argument as befitting the 'pitiful and comic and cruel misapprehension of the world': 'You are the Christjaw it's obvious' (recalling Ursula's wonder at the labyrinthine logic she strenuously discerns and imputes to 'the Christmind'); and it follows that only she, 'who prays to nothing', is fit to know him in his cosmic ironies, which drive her to fashion her own prayer:

GOD WHO IS NOT NOW AND NEVER WAS
GOD WHO IS A PLACE OF PAIN A TREELESS LANDSCAPE A BARREN
FIELD A SCORCHED PLACE A PARK OF SUICIDE A BROKEN TABLE A
POISONED DISH HEAR ME LIKE THIS HEAR ME
AS A WALL HEARS AND LIKE A WALL SENDS BACK MY CRY A
WALL OF SUNBAKED CONCRETE IN A LAND OF ROCKS I
HUDACONCEVA DENY THE MEANING PURPOSE BEAUTY OF MY
LIFE AND OF ALL OTHERS I CRY MY SOLITUDE I CRY MY
BITTERNESS I PRAY IT I PRAY IT MY PRAYER RUNS ROUND THE
BARE BOWL OF THE WORLD LIKE A TONGUE ... Amen

The jawbone protests when she leaves it briefly in an inn ('AM I LITTER / AM I JUNK'), but she tells the persistently loquacious relic how she is elevating the terms of her perception of life, and her self-dedication to its principles: 'How painful my solitude / I do not wish it to be relieved by you'. She is reluctant to continue with their 'intimacy' (the voice seductively but rather desperately ripostes by imputing to her a 'GREAT LOVE' which is distinguished by its fear and 'KNOWN BY ITS RELUCTANCE TO CONSENT'); instead she proclaims herself 'GREATER THAN THE CHRIST-JAW ... I PRAY TO NOTHING AND NOTHING CAN BETRAY ME THERE-FORE'. In pity for her own loneliness and courage, she briefly sobs, but persists in the 'faith in Death', 'which fills my gaze and straightens me'. Whereas for Christ, Death was a 'doorway' to the 'gardens of Eternity', Hudaconceva insists (in terms which to some extent reflect those of the philosopher Kierkegaard) on her own self-conscious existential commitment to what is consciously beyond human reason or even belief ('I pray to the wall / The wall speaks back / My own words / I SPEAK ME TO ME'). When she rejects the rationale of her trek back to the abbey, the pathetically self-pitying jawbone clamours for its return, divulging 'I am Lazarus I envied Christ and pretended to be him'. Deciding on an arbitrary death, of which the arbitrariness is 'however OF HER OWN MAKING', Hudaconceva lies down in the earth entwined with what she calls the 'remnants of Jesus Christ' and ignores the wails of the jawbone (rebuking it 'PLEASE DO NOT STEAL MY DEATH FROM ME'). In a further irony, the grieving nuns who discover her body impute her 'EXEMPLARY' in her preservation of the relic, refusing to admit the uselessness of prayer or the wrongness of her action, and thereby subsuming Hudaconceva into canonization and Christian design, so that she herself becomes a future relic, in which the desperate hopes of others are invested.

Two Skulls surpasses *The Seduction of Almighty God* and *The Moving and the Still* in its dramatic complexity and investigation of the boundaries and overlappings of faith, flattery and meaning. The central dramatic premise of a recurrent conversation, which proposes and questions an inscrutable identity and power, with an unseen seductive interlocutur, is a particularly ingenious use of the combination of *intimacy* and *unverifiability* which characterizes events and identities which unfold through the medium of radio drama. Hudaconceva is a more engaging protagonist than Loftus or Hoik, because more imaginatively passionate, resourceful, distinctive and pitiful in her challenges to God and the world, closer to Barker's Ursula in these respects; and her story, like many of Barker's strongest narratives, similarly suggests the reversibility of all forms and terms of power, whilst itself being irreducible to a single authoritative perspective of interpretation. It depicts the 'pitiful and comic and cruel' misapprehensions of the world, even in the case of one who seeks to repudiate it: if Hudaconceva struggles to raise herself and her supremely hope-less form of prayer to the level of a work of art, exhibited primarily to herself, her disruptive effect makes her a particularly urgent and enticing case for social and religious reclamation by the deceptive and self-deceiving forces which surround her. She is lured, not by human love, but by the promise of a divine preference and intimacy, only to discover the inadequacy – even the illusion and (self-)deceit – of such intimacy. She therefore decides to make her own defiant, consciously absurd, ultimately inscrutable terms of meaning through fixation, not on life and the world, but on Death, and the strange liberations of its acceptance. She places herself beyond the ongoing process of human valuation, which is of course driven to interpret and subsume her actions all the more energetically. In its resonant dramatization of Cioran's *aperçu* that 'All forms of divinity are autobiographical' – 'Not only do they come out of us, we are also *mirrored* in them'[46] – *Two Skulls* is one of Barker's best radio plays, and worthy of wider production.

Acts (Chapter One) was written in 2004, is as yet unpublished and awaits a professional production, but is one of Barker's strangest and most haunting plays, not least in its dramatic use of a character who seems even more, and literally, an intruder on the earth: an angel, called Coff, who appears to a woman, Oxford, at the moment her husband leaves her. The neo-Biblical title suggests a pre-eminently theatrical alternative or supplement to Old Testament books (such as Numbers, Psalms and Proverbs), and may also alert the audience to how the narrative will offer a surprising development of the Biblical advice, in Hebrews 13.1., 'be not forgetful to entertain strangers; for thereby some have entertained angels unawares' (in terms very different to David Edgar's 1985 community and Royal National Theatre play *Entertaining Strangers*).

Acts (Chapter One) opens with a pithy, memorably powerful trope of inter-personal extremity: Stone, the lord of a manor, repeats the phrase 'I love you with all my heart' four times, each time presaging a silence of worse pain; when Oxford *'yields nothing'*, he leaves the house and estate; their ambitious steward Floy immediately but casually plies her with seductive insistences, whilst maintaining an air of orderliness. Oxford rages bitterly in hatred of her husband, and vengefully considers admitting Floy's attentions (so that 'THE MOON CAN SEE A HURT WIFE IN HER ECSTASY'); her mother China listens and, most irritatingly, deterministically subsumes the responsibility for not only Oxford's distraught state but her entire personality – 'What you are darling it's all me' – effectively denying Oxford original-ity or autonomous agency. However, Oxford *'closes her eyes'* and *'is altered'* when *'she senses a stranger'*: Coff simply states 'I walked in', whilst sympa-thetically observing the gravity of the situation, of 'seeking a new life or poss-ibly a means to end an old one'. He tells Oxford 'you need me perhaps the scale of this need is not yet obvious to you'; however this is not, as she first suspects, a sexual proposition akin to Floy's. As Coff observes, seductively but not sexually, he inspires in her a trust 'so profound' as to be shocking, whilst retaining his manners and composure at her outbursts ('Please don't apologize it is perfectly natural to quarrel at least initially with something so welcome to us it appears almost to parody itself') yet darkly hinting 'I have to warn you however that it cannot always be like this'. Oxford, bewildered not least at herself, accepts him into the household; Floy importunes her in the darkness with a demanding insistence.

Stone has sought refuge in an abbey, where the old priest, Gibbon, encourages him to articulate his contradictory feelings towards his wife. Gibbon addresses the details and tumultuous complexities of sexual resent-ment; Stone acknowledges 'sensations of desire violence curiosity' but seems the more existentially philosophical and deductive in his forcible, but possibly evolutionary, detachment from his former self ('I will either die or become what I have not yet been I believe all men have many lives but choose to smother some in order to live others'), even as he admits himself still enthralled to Oxford and currently lost 'IN THE MOST SICKLY SERVITUDE'.

Oxford tries to dismiss Floy, who stands his ground, territorially, in rela-tion to both house and mistress; the appearance of Coff has the surprising (but characteristic) effect of releasing the tension between them. Oxford is *'wounded'* when Coff declines her offer for him to take her husband's place in her bed; he insists and maintains that he desires nothing, even as he acknowledges that they do have a non-sexual, 'much greater', but unspecified intimacy which Floy will find intolerable. However, Coff points out their continuing dependence on Floy ('What do we know of farming an estate'); his frank directness characteristically leaves Oxford *'torn between laughter and rage'*. Coff is also disarming towards China, who challenges him to value the 'property' of Oxford's nakedness, and of her own. Coff calmly replies that the value of Oxford's nakedness can be assessed 'only by him who suffers the urge to possess it', and testifies to the instability of sexual power ('neither of you naked out-values the other'), which he insists China also commands in her own way, at the age of sixty. When Oxford laughs in superiority at her mother's brief erotic disarray at this proposition, Coff is uniquely abrupt and menacing towards her. He shamelessly makes an assignation with China ('Inhibition what is it a fence only a fence'), which defuses the sexual competition which Floy would wage with him for Oxford (and her seven thousand acre estate).

Stone, who has become a monk, is nevertheless racked with feverish sexual imaginings of his wife, but determines 'God is in this also'. Gibbon urges him to talk more of her, pre-emptively voicing Stone's possible suspicions of his motives as a vicarious delight in inflammatory confession: 'YOU ARE THINKING THIS MONK IS A DOG OF SQUALOR TASTING MY WIFE THROUGH MY OWN MOUTH'. Gibbon insists his interest is theoretical, but that it focuses on the hinge which sexuality constitutes between the physical and metaphysical, and hence provokes the re-evaluation of the ethical: first he proclaims the possible reinvestment of taboo: 'say your acts with her starting with the worst by worst I mean the best with the loved one all values are reversed'; 'the body of the loved one does it not abolish dirt God made us filthy but the loved one naked can make filth an empty word'. Gibbon is thus in his own way seductive, if only from a (satanic?) point of exclusion, extending his speculations: 'like you I am a mess of contradictions if we resolved them all we would ourselves become God would we not unless that is we say of God an altogether different starting point He is Supremely Contradictory'. Stone threatens Gibbon and commands him to lick his hand like a dog; recovering his placidity, Stone guesses that Gibbon's curiosity is that of 'a killer of women': 'What else could explain your reluctance to be where you crave to be?'.

China professes concern about Oxford's conduct and urges reconciliation with Stone. Oxford retaliates by alluding to China's bruisings from energetic lovemaking with Coff and challenges her to terminate their relationship. Gibbon observes Oxford, and Coff subsequently confronts the priest at his church to warn him off; Gibbon explains that he was scrutinizing her at Stone's behest, and admits 'I killed a woman once'. Gibbon maintains that this woman 'asked and I obliged' and so 'I felt it neither sin nor crime but ecstasy'; moreover, he hints to Oxford about his 'inevitable intimacy' with her, which Stone imagines and aims to facilitate. When Coff (thoughtful, on a garden swing) later detects the smell of the priest on Oxford, he deduces 'The old man knows he is the sacrifice what he does not know is the identity of his killer'; he kisses Oxford for the first time and orders her to 'pray'. Coff intuits that if Stone 'looked for God in the abbey it can only have been because he had lost God in you'; Oxford asserts 'God was not in me to be found ... Nor in any woman'. This drives Coff into a fierce empathy with Stone ('who found mystery in the body of a wife who could not find herself mysterious'), whom he finds deserving of pity, even as he proposes that he and Oxford 'fuck now', a challenge which only serves to compound the resentment of their standoff. Stone reappears, an apocalyptic figure who feels himself purged of 'contradictions' and there-fore 'perfect', since his dismissal of Gibbon as his spiritual adviser. After Stone's brief meeting with Oxford, Coff observes 'if it's true I am an angel it might also be true that Stone's God', as Stone has indeed eradicated all con-tradiction from himself. In the chapel of Oxford's house, Stone speaks what he has composed as 'the world's worst prayer', to which Gibbon must say 'Amen' before Stone cuts his throat: 'Father / Because all you have made appals my sight blind me / Because all you have said is insufficient deafen me / Because to speak is to threaten and I am tired from threatening take my speech from me / And when I am this because I can still kiss take my life from me / Let me be a bone and after the bone nothing / Father / For none of this do I make criticism of Thee'. The killing performed, Stone leaves Oxford to Coff. The final scene shows Coff and Oxford either side of a bed, motionless and then '*with decision*' undressing, and choosing to ignore the worldly distractions represented by a windblown casement.

Acts (Chapter One) is eerie, surprising, poignant and evocative in its own peculiar ways. It does not attempt the exhaustive reach of *Rome* or *The Ecstatic Bible*, nor the traumatic archaeology of a historical landscape as in *Found in the Ground*; it does not feature the exhilarating up-against-time narrative drive of *(Uncle) Vanya* or *Hated Nightfall*, the wry unpredictable disruptions of international strategies as in *The Europeans* or *Animals in Paradise*, the breathlessly intense focus of *Isonzo* and *N/A*, nor *Gertrude*'s combinations of the above. As my précis suggests, its narrative events can be summarized (more easily than those in *Found in the Ground*, for example), but not as readily as those of *Ursula*, where the desperate insistences of the *story* still do not entirely account for the unsettling resonances of Barker's *play* (or, indeed, an effective production). *Acts (Chapter One)* does not depend on a narrative structure like that of *Ursula*, *The Castle* or *Gertrude* in which one pressure necessitates a counter-pressure and events acquire a terrifying, apparently irrevocable momentum. So what does it do, and how?

Acts (Chapter One) is a principally character-driven play which nevertheless suggests an external force of design – which alternately may and may not be associated with Coff, who proves knowledgeably insightful, but not omniscient. Coff is an unusual character in that he is strategically effective primarily through remaining even-tempered, yet inscrutable (Starhemberg in Barker's *The Europeans* also demonstrates this unnerving self-composure). The setting of the manor house, estate and chapel may seem socially selective but are also atmospherically distinct (rather than abstracted or rarified); the incursion of a potentially supernatural force into the twilights of an aristocratic languor may suggest a medieval setting, or else initially recall 'classic English drama' of an earlier period, such as Eliot's *The Family Reunion* or Priestley's *An Inspector Calls*. A more precise analogue may be John Whiting's *Saint's Day* (1951), which also, in its own different ways, destabilizes the Anglican landscape of village, church and big house through incursions of an irrationality borne of 'volatility and sudden turnabouts in apparently ordinary situations and without formalising them into a new dramatic system, which makes them even more surprising and ominous'[47]. Like Whiting's play, *Acts (Chapter One)* depicts a landscape charged from the outset with psychological (and, moreover, sexual) anxiety and desperation rather than hierarchical security, becoming increasingly apocalyptic. Stone goes from an archetypally self-repressed Englishman ('his manners were a castle lovely manners but he hid inside them' – suggesting a parallel with Stucley's hysterical extrapolation from personal injury as amplified and explored in *The Castle*) to an ominous harbinger of death (recalling the transformation of Procathren in *Saint's Day*) and a finally eloquent pursuit of transcendence (Stone senses in 'the perished purse that was my heart' still one 'small coin of regret', when 'the true saint has none ... I have some way to go'). Oxford is engagingly and expressively off-balance, not only in her combination of 'her pettiness her grandeur her loathing of herself her insufferable pride' but in her constantly surprising susceptibility to others; and, as in *Gertrude*, the remaining characters who may initially seem stock (even comic) types all demonstrate surprising depths of feeling, possibility or depravity. It is through these reversals, and a characteristically Barkerian wry humour and possibilities of individual redemption, that *Acts (Chapter One)* eludes the irony and nihilism of *Saint's Day*. The ferocity of interpersonal and internal struggles is relentless despite Coff's relative composure, and even he finds himself in rage and tears; and the classicism which the play's setting initially seems to invoke is shot through with a startling sexual explicitness (Oxford's report of Gibbon urinating on her, Coff's methods of reviving China's dormant sexuality: 'Acts of recklessness she calls sin I applaud her and pin bruises to her like dark medals'). However, Stone's aggression is finally focussed on eradicating the implicit murderousness of Gibbon; he pursues his own chosen terms of redemption, as Oxford and Coff finally consider their own, with each other, in what may be a final challenge to (and acceptance, or dissolution, of?) the law that forbids an angel wooing human 'clay', lest he lose his wings at the dawning of the day. Thus, *Acts (Chapter One)* shows characters who are surprisingly engaging in their choices of terms of faith and their bids to step outside personal boundaries, challenging divine order in such a way that even an angel might be seduced, and the terms of the game of life resurrected.

9
Servility and Servitude

An Eloquence, The Blood of a Wife, A Rich Woman's Poetry, Stalingrad, 13 Objects, The Dying of Today, Dead Hands, Christ's Dog

> How can freedom know itself unless freedom is confined?
>
> *Dead Hands* (*DH*, 56)

Cascan's insistence in *Gertrude – The Cry* that his devotion to his queen constitutes an 'ECSTASY', 'a passion and a faith' is the most fulsome expression of the brilliance of the servant, beyond even Barker's play of that title. Cascan responds to the beauty of the searching and pain in Gertrude's life, inspiring in him a passion (the pain of wanting) to offer Gertrude at least occasional relief. Claudius similarly discovers a 'religion' in Gertrude's agonized reach of self-invention, though Cascan queries how far Claudius will be able to 'meet' her. Cascan's form of **servitude** (like that of Shoulder, Practice in *Animals in Paradise*, Just in *Minna*, Kidney in *The Fence*) is shown to be more honourable than some of the higher-status posturing of their social superiors, and, in sacrifice, an important resistance to those who would attack the subjects of their devotion (and render them into objects). These aggressors represent the **servile**, those who would tear down everything to a facelessness to match their own. Lingis, following Nietzsche, offers a useful exposition: servility is characterized by dependence: on one's fellows, the institutions, and the technological equipment in place; whereas nobility 'realizes traits, behaviours and works which do not depend on support from the surrounding social and technological environment'[1]. Under the reductive terms of the servile,

> health means closure from the contagions of the world, beauty means enclosure in the uniforms of serviceability in the herd, goodness means productivity, and happiness means contentment with the content appropriated. [Their notion of goodness] is a belligerent term ... a demand put

on others. Their forces, disengaged from the substances of things, recoil upon themselves and turn a surface of closure to the world.[2]

This aptly captures both the ultimately tenuous and self-undermining quality of ideological and moral closure, and the possibility of dramatic and demonstrable reversibility and disclosure, as frequently depicted in Barker's work. Several plays in this period of his writing examine the distinctions between servility and servitude, in the shadow of the recurrences of fear and hope: oscillations which are presented, to and in the imagination, in forms which dissolve conventional restrictions (and therein resemble dreams and nightmares)[3].

The as yet unproduced filmscript *An Eloquence* (written 1997) is, in terms of dialogue, almost wordless: a measured, gradually unfolding and evocative visual meditation on solitude, duty and abandonment set on an unspecified part of the Roman frontier in 210 AD. It dramatizes the perspective of a soldier attempting a consistent and orderly servitude, notwithstanding recurrent evidence that he represents an enclosing and destructive servility, through the increasingly apparent myopia of the imposed rationalism which he (somewhat eccentrically) serves. He takes up office at a signal tower in a bay where a native Woman is staked out in the sand and casually punished, and an Old Man smooths the sand with a length of wood, his *'methodical manner'* indicating *'this is the labour of a lifetime'*. The signal flag is incomprehensible to the Soldier; the tower is now occupied by a mocking Greek, with whom the Woman transgressively consorts. The Soldier takes up charge and residency of the tower, but finds the combination of signals flying from a distant tower impossible to interpret, let alone relay. The script draws the audience into an inhabitation of the Soldier's perspective, studying the landscape with a mixture of bewilderment and attempted self-reassurance; in the process, he crucially views those who people the landscape *externally*, to an unusual degree (the Soldier finds them incomprehensible in word and deed). When the Old Man's wife is killed, the Soldier has to shake him physically from his sand-smoothing task to encompass the experience and prepare her grave, which the Old Man determines to occupy himself, also, in fatal resignation. This contrasts with the Soldier's focus on a persistent routine of attempted communication (though not with those most proximate), which is further beleaguered by the discovery that the Woman has stolen at least one signal flag for use as a skirt. Exasperation and terror drive the Soldier to kill one of two hooded conversationalists who pass the tower at night. A Roman patrol hangs the Greek, tortures and kills the Woman and punishes the Soldier for neglect of his duties. The surviving conversationalist continues his theological debate (in the only spoken language in the script which is comprehensible to the Soldier and the audience), taking both sides in debating the essential human pain of solitude, in a Godless world where men can offer 'nothing more substantial than the painful and possibly eccentric experience of their own brief lives' (and how 'fraternity, conjugal love, ecstasy, oblivion', only disclose a central experience of absence, 'the void'). However, the conversationalist remains 'undecided' as to whether 'the existence of God even as a proposition is a characteristic of an infantile stage of human development' when it may be that not 'all expressions of a will to submission are themselves evidence of immaturity': as the Soldier's tasks demonstrate in a secular military context, they may be bids to make and transmit meaning.

The Soldier is wounded by an unseen archer; he recovers, but a Young Replacement is sent to continue his duties. The Soldier is displaced from eminence within his own small kingdom of the signal tower, and performs a resentful passivity. It transpires that the Soldier nevertheless has a greater comprehension of flag signals than the Replacement; the Replacement is exasperated, the Soldier's coarseness is *'designed to offend, to assert his absolute domination and title to the room and all it contains'*. His limited authority is secured when the Replacement is killed by an arrow from an unseen assailant. The *'ecstasy and relief'* of the Soldier is interrupted by a visit from a boat, bearing a Governor, who *'proceeds to enunciate with a serious manner the strategic and political considerations of the authorities'* but in a language the soldier does not understand. When the Governor departs, the Soldier realizes that the Governor has foisted the custody of a child on him: a figure of uncomprehending helplessness, to which the Soldier must nevertheless become dutifully subservient. The Soldier in turn foists the child upon some passing refugees, but then experiences regret at this action. His contrary impulses bring him to *'a sense of absurdity'* and he prepares his own grave. Insurgents approach, with the baby, whose cry shatters the Soldier's terminal solitude: *'The perfect organization of his death at the hands of an enemy still to arrive seems threatened by the reappearance of the baby'*. The soldier is killed; his hand opens and shuts, like that of the baby, as he tries, in death, to stir a watching hare, which remains *'undisturbed'*; *'Then the hand turns, palm upwards'*.

The unfolding visual and emotional details of *An Eloquence* vividly substantiate its modest but significantly indefinite title: in counterpoint to the received, privileged truth and grand narratives of imperial history, the narrative offers a resonant account and series of images of experience at the edge and limits of an empire's power: an experiential encounter with indeterminacy, and a predominantly visual example of Barker's feeling for dramatizing 'anti-history'. The script skilfully orchestrates details which engage the audience with the Soldier's dutifully methodical hopes, his repeated confrontation with indeterminate images and forces, his *'exaggerated self-congratulation'* and *'wordless despair'*. It suggests how abject and *'obscene'* human forms compulsively return to provide reminders of the consequences of power, its circumscriptions and exclusions (the corpses of the women, re-emerging in skeletal transformations from sand and sea[4]). It shows how a faithful (though beaten, foist upon and uncomprehended) servant can nevertheless be distraught when the small territorial enclave that constitutes his shrunken world is threatened by a potential successor (a scenario which may remind some of Beckett's *Endgame*, with the crucial difference that Barker locates the existential nausea of his protagonist in a historical, as well as a metaphysical, context). Barker's script is classical in the sense of being timeless in its resonances, despite the specific historical location of events, with the signal tower forming another of his memorable images of the collapsing defenses of the fortified but fragile self (compare *The Castle* and *A Hard Heart*) which reveal the stress points of subjectivity in terms of 'the flayed skin'[5]. The political reverberations of its visually poetic imagery are even more immediate since the script was written, not least in its confrontation between the anxious self-reassurances of haplessly dutiful (and ultimately forsaken) representatives of a significantly self-distancing global power and its undefined (because indefinable) enemies. This metaphorical prescience alone makes the film particularly deserving of resources for production, though its discomforting power may equally preclude the support of some funding institutions.

The Blood of a Wife, another unproduced filmscript written by Barker in the same year (1997), offers another series of haunting visual images which are even more defiant of single interpretation. In the closing stages of a war, characters pursue strange hallucinatory encounters, surrounded by the ominous forms of escaped circus animals (performing dogs, a bear and, most incongruously, a kangaroo who has taken refuge in a church) and museum statues which are being transported for safekeeping. A Boy watches a wilfully elusive yet provocative Woman, who frequently claims to have been 'attacked'. Her husband ascribes these attacks to her irritation of the world, its forces of servility which strive to expel her as intolerable; he is affectionate, but their marriage seems a matter of accommodation. The Boy's father, a museum curator, defends the collection of works of art, however futile its attempted defiance of historical oblivion: 'I like futile acts...It's the significant acts which bother me'. The Boy reveals to the Woman a vision, reflected in the liquid of a copper pan, of a woman in flight from hounds (an image from Boccaccio's *The Decameron*), reflecting the posture of a statue with which he is fixated (it has the almost carnal allure of works such as Bernini's Proserpina and Daphne). At a dinner party, a male guest, termed simply the Man, speaks from his own passionate sense of mortality; his words have resonances through several of Barker's works and propositions:

> We all have secrets ... They are necesssary. They are the condition of all freedom. On the other hand, they kill. They both make life possible and extinguish life ... And these secrets are more beautiful the greater power they possess to wound and injure those who live outside the secret.

> Is that not terrible, but then beauty is terrible is it not? ... this secret is
> therefore a punishment, a weapon which hangs suspended over others,
> and possibly a revenge ... and we must all revenge ourselves on the
> world obviously, especially on those who love us, they particularly must
> be singled out for our revenge, is that not odd, is that not terrible but
> also the truth of the secret?

The woman is desperately protective of her unborn child – of which,
perhaps, the Man, rather than the Husband, is the father? She tells the Man
how 'it is obviously necessary that the truth is sacred', but also that 'the lie
might be', 'Some need truth, and others, lies...'. The Husband recognizes
that 'There is a purity in love which is indifferent to the truth...perhaps
even...it abhors it'; the Woman's mother goes so far as to say 'always the
wife lies ... This lying the husband has to love'. The Boy menaces her pregnant stomach with a knife, but she escapes; after a bombing raid which fills
the museum with shattered glass and upturns the circus trailers, he frees
the performing dogs, which pursue the woman in reflection of the classical
image; the Husband rescues her, assuring (himself?) 'I am your husband.
And my child is in your womb'. The Boy is caught and castrated by soldiers. Her child safe, the Woman confronts the Husband with a look
'ambiguous in its combination of fear and self-assertion'; the image cuts to *'the
classical figure of the HUNTED WOMAN, closely pursued by hounds'*, cascading
into the museum which is heaped with broken glass, indicating *'the end of
the chase, her entering a cul-de-sac of a building having doomed her to destruction'*; then back to the Woman's decision to sit, as the Husband holds her
look: *'The chair, moved abruptly over the tiled floor, emits a painful sound'* – a
telling instance of an unremarkable domestic object's ability to suggest the
grindings of the will[6].

The dreamlike quality of events in *The Blood of a Wife* makes it difficult
to summarize, let alone exposit; in this respect, its style recalls Barker's
unproduced filmscript from 1985, *The Blow*[7]. Its images of accommodation
and resentment are hauntingly elaborated by the (occasionally Lawrentian)
imagery of caged and released animals, which reflect the characters'
passions, and their terms of restraint.

The figure of the aristocratically sexual woman, who provokes and irri-
tates the terms of servility, provides a more central focus in Barker's stage
play *A Rich Woman's Poetry* (written 1999, as yet unperformed). Dexter is a
duchess and titled patron of the arts: she is also mad by common standards, but self-knowing: she remarks 'whereas I am mad I am not foolish'.
She is a compulsive disrober and masochist who nevertheless achieves a
performative and existential grandeur compared to those around her,
whose servility is not merely economic, but spiritual. In some ways, Dexter
foreshadows the greater protagonists Gertrude and Oxford, not least in her
strain at enclosure in a bad marriage, her predilection for blue shoes, and

her pleasure and strategic impact in nakedness (as another character observes, Dexter 'has an altogether mischievous relationship with nudity ... the body in her case acquires the characteristics of a weapon she loads the thing she aims she fires it at you'). In other ways, the play offers a reversal of Barker's *Scenes from an Execution*, in which the patron is the existential seeker, impatient with material encumberances, and the artists and their agents are resentful, superior, disrespectful and ideologically reductive. Dexter has agreed to fund a glass sculpture by the female artist Trolley: its subject is 'Masculinity', and it is designed to be damaged in transit ('Masculinity is made of glass precisely to be broken': it 'shifts', 'can't keep still', 'is shattered / Obviously', then stitched up with wire, representing the male 'so-called powers of analysis', 'discrimination' and 'moral integrity'). Trolley's life model, Cartmel, suspects a reductivity in the project, in its insistence on a rendering of complexity into theory. Cartmel is more inclined to the contemplation and potential discoveries of death ('I only know death is the key the key to life strange at the very end you find the key'), which, he acknowledges, in itself makes him anomalous as 'the model for a thing called masculinity'. He admits he does not know what masculinity might be, but knows it is not 'barbarity', whereas the postures and theories of some artists, agents, theorists, critics might be just that: forms of closure which 'serve to immunize' them 'against shame or embarrassment'. Both Cartmel and Dexter manage subversive friendships with Trolley's son, called simply Boy. Trolley insists, in her ideological feminist certainty, that every action that Cartmel does or considers, clothed or not, is an attempt on her 'authority', 'a disputation of this space' and thus 'the entirely predictable manoeuvre of the male'. Dexter contrastingly performs the unmanageability of the body in intimacy when she asks the Boy to cut her, *'carefully'*, then thanks him. Dexter tells her loyal and faithful maid (whose careful attentiveness contrasts with the barely concealed contempt of her financial "suitors"), 'I don't love the boy / I am creating him', in different terms to Trolley's bids for all-encompassing theoretical control. In fact several characters struggle for the soul of the Boy: Dexter's husband implores the Boy to help him impose his own form of repressive control, to murder Dexter. Contrary to Trolley's definite conviction 'Masculinity is the fear of women', the Boy works towards his own notion that 'The masculine is rarely where it seems to be and the reason for this is that paradoxically it does not have its origin in men at all'.

Dexter, by her very terms of being, drives the surrounding characters to wilder extremes: her husband blinds himself, preferring this to the 'constant revelation' of her infidelity. Dexter tells Cartmel that ecstasy is 'paid for with the pain of others', 'not just a melancholy consequence but the condition of this ecstasy the ecstasy is launched on pain as a ship is floated on the sea'; and she deduces the transformation (hence, overturning of determinism) in intimacy, that Cartmel 'brings some rag of maleness to me' which she tends and gives back 'as masculinity'. Trolley proceeds to a new work of art, spurred on by her offence at Dexter's 'ecstasy'; Trolley strips and plunges into a wire cage of broken glass (stage directions suggest *'Instant lighting effects isolate the sculpture and the entire cage and TROLLEY herself is washed in blood by a revolving hose'*). The Boy recites a prepared commentary ('Sandra Trolley swims the experience of Masculinity a pool of broken glass that lacerates her body and drenches her in blood the blood of animals the blood of slaughtered innocence the blood of violated integrity … obliging us to become however reluctantly connoisseurs of trauma'). Dexter challenges the work: 'How is that Masculinity? … Masculinity is the gift of women … Not the cruelty of men'. When Cartmel fatally stabs Dexter, the Boy embraces her, remains kissing her face whilst others drift past and away, with insults and laughter.

A Rich Woman's Poetry features a memorable female protagonist in Dexter but otherwise falls short of the achievements of the subsequent explorations of sexual transgression, *Gertrude – The Cry* and *Acts (Chapter One)*, for which it partly prepares (a feature of Barker's *ouevre* as a whole, in which some texts are identifiably exploratory works, bridges towards the full flowering of a theme). The characters which surround Dexter lack comparable stature and complexity, and the modern setting may limit the play to seeming a satire on contemporary types (surprisingly reminiscent of Barker's *Stripwell*, not least in its 'big house' setting) – albeit a mischievous and profoundly moral satire. Problematically, Cartmel's final decision to stab Dexter is vague; unless it is interpreted (rather strenuously) as a botched act of love or a charitable deliverance from the demeaning resentment of her husband, this outbreak of aggression might be drawn into support for Trolley's characterization of masculinity. Nevertheless, *A Rich Woman's Poetry* provides an incisive, and sometimes moving, pursuit of ideas surrounding transgressive sexual freedom versus gender ideology as represented in artistic form, and the terms of its support and administration; at best, it eloquently dramatizes the limits of ideas and theory as opposed to the complications of compulsive experience, as when Trolley feels obliged to diminish the implications of Dexter's unruly, restlessly active sexuality with her own ludicrously Messianic postures of gender-fundamentalist martyrdom. This satire on the modern vogue for Explicit Body Performance[8], such as that practised by Maria Abramovich, distinguishes Barker's theatrical practice (and use of nakedness, as exemplified by Dexter) from the worst of such work, which proclaims and indulges ideological and moral superiority whilst simultaneously purporting (in Trolley's words) 'to take refuge in an aesthetic beyond the moral beyond the ethical beyond beyond the parameters of pity'. Crucially, Barker's theatre and its terms of enquiry are actively and complexly moral and ethical (and concerned with pity, as exemplified by the Boy's final embrace of Dexter) rather than self-righteously and fetishistically re-presentative of postures of victimization.

'The parameters of pity' provides a keynote from which to consider Barker's opera libretto *Stalingrad: The Envy of the Painless* (written 2000, staged in Denmark 2002, music by Kim Helweg), in which Albert Ragusa (whose name combines that of two character from *Gertrude – The Cry*), a maimed and blinded war veteran of truly and catastrophically 'active service' in Stalingrad, arrives at his 83rd birthday in the East German war hospital where he has lived since 1945. The hospital, formerly an eighteenth century château set in a park, was the scene of an intimate meeting between the young Ragusa and a young woman, Bischof, in 1941, when, illuminated by searchlights, she undressed for him and posed between two statues of Augustus and Voltaire. Ragusa assumed she had fled to the West, whereas in fact she had waited for his return but was raped by Russian soldiers and drowned in the same lake beside which Ragusa takes his daily exercise. On the occasion of his birthday, an irreverent party is arranged for Ragusa, involving a kissogram stripper, who provokes him then resents the intimacy of his kiss; in response, some loutish waiters tip the old man (whose scarred face makes them impatient to 'relieve' him 'from the burden of self-consciousness') into the lake; in further representation of destructive servility, they also double as Bischof's attackers in the wartime scenes. As Ragusa sinks, drowning, to the bottom, he encounters the ghost of Bischof and learns of her experiences; he summons up the spirit of his idealistic youth to judge his life and consummate at last the sexual union with Bischof that he craved. One of the louts, haunted by his violence towards the weak old man, contemplates the reason for his cruelty.

Stalingrad thus weaves poignant and savage juxtapositions in dramatization of a major theme in Barker's poetry, historical relativism: specifically, how morally infantile philistines of the present day seek, but fail, to belittle and dismiss the past. Ragusa the Boy laments 'It was such bad luck to be born in 1920' but Ragusa wonders 'Perhaps / Or was it luxury', contrary to all notions of social promise beyond that of 'honourable service'. The erotic delicacy of his tryst with Bischof is contrasted with the spiritually servile and meanly resentful waiters and stripper: the zenith of whose sensibilities are demands for the 'posh things' of consumerism, the paradise of painlessness, allied to the smug self-righteousness 'AT LEAST … WE AREN'T THE VICTIMS OF PROPAGANDA' – except for the one who realizes his 'jealousy' of Ragusa's 'agony' ('One of your hours … Might be worth whole days of my existence and one kiss more than five hundred fucks'). Notwithstanding – or because of – his catastrophic life, Ragusa repudiates modern social anaesthesia as he sinks in the lake: 'The bottom / Is the only place from which to see the sun … Of course no one (*he utters these words silently*) / DARES / SAY / SO'. He thus reverses the dominant terms to ask: who has the privilege? Who has the moral authority? Barker's libretto succinctly and poetically suggests the error of received moral wisdom which would shrink from and diminish pain, and how this ultimately issues in infantile aggression: the envy of the painless.

Barker's dramatic compendium *13 Objects: Studies in Servitude* (written 2002, staged by The Wrestling School 2003) might take its keynote from Girard's suggestion in *Violence and the Sacred* that '*desire* is produced through competition with an other for the possession of some object that comes, in the desiring struggle, to emblematize desire itself'[9] (compare the importance which the characters attach to objects in *A House of Correction*: the empty lipstick case and photograph). Barker's programme notes expressed the premises of the compendium with a direct precision:

> Everyone thinks his life should be better than it is. Everyone thinks life itself should be better than it is. The ancient torment caused by the mismatch of hope and reality is the theme of these short plays, where the sense of disappointment, so profound as to become almost a death-wish, is stimulated by mundane objects. If the object by definition is lifeless, it is given life by our imaginative relation to it. In this sense, it *proposes*, it is not inert. On the surface of these common things a spiritual struggle is enacted. They become the focus for desire and loss, desire for what might have been, and despair at what is. Whether they are relics of the dead, symbols of sacrifice or pathetic items imbued with love, they carry immense significance, and serve as a pretext by which the world is judged. We are enslaved by our hopes...[10]

In 'A Lonely Spade', a philosophical officer, proclaiming that dignity 'comes only as an intimate of death', tests the will of two prisoners, and possibly his own, by ordering them to dig a grave, but tantalizes them by changing the rules: first he says the grave will be occupied by the less enthusiastic digger; then that it will enclose the 'scum' who was sufficiently servile to dig his 'comrade's grave'. The officer's speculative experimentation is existential: he wonders how one might 'invent new and untarnished forms in which to enter death?'. When the digging prisoner relinquishes hope and turns the spade over to the officer to dig his grave, surrendering not only the hope of survival but the emblem of desire itself, the officer declares 'how beautiful you are today', and walks away: the digging prisoner flees with '*agonized disbelief*', the one in the grave seems bewildered by his own sudden neglect.

In 'Cruel Cup Kind Saucer', a woman fights her shame by returning to a café which she frequented with a lost lover, a site distinguished only by 'My / Struggle / With / Myself': she infuses compulsive calculation with erotic melancholy, speculating, from the sixty cups and sixty saucers at the counter, how many cups 'will bear the intimate traces of his lips' before they too become representative of irrevocable loss, either through breakage or replacement, obviating her testimony of their history. She hides her tears behind a saucer, then exposes her consternation; in discreet pity of her contradictions, a waiter proffers a handkerchief. 'Cruel Cup Kind

Saucer' may be Barker's most beautiful compendium play in the way it presents a full flourishing of a character's distinctive contradictions within such a succinct form.

Other plays in the compendium present human vulnerability in terms of a wry comedy: the comedy which, in the words of Simon Shepherd, 'comes from the realisation that objects have their own logic, and that human activity is vulnerable to this logic' when this vulnerability 'derives less from human passivity than human desire and determination'[11], giving rise to an idealism which subordinates, or even necessarily defeats, its own agents. 'Cracked Lens' depicts a youth whose self-defeating idealism locates Hamlet's dilemma in bathetic terms: 'the very simplest encounters with the world pose … agonizing dilemmas': the gift of a camera obliges him to focus on the banal, and selectively frame it as the miraculous; he prefers to wield, and associate himself with, the camera without film. 'Not to Escape Now' shows how a ring can imply property, and invite the excitement of theft as well as blindingly wilful idealization. 'Blind Prejudice' depicts subordination of self and others to the conformity of a generalized and idealized myopia, in comic terms. 'The Hermit's War with God' shows the indefatigable appetite of disciples for subordination and fetishistic iconography. 'The Talk of a Toy' dramatizes wilfulness and pleasure in the domination of others, as an unusually (and artificially) eloquent child coerces others to repair her loss of a plaything[12].

Several of the plays in the compendium focus on the erotic power of objects, in ways which Bataille's speculations may illuminate, when he suggests that eroticism can be a 'fusion which shifts interest away from and beyond the person and his limits', and thus be 'expressed by an object': 'We are faced with the paradox of an object which implies the abolition of the limits of all objects, of an *erotic object*'[13]. 'South of That Place Near' shows how a woman (in a reversal of 'Cruel Cup Kind Saucer') can prefer the erotic mystery of an indeterminate object (a summoning postcard, its source obscured) and a thwarted hope to the obligation of joining her lover (indeed, finding herself so 'RELIEVED' she does not have to go, obediently). 'Navy Blue' shows two lovers who encounter a woman's discarded blue shoes (as if they stumbled on the site of Dexter's assignation with Cartmel, or Gertrude's with Claudius); the man becomes aroused by the implied narrative and infatuated with the absent woman capable of such an 'ecstatic gesture'; the woman insists 'We all do that', leaving artful scatterings to excite the imagination, but the man is unconvinced by her, preferring the object which proposes the abolition of limits ('Catastrophe … How vastly superior this is to the tedium of knowledge', *OP2*, 276) and the image of the mysterious woman who 'retains forever an incorruptible distinction'. 'Listen I'll Beat You' stages the meeting of an itinerant drummer, whose rhythms provoke 'ecstasy', and a female recluse who attempts her own form of coercion with a gun. When he drums, she finds her habitual preference for silence overwhelmed, and her body animated; however, he denies the erotic consummation of the appetite he has kindled in her (claiming 'I open doors but don't go through them'). She shoots him, concluding that 'If you can't keep the drummer you can keep the drum', though this leaves the drum unbeaten, 'Undrummed'. This revisits the premise of *The Twelfth Battle of Isonzo*, in which the young woman enters the place where the old man contemplates his departed lovers as if trophies (realised in Barker's 2001/2 production by the shoes suspended in cages) to embark upon a duel which will constantly question who is, ultimately, whose trophy, and on what terms. The woman in 'Listen I'll Beat You' reverses the power and expresses her rage at humiliation by his cavalier wanderlust, but his instrument is a hollow trophy, an apparently delimiting erotic object which in fact proves inanimate, and crucially limited, when mute. Their wieldings of drum and gun are both forms of coercion, which excite, but the respective entrenchments, involved in coercion, significantly disappoint.

Barker's 2003 production of *13 Objects* for The Wrestling School provided the most elegant staging of one of his compendia to date, opening with an exordium based around a 'Wedding Machine', a contraption, suggesting idealization, best described by its designer:

> a simple fulcrum rises and falls, lifted on a pulley by a seated, visible actor, one end plunging a collection of objects – lilies, a clock, a trowel – into a tank of fluid, and the other simultaneously lifting a framed wedding photograph high, away from a half-naked bride who is contemplating it...she rises to follow it and turns to accuse another seated bride...as her gesture is repeated a shreik comes from the sound system (pre-recorded of course) [at the same time]...[14]

Performers made artfully swift transitions from one play to another (for example, the exordium featured the waiter cleaning a café with a vacuum; he then became the victim of the Officer in 'A Lonely Spade'; when released by him, the waiter ran into the café, which was revealed behind a barrier of corrugated iron, opening to disclose also the solitary customer for 'Cruel Cup Kind Saucer'). The production was able to field a strong and experienced ensemble, with four performers from 2002's production of *Gertrude – The Cry* – Wicks, O'Callaghan, Avoth and Jason Morell (Cascan) – joined by Sarah Belcher (from *Ursula*, *The Ecstatic Bible*, *Knowledge and a Girl* and *He Stumbled*). Morell excelled in 'The Talk of a Toy', Avoth in 'Cracked Lens', Wicks in 'Cruel Cup'; as did both Avoth and Wicks in 'Listen I'll Beat You'. Impressively realized in splendid detail, *13 Objects* remained a strong example of a small-scale Wrestling School show rather than one of their great expansive works; its wryly ironic analyses of misplaced hopes made it, on the surface, more engaging to prevalent British theatrical tastes. It was also a notable demonstration of how Barker's compendia (including *Wounds to the*

Face, Twelve Encounters) present and constitute a peculiarly Barkerian sense of comedy, and comedies: ironic reversals which mock any sense of natural order or tenable theory, presenting rather the mismatch of ideals and reality, and the vulnerabilities of human desire and determination (from which none of us are entirely immune).

The old adage 'A man who is laughing has not heard the news' might provide a starting point for considering Barker's short play *The Dying of Today* (written 2004, performed in Caen, France, 2007), which considers how the first signs of the 'inexorable collapse' of one's personal world are small ones, 'distinguished not so much by menace as uncanniness'. Dneister visits The Barber and, seated in the chair, proposes to unfold 'bad news', with an unusual sense of purpose, cool appreciation and even 'ecstasy'. Dneister professes to find men 'more beautiful flung down than standing up' and welcomes the 'gratifying intensity' with which he is heeded, as a Cassandra whose words cleanse away trivia and ephemera. He is thus an unusually self-conscious student of the activity centrally dramatized in *A House of Correction*: the human imagination's striving to face the wound of its own terror before the death of meaning. In a new refinement of 'intimacy', he even bids the Barber sit in the chair so that he can watch how every detail and effect of his news registers on the Barber's face, in scrutiny of revelation and his own apotheosis. Dneister introduces the topic of the Barber's son ('no longer in his infancy a subject for apprehension therefore rather than pleasure'), and his foreboding rhetorical flourishes drive the Barber to determine 'I'll tell *you* ... What I don't know ... My boy is dead'. Indeed, the Barber grapples with this possibility in particularly dire terms, in an active exercise of imaginative will ('The discipline of bad news says there is a much worse fate than death'), and mindfulness of the particular poignance of the soldier:

> Beautiful is the army so my mother said because of our tunics because of our polished weapons I asked her those are beautiful certainly she said but clowns or actors might put on uniforms the beauty is the fact so many of you will be dead it is death that makes the soldier beautiful

Dneister marvels at the Barber's words, 'only grief could lend you such authority', but maintains an analytical posture, even as he observes 'You have made such a good start especially in regard to placing private melancholy in the wider context of what we both know or rather I *know* and you *suspect* to be a national catastrophe', namely impending invasion by the enemy. He further accedes, 'The barber has already far surpassed me' in his telling and instinct for the outcome; 'I am by comparison a dilettante'. The Barber rages at Dneister for his cold-blooded lack of feelings; however, Dneister has in his own way been transformed by the encounter: 'oh dear I have broken the first rule never to be animated and the second always catch the first boat out'. Moreover, the Barber's resolve to stand and meet the enemy stoically defeats

Dneister's appetite for inconsolable weeping; Dneister protests 'The victim dignifies only so long as he remains a victim … it was a privilege to witness you and now it is humiliating'. The Barber even manages to be gracious in acceptance of their different capacities and destinies. He outperforms Dneister and achieves tragic status in his servitude, raising himself above that of the would-be godlike connoisseur of catastrophe. In a way, Dneister is an extreme parody of the human fascination for pain which Barker has argued is integral to tragedy[15] and Theatre of Catastrophe. However, Dneister's would-be divine perspective renders him inhuman; like Shakespeare's Iago, he represents intelligence devoid of emotional sympathy, and so his "play" with others is manipulative rather than creative: his insistence on the *comprehensibility* of pain, principally to himself, does indeed diminish tragedy to 'bad news'. Those like the Barber who live through and endure catastrophe may therefore be more alive, if not always enviably, than those who observe it from a distance and speculate in it like a currency; because their will can subvert imposed terms and challenge notions of human powerlessness.

Dead Hands (directed by Barker for The Wrestling School in 2004) develops from *A House of Correction* and *The Dying of Today* in its explorations and interrogations of the balance between 'spontaneity and utter calculation'. When Eff arrives too late to attend the death of his father (who initiated and trained his capacity for 'incorrigible speculation'), he confronts the (dead) body of his father but also the (quick) body of his father's mistress, Sopron, her sudden desirability to him, and the prospect of her further intimacy with his brother Istvan. Barker's programme notes for The Wrestling School's production (in October 2004) crystallize themes and resonances:

> The erotic energy that surrounds the body of the widow…its sudden availability…might we say here as in so many instances that nothing in Death is what it seems? What arrives in the form of *loss* might turn into *longing*… of the many paradoxes of Death, the most shocking is the eruption of a reckless *joie de vivre* created by the spectacle of the cadavre…an exhortation to *live while you can*…
> The funeral…its lather of chaotic and contradictory emotion…only the wedding is comparable…is there not Eros in the bereaved? Is there not Death in the bride?
> If we can bequeath our material property might we not bequeath our emotional property also?
> In sex as in politics, the sense of *having no choice* might be experienced as both catastrophe and relief…
> The heir beholds the cadaver…through loss, a growing sense of freedom…or has the character of the dead man merely migrated into the heir?

As Helen Iball observes, Barker's play 'literalises and sexualises' the platitude that the characters are 'united in their grief'; and Iball goes on to argue that

'*Dead Hands* is a chamber piece that relies for its concentrated impact upon ... the disorientation of social realist domesticity'[16]. I would add, *Dead Hands* develops from *The Twelfth Battle of Isonzo*, in its stagecraft as ballet of power: *Isonzo* can be analyzed as *pas de deux* for two performers and a single (finally upturned) chair (in which even prostration can yield radical reversals, such as the "dead" Isonzo's continued speech); *Dead Hands* initially appears to be a *pas de trois* for performers and chair, but also problematizes and extends this notional triangularity: Eff imagines Istvan with Sopron before his father, 'The two of you / A dying man / A single chair' (*DH*, 30), in a relegation of one party to that of voyeur, an arrangement which is arguably subsequently replicated between the two brothers in relation to Sopron, and each other. But just as the 'prone' Isonzo refuses to be relegated to the role of mute stage prop in the closing stages of the earlier play (for all that, in Barker's 2001/2 production, he was ultimately denied the mobility to prevent Tenna's escape from his room, when she had assumed 'the erotic energy that surrounds the body of the widow'), in *Dead Hands* the dead father continually threatens to make the play a *pas de quatre*. The title refers not only to Sopron's startling, prematurely aged hands, whose repulsiveness is 'tempered and to some extent abolished' by the power of her general (if paradoxical) beauty, but which remain memorable as the 'girl-sized talons of a pharoah's queen' (*DH*, 16)[17]. It also suggests the father's influence, or even control, of the interactions of his mistress and his sons from beyond the grave. In this respect *Dead Hands* reverses a central premise of Shakespeare's *The Merchant of Venice*, dramatizing a scenario which is ambiguous in its promises of both liberation and constraint, but here from the male perspective: Eff suspects himself the compliant dupe in a proposition which is, at root, 'the enactment of some sordid and corrupt conspiracy between a dead man and his mistress an erotic contract which would have reduced me to the status of an instrument' (*DH*, 14), like Portia's suitors. Sopron enacts, and, contrary to her characteristic wilfulness (expressed by her cruel aspect, what she calls her 'stone' nature and the seductive claim 'I prefer the darker ways of love'), perhaps artfully refines (like Portia) the challenge of the performance scenario of servitude imposed upon her by an older man, dominant beyond death:

> All my gestures / Every one / Hang in the air / Like smoke ... be smoke always he said to me by which he meant take on the shape of others surely be formed by them concede be moved as smoke is by the air (*DH*, 12)

Eff is transfixed by the imaginative prospect of intimacy with Sopron, in terms which are repeated by all three characters (with only minor variations) like a musical motif: the desire to take Sopron's 'whole cunt in my mouth the flesh the fluid and the hair': this motif thus provides a poeticization of desire in terms which the characters successively use and unconsciously share (or somehow transmit?), an example of how Barker's drama and language, like Shakespeare's, 'aim for a distillation of life, not an imitation of it', in the words of Ian McDiarmid[18]. Eff has imagined Sopron 'Marching in here naked but for shoes' and thus found himself 'Passively anticipating' an action of 'exquisite eloquence' which 'she almost certainly has never contemplated let alone possesses the courage to perform' (*DH*, 10); but in fact, she promptly and repeatedly performs this very action, and offers herself using his very words, contributing to the dream-like or hallucinatory rapturous atmosphere. Eff strives to maintain a dispassionately analytic philosophical distance, but concludes that an ominous formality is at work, around and upon him: 'It is as if in selecting Sopron to be his mistress my father knew the agonies I should experience in my encounter with his corpse' (*DH*, 33).

There is a strong sense that the characters in *Dead Hands* are, quite as much as the protagonists in the plays considered in my last chapter, 'Wrestling with God' – or at the very least with a deceased, but unnervingly knowing, imitator of Him[19]. Istvan is the less sensitive or speculative brother (Eff assesses his limitations: 'I sense you know nothing of ecstasy but correspondingly nothing of melancholy either', *DH*, 68), who remained more physically proximate to his father but crucially effected a final recoil, as he confesses: 'I / Fled / My / Father … and hid myself until he died' (*DH*, 18). Eff tries to assuage his self-recrimination, proposing: 'whilst your action was strictly speaking a betrayal of a doomed and frightened man in actual fact you rendered him a critical and final service': 'If the world were not revealed for all its affectations as a fatuous and sordid place how could we ever bring ourselves to leave it' (*DH*, 22–3). Istvan pursues the resonances of this sophistry, of "altruistic egoism", to a point of its evident fragility, which Eff prefers not to acknowledge, in terms which challenge the powers of response of both his brother and the audience:

```
ISTVAN:  ...My cruelty was kindness
EFF:     Certainly
ISTVAN:  My selfishness was generosity
EFF:     In effect
ISTVAN:  Yes
         Yes
         Of course one must be cautious not to extend this principle to
         any act of
EFF:     Why not?
         (Pause.)
ISTVAN:  Because
         (Pause.)
         I don't know                                    (DH, 24)
```

Eff proclaims Istvan's diminished responsibility, 'in your inveterate egoism you were merely the agent of another', even as Sopron seems to manifest the force or agency around and between them when she walks in, *'exquisitely dressed in mourning'* and *'takes the chair'*, ordering Istvan into (consciously imperfect and, in performance, consequently somewhat comic) postures and *habillements* of mourning[20], expressing impatience with his childlike dependency then smothering him in kisses before she exits as mysteriously as she entered. Eff, philosophically Protean to the point of being manifestly (and desperately) self-contradicting, seeks to re-assure Istvan, 'Shame is apparently essential to our sense of order and pro-priety' (*DH*, 22), but finds his own powers of reserve and resistance strained by Sopron's ritualistic repetition of her formal gesture of mourning, which is infused by the provocative eroticism of her nakedness (and, like the verbal trope of the 'cunt' speech, repeated, here as the physical equivalent of a musical motif within a consistently orchestrated, but inscrutably unpredictable, progression). Eff, in his injured pride, wishes to avoid 'the craven disposition of a servitude' which he ascribes to his brother but senses himself overwhelmed, prepared to marry Sopron if she appears naked a third time, and potentially exposed to desire, in its full declaration of war on all aspects of the self, as it was previously known ('I am flinging my entire life at her feet to tread to desecrate let her dead hands claw ribbons out of my face ... I want to stand in the ruins of my face', *DH*, 39). When Sopron sits *'smartly'* on the chair (with which Eff earlier collided and overturned) and *'composes herself'*, Eff tells her 'I love you', in the know-ledge that 'if the love destroys me I don't care'. But is Eff being destroyed, or fashioned? He glimpses anew the possibility that his late father may be orchestrating his own replacement in Sopron's attentions. This could be as a conscious and progressive supplanting, which would confer a purpose and virility on Eff through the assumption (and mythically, the sacrificial ingestion) of his power, a prospect Eff entertains as the exchange of one form of authority for another: 'Father funny word I think it must mean sacrifice but in what way a sacrifice? ... Surely a willing one an ecstasy if sacrifice can be yes to be murdered by the son is ecstasy' (*DH*, 44). But, at worst, it could be reducing Eff to a slavish and deficient mimicry of the 'servitude endured' by his father, and haunted by the nagging intimation that 'no woman speaks the truth when interrogated on the subject of her former lovers'. Eff's desperately repeated statement 'I am my father's son' permits both interpretations (not least by himself). Whilst Sopron repre-sents 'the apotheosis of the inauthentic' ('she knows the secret of desire and how to keep desire from expiring in the blankets of a fond and tepid domesticity', *DH*, 62), Eff is dogged by the sense that his capitulation to her will also be to a larger scheme, the service of which would diminish his originality, even his existential authenticity, rendering him essentially face-less and replaceable.

When Eff does indeed consummate his desire with Sopron, it is with a sense of pleasurable formal completion, but Sopron immediately proclaims she has 'a train to catch': 'My sense of order and propriety which both amuses and enrages you forbids me to attend the funeral' (*DH*, 51). Eff denies the existence of the station, Sopron maintains her possession of a ticket. This collapse of logic in the face of duel convictions[21] suggests to Eff what is at stake: 'The ticket represents something injurious to me your free will for example' (*DH*, 52-3) – in terms which threaten his own. Eff fluctuates in voluble self-contradiction, praising Sopron's rightness in leaving then begging her to stay; Sopron upbraids him, on the terms of his performance of masculinity: 'The dignity of man consists in his not yielding to his feelings however natural those feelings are that is the paradox of dignity' (*DH*, 55-6); she characteristically concludes, 'Let us from this moment on abstain from all those manifestations of emotional excitement that are commonly described as natural' (*DH*, 56).

Eff swings between sympathy for his father, embroiled, yet self-sacrificing, in a fatal relationship with a woman whose power even now threatens to engulf his son similarly, and rage at the 'Monster' who could knowingly expose his 'little boy to such a vile vile journey', or even design its route. He also realizes that, under current terms, he will not pass on his own costly knowledge to a son because Sopron is beyond child-bearing age; and also that he cannot share her. Istvan, who is 'not ambitious', proclaims the abandonment of his own hitherto 'relatively innocent' intimacy with Sopron. But the duration and quality of silence which accompanies Istvan's unseen and offstage farewell to Sopron drives Eff to suspect 'He has her whole cunt in his mouth'. Eff realizes further: 'Inheriting the property of one's father one inherits both the pride and agony of it'. Sopron entertains his suspicions: 'I never lie / He begged me to / And so will you / I never lie however' (*DH*, 71). Eff occupies the central chair, reflecting on the fateful hinge of his own decision to undertake his journey and accept its consequences. Importantly (again like *Isonzo*), Barker's own production added significant directions not in the published text: the seated Eff placed his hand on Sopron's face but turned this into a decisive gesture of repudiation, reflecting 'the oscillations that accompany great sexual encounters – the desire to escape the enslavement of them, followed by the discovery that one cannot breathe without the presence of the loved one'[22]. Here, Eff seemed to have determined (if only fleetingly) to reject the sexual 'legacy' of his father: but what Eff intended as an uncompromising independence and defiant authenticity could also be viewed simultaneously as a cruel subjugation of Sopron. This dramatic ambiguity is more disturbing and more resonant than the apparent resignation suggested in the text. It also returns, in a new light, to Barker's question: 'through loss, a growing sense of freedom...or has the character of the dead man merely migrated into the heir?'. As within the play, the questions extend infinitely, *beyond* the play.

For a mid-sized 'chamber work', *Dead Hands* is, like the formally and thematically related *Isonzo*, constantly surprising in its intricacies, depths and resounding resonances. Both works consider sexual relationships as both the opening and the closure of possibilities, where a drive towards breakthrough into a vivifying self-inventive originality is pitched against the claims of personal and social determinism, and the lure of duration through self-accommodation which will inevitably partake of self-subjugation. The characters are compelling in the 'agony and pride' which their actions and demands require. Appropriately, Barker's production located the events in a delimited black space (no walls or decor were visible) in which the characters were nevertheless surrounded by invisibly suspended mirrors. Correspondingly, the corpse of Eff's father (a dummy, rather than the coffin suggested in the text) lay "in state" on a glass panel off-centre, giving a similar and appropriate sense of a presence in hovering suspension. Justin Avoth's notable confidence and authority with the linguistic and sexual demands of Barker's work in the previous two productions made him a deserving and powerful choice for the role of Eff; Avoth scaled and navigated the mountainous page-long speeches with a precision and wit, in a *tour de force* characterisation and performance of a man constantly overwhelmed, either by others or by his own calculations. Chris Moran demonstrated a delicately comic touch in the low-status role of Istvan; and Biddy Wells caught the striking balance of stylish poise, erotic allure and faintly decadent menace required for Sopron.

Christ's Dog (written 2004, staged 2005, Vienna, Austria[23]) presents Barker's (re-)imaginings of the last days of the eighteenth-century Italian seducer and dandy Giacomo Casanova[24], renamed Lazar to avoid some of the more crass cultural associations. Lazar, aged 65, presides over his two wives, Buda and Sisi, with a supremely introspective, and paradoxically disarming, narcissistic egoism. As they sit in a ploughed field, on which snow falls, the women become the subjects of the envious wrath of a peasant farmer, Bubo, who, through raping Sisi, becomes infatuated with her. However, Lazar displaces the aggressively servile Bubo from the bed on which he has spent his entire wealth. Having thought to have thieved Sisi, Bubo finds himself additionally dogged by Lazar, who occupies the centre of his dwelling with disquisitions on his own forthcoming death, and Buda, the other component of Lazar's entourage. Thus, even whilst Lazar proclaims how 'death consists precisely in the abolition' of the resources one has 'so assiduously accumulated in his life', he continues to dominate those around him, particularly the wives and mistresses whom he draws into servitude. Buda resentfully claims that Lazar 'swallowed' her life, admitting 'the more preposterous [his] ideas the more they appealed to me'. Lazar is another of Barker's compulsive irritants (a servant observes 'you seem to make the world disclose itself more brutally than some would wish'). However, his charisma still draws his former mistress, Canal, who journeys to his bedside when Buda tells her Lazar is dying. The assembly of the jealous and competitive mistresses, resistant yet susceptible to their common focus, recalls Lvov's summoning of the disciples in *The Last Supper*; and Lazar's requirement of a supporting cast for his self-quoting, self-mythologizing dramatization of his own (surprisingly long and shockingly unburdened) crawl to death additionally provides, of course, a mock-heroic parallel to the passion of Christ (bathetically undercut by Canal, 'Lazar you are not dying are you you are merely talking of yourself in the past tense', notwithstanding her own devotion to seeking out a lipstick of the proper 'moral and historical significance' with which to please and taunt him).

Lazar's son, Tax, introduces the apocryphal anecdote of Christ's dog, an animal who literally re-presents a paradoxical, surprising, and possibly unwitting, servitude: pure-bred but embarrassing, highly strung and ludicrous, mischievous yet disarmingly loyal, 'it laid at the feet of Him who was pure in heart the evidence of its own impurity'; 'the dog was sent to try Him as all life tried Him' as they trod the same road to Golgotha. Thus, Lazar, in his profoundly self-conscious egoism, provides the mocking but compulsively persistent deject which borders conventional Christian behaviour and values; he may even be required, or called forth, by these values in order for them to establish their perameters of distinction from the impurity he represents ('Never Christ without a dog never virtue without mockery').

When he learns that Bubo has strangled Sisi, Lazar is briefly and uncharacteristically animated with anger, but this degenerates into a drunken mutual tolerance; Lazar senses that his wife's murderer needs him, Bubo's 'dog's eyes' expressing a craven loyalty that compulsively courts punishment: like that of Lazar's mistresses, and like Lazar's for life itself. Rather than Lazar, it is the younger characters who die first: after Sisi, Tax. The surgeon who self-consciously failed to save Lazar's son takes up and concludes the story of Christ's dog: 'he beat the dog this had no effect still the dog loved him in the end he killed the dog ... because the dog was love'.

Bubo tries to convert his servility into servitude, and is challenged by his adopted 'master' Lazar to help him find death and the relief of 'silence' from the tinnitus which rages increasingly in his head. Lazar tries to provoke Bubo's hatred to a point where Bubo will hang him, but the attempt fails when they topple in the snow, and then Bubo collapses in death[25]. Unaware of this, Lazar tells him of his sudden realization that Death is, contrary to 'all these images of old men with scythes and hourglasses', a woman: to be precise, 'AN UNSEDUCIBLE WHORE' who provokes yet mockingly determines time and duration, unbiddable to any rhythm but her own.

The play shifts gear for its bravura final scene, in which Lazar is delivered to a grand dinner party, as if himself a dish; he proceeds to regale the guests with an obscene song. *'They are neither shocked nor amused'*, but he persists, through derision and jeering: *'Just as LAZAR resembles a parrot in his manner, so they adopt the attitude of zoo-visitors'*, then he becomes *'irregu-*

lar in his recital and ignored'. In the failing light of the party's decay, one guest recalls the story of Christ's dog, 'an irritation certainly but often what is irritable is necessary', and insists that, whilst the 'perfection' of Christ's teaching 'required the contrast of the imperfect to be understood just as an alabaster statue shows better on a dark ground', Christ 'WAS / NOT / MOCKED'. This spurs Lazar into a defiant reprise. A female guest exhibits herself to Lazar, deliberately depersonalized by the invisibility of her face, and identifies a further facet of Lazar's unforgivable provocation, his scrutiny which makes women feel 'more than naked ... stripped to the bone' and refuses to 'save them from themselves': 'NO WOMAN LIKES MEN WHO LIKE WOMEN SO MUCH'. Ironically, Lazar is now deafened by tinnitus and cannot hear; he proclaims himself 'THE TRUE SERVANT OF THE FAITH' who despairs of the surrounding slavish diminishment of sexuality, which constitutes the passion in which he discovers nobility and religion: 'And cunt was sometimes pleased and sometimes turned black looks on him for that is the way of God that is the sign of Him'. The sounds of a cacophany locate the audience sonically inside Lazar's head, up to the point of an abrupt cessation, which signals his death[26]. Buda and Canal attend his deliverance from this final rapture, which may have been imaginary, but is undiminished in its vividness and instinctive insights, characteristically 'wreathed in contradiction'.

Buda nevertheless testifies to her status as 'happily married'. Thus she, like many other characters considered in this chapter, shows how, in the varying ways of 'sex as in politics, the sense of *having no choice* might be experienced as both catastrophe and relief...'[27].

10
The Boundary and Beyond

The Fence in its Thousandth Year

Many of Barker's characters situate themselves (self-consciously) at limits (physically, imaginary and emotionally). This experience, of being at a limit, is painful but also opens up possibilities: it indicates that in fact there must be something 'beyond' the limit (otherwise it would not be a limit), and that the limit is not a final authoritative end[1]. Barker's work frequently addresses, depicts and exemplifies the phenomenon of *singularity* within sociality: that which occurs *at the limit*; that which challenges and redesignates conventional terms of time and place by occurring specifically and exceptionally, eluding appropriation. Jean-Luc Nancy suggests singularity is 'bound up with a sudden appearance' that also implies a disappearance[2] (as does theatrical performance, we might add): singularity implies its own limit but presents that limit as self-consciously *'its own'*[3], an intrinsic quality, and feature of identity, rather than something imposed externally. Nancy claims 'The border is an extremity where structure comes to a stop'; and that, whilst being nothing tangible, a limit proposes structure in order to respond to, and denote, a sense of 'excess'[4]. Thus, the border is 'the part or dimension that exposes or has exposed the singular', configuring something previously and otherwise indefinable (an Other or a Nothing); in exposing and being exposed, the configuration of a border achieves some degree of containment; yet, on the border, 'the limit [both] loses and recovers itself in the same movement'[5]; it responds to (and delineates) the force of an impulse (or gesture) of self-overcoming. This impulse gathers a sense of the excessive, and configures a (potentially catastrophic) singularity which would uproot itself from standard continuous terms of time and sense.

The frequency (in both senses, of recurrence and medium) of this singularity is the self-conscious territory of Barker's play *The Fence in its Thousandth Year* (written 2003, staged in Barker's production for The Wrestling School in 2005), in which challenges to "go beyond" are the theme and subject of the play, as well as the proposition of its theatrical actions and effects. The Wrestling School's publicity and funding applications alluded to physical frontiers such as The Great Wall of China, The Berlin Wall and, most recently, the manifestation of a fence in Gaza to separate Palestinian and Jewish communities. Thus the resonances of the play could be cited to justify its production in terms of the dominant British neo-utilitarian artistic criterion of topicality, whereas *The Fence*, like *The Castle*, is profoundly metaphorical in its entwined explorations of public and personal identity in ways that are not restricted or limited to the specifics of time and place. It is also a supremely challenging meditation on forms of freedom, in terms identified in Barker's programme notes:

> We like to imagine a world without frontiers. This flatters our humanity. It is a vanity of the Western mind. But the barrier merely moves. We dismantle one to erect another. We identify another alien *inside* the fence. It is possible we are human *because* of the frontier. Does it not define us?[6]

From the outset, *The Fence* is a good example of how many of Barker's plays manage to be simultaneously narratively original, tightly and suspensefully plotted, and provocatively enigmatic. The opening scene is wordless: Algeria, an elegant duchess, promenades a frontier fence, then presents herself to a rush of dim figures on the other side, with whom she copulates. These are definite and decisive actions that nevertheless have no ready or obvious meaning, posing a conundrum that is not immediately explained and cannot be fathomed. Then the dramatic focus moves to a cemetery, and a funeral service that positions the secret events round the fence in the context of a slowly emergent Oedipal mystery, a personally and nationally crucial Sphinx-riddle that hinges on the transgressive interpenetrations of eroticism, coercion and death. The precociously intelligent, blind and supersensitive youth Photo delights in his intimacy with Algeria, his bereaved aunt. Her enticing widowhood drives a suitor, Doorway, to engage in rivalry with Photo, whom Doorway dismissively consigns to 'suicide' by proposing the crucially greater strength of his own will in this highly personal, yet decisively social, context.

Algeria is (like Gertrude) an aristocratic woman of demanding, volatile and potentially fatal eroticism and singular passion, literally and metaphorically living at the limit: she acknowledges 'I am a war' (in which men 'fall'), along with her need to challenge and test her power ('And yet I must be married without marriage I', *FTY*, 9). She suddenly accepts Doorway, consciously calling his bluff in the knowledge that the 'poetry' of his courtship will give way to dread. Significantly, Algeria's recurrent action involves exposing herself and provoking exposures, to 'lose and recover' herself in the same moment through a distinctive and persistent self-overcoming (as when she thrusts a handkerchief into her own mouth to stifle her gasps). She discloses to Doorway that her incest with her ostensible "nephew" is intensified by his actually being her son; thus, Photo speaks more prophetically than he knows in proclaiming his own devotion to the 'excess' in which 'terrible truths appear' (*FTY*, 12); appropriately, he also identifies himself with the Fence, dining at the paradoxical perimeter in the knowledge that 'the fence is not beautiful nevertheless it is the condition of beauty' (*FTY*, 17) – not least in the way it endows the surrounding landscape with 'that quality of intangibility that is the critical element of the picturesque' (*FTY*, 16) and provokes its own violation and reconfiguration by successive generations. The Fence is the site of ritual performative provocations, by which the citizens of Algeria's country designate themselves 'farmers' in opposition to their neighbours, 'thieves'; a ritual paralleled, but also partly subverted, by Algeria's nocturnal assignations involving repeated semi-anonymous copulations through the wire, which (re)charges the ostensible binary opposition with an eroticism which makes it a site of problematizing singularity. This informs the observation of Algeria's servant Kidney: 'Long is the fence ... HOW DEEP IS IT THOUGH' (*FTY*, 29).

The Fence thus develops more substantially from *The Gaoler's Ache* in its heretically sympathetic depiction of an incestuous relationship, which renders the female instigator both reviled as taboo and seductive in her promise of a narcissistic vertigo, which Baudrillard conjectures may be associated (mesmerically) with death:

> The great stories of seduction ... are stories of incest, and always end in death. What are we to conclude, if not that death itself awaits us in the age-old temptation of incest, including *in the incestuous relation we maintain with our own image*? We are seduced by the latter because it consoles us with the imminent death of our sacrilegious existence. Our mortal self-absorption with our image consoles us for the irreversibility of our having been born and having to reproduce. It is by this sensual, incestuous transaction with our image, our double, and our death, that we gain our power of seduction.[7]

Doorway finds that marriage to Algeria brings him into confrontation with his own limits, where emotions threaten to turn into their opposites (adoration into violence and *vice versa*), even as he refuses to be 'killed', or driven to choose death, by her wilfulness (which outstrips his own initially confident performance). A parallel experience occurs for Algeria's friend and confidant Istoria, who is also made to confront the conflicting emotions at her own limits when a man throws a baby to her over the Fence. Kidney urges her to throw it back, 'overcome your feelings what else is bravery' (*FTY*, 27), but, here as elsewhere, possession proves powerful and paradoxical, and the state adopts the child, who is lodged in a 'baby park' containing many such foreign orphans. Photo senses a disturbing kinship with this child, who is diagnosed as also blind; he names her Camera. Doorway threatens Photo with disclosure of his true nature, and that of his relationship with Algeria; Photo summons the bravado to propose that Doorway is 'INSTRUMENTAL' to their own preferred terms of death. He has broached this topic earlier with Algeria, in a speech in which he envisions the pre-emptive design of their own deaths (in a manner crucially redolent of that of Antony and Cleopatra rather than that of Clara Petacci and Benito Mussolini):

> how shall we die it would be so pitiful and abject if we were taken unawares and strung up from a lamp-post like slaughtered dictators heads down your hair in a drain and our underwear on show all stained all filthy with our own filth no let us eschew the conventional in this as in every other instance and create our own deaths in dying beautifully one lives forever (*FTY*, 21–2)

An ominous blind thief, Youterus, dogs Algeria and confronts Photo at the zoo (which Youterus provocatively characterizes as a place of necessarily artificial civility, 'we should all be in a zoo I sometimes think then we would not savage one another … some say zoos are not natural does everything have to be natural I am wearing trousers are trousers natural I despise that argument', *FTY*, 39). Youterus intimates, in recollection of his own fateful union with Algeria, that 'the boy' Photo is his congenitally blind offspring, further entangling the lines and claims of definition and identity (driving Kidney to protest, 'But you're a Thief…'): a note of particular enigmatic suspense at which the play breaks for interval, and (as often in Barker's drama) thus assists the effect of some passage of time in the narrative.

Once broached, the Fence has fallen into disrepair, and is '*rotted and thick with clinging litter*'. The women's ritual has transformed into a formalized emulation of Algeria: they fling up their skirts, then laugh and continue along or through the fence. Whilst in one sense this reflects Algeria's status as a 'legend', and licenses an innate disloyalty in the younger women, her action is diminished through generalisation, into mere sexual mutiny in defiance of elders or even the taunting but hollow postures of a new vogue in adolescent conformity, which Photo terms the spiritual condition of this 'miserably fenceless' and 'unsecret place' (*FTY*, 48). Nevertheless, the 'legend' of Algeria, when related, permits him to deduce a less anodyne scenario which led to his own birth, and Algeria's status as his mother. His initial homicidal rage turns to an appreciation of the aesthetic symmetry and irony of his habitual dining by the Fence ('My starched table cloth a mocking of the marriage bed') and transforms into a paradoxical passion for his compulsively sexual mother: 'it is impossible to like you and impossible to dislike you … I ADORE YOUR TERRIBLE LYING I

ADORE YOUR TERRIBLE TRUTH' (*FTY*, 50). He commands her to 'UNDRESS' even as she describes the agony of his birth, laughing wildly in their joint transgression.

However, Algeria also persists in her sexual relationship with Doorway; when (in another echo of Gertrude, with Hamlet) she claims 'I don't know why', Photo accepts and appreciates this compulsiveness: 'half your beauty is your don't know why' (as he observes later, 'she only acts consequences are for others'); he also tells Doorway he has 'become the subject of her mischief', or worse: 'I think in her case betrayal is a perverse faith which like all faith begins in ecstasy but ends in obligation' (*FTY*, 53). But Photo is wounded by Algeria's (re?)discovery of desirability in Doorway, her shift in prime status from his 'mother' to Doorway's 'wife'. The sense of degeneration is national as well as personal: Kidney finds Algeria's dogs have fled (a rupturing of a bond and sign of entropy, as when the dogs are dispatched at the end of *Found in the Ground*). The young woman Lou (like Ragusa in *Gertrude – The Cry*) becomes the representative of a vengeful moral orthodoxy, interpreting the wishes of her unborn baby to proclaim the new dictates and prescriptive "knowledge" which springs from miscegenation and outlaws women like Algeria (compare the similarly totalitarian regime of 'Mum's Troy' in Barker's *The Bite of the Night*, in which babies are ascribed the authority of adults in a state of 'moral excitement', to be interpreted solely by the self-appointed governors); though, as Kidney points out, this is a "knowledge" unworthy of the name – indeed, it legislates against the painful knowledge of singularity, through the attempted imposition of mere authoritarian ideology.

A new '*pristine*' fence descends, that of an institution which imprisons Algeria, enabling her captors to identify themselves as sane and/or human in opposition to her. Photo has been infantilised, his voice '*curious*' within an '*ominous*' pram to which he is consigned[8]. With the insight that 'MADNESS ALSO IS A FENCE' (*FTY*, 59), Algeria couples once more with Youterus, through the wire of the asylum's courtyard, and, after an initial apprehensive agitation, Photo expresses the perspective of her newly conceived child. Ignorant of this new conception, Istoria expresses a sense of despair that Algeria and Photo have missed the chance to choose their deaths and are now doomed 'to walk a road of infinite humiliation', a *via dolorosa* as exhibits of their triumphalist enemies. Doorway visits Algeria, but he is in fatal decline; his characteristic inability to act, the weakness of a Chekhovian romantic, is now progressed into a terminal withering of the will. Webster's eponymous tragic heroine in *The Duchess of Malfi* was tormented by enclosure with madmen to break her spirit; in a similar but intensified scene, Barker's Algeria is paraded before doctors and warders as the lunatic, yet she summons up a fierce expression of existential defiance of her position in 'the museum of your lies', discomforting her initially patronizing audience by insisting 'I am a duchess' and discovering a self-expression whose vehemence is intensified by its partial (but problematized) containment ('I find with poetry wire helps'). In the antiseptic institution, Algeria gives birth to her (again, congenitally blind) child, Photo shaking a rattle from his pram in encouragement through her pain, and Kidney holds fast to his promise not to desert Algeria, by appearing and throttling Youterus. The blind girl Camera appears, now grown, to utter her contempt for Youterus and his feckless selfishness in throwing her over the Fence; Camera relieves Algeria of the new baby, whom she accepts and names Film. Disingenuously proclaiming her baby "stolen", Algeria eludes the perfunctory attentions of hospital cleaners, and, *'with a profound effort of will'*, she pushes away the pram, in which Photo remains *'incorrigible'*.

Istoria finds them on a hillside by the frontier, Algeria at a point of terminal exhaustion, Photo unwilling or incapable to be roused from his regressive posture, even as he utters the gnomic (gnostic?) words:

> In the first instance desire represents itself as freedom … The freedom of the vortex … Freedom *from* therefore not freedom *to* … Freedom from will freedom from opinion freedom from choice paradoxically … Inevitably this paradox degenerates desire turns inexorably into responsibility the sense of debt the paralysing horror of reciprocity and here is the conundrum surely … Is this reciprocity a synonym for the extinction of desire? I don't know (*FTY*, 70)

As Istoria laments her friend Algeria, the sound of surveyors tapping pegs finally entices Photo from the pram: *'he staggers on unused limbs'* as they make plans to renew the Fence. Istoria seizes the opportunity to release the brake on the vacated pram and send it rolling away. Algeria is roused, and disconcerted, but she and Photo join in a verbal elaboration of the sense of heretical relief and cruel release from this vehicle of infantilizing regression and paralysing responsibility[9]. Istoria releases Algeria's dress, presenting her as monumental, proud and statuesque in her nakedness, as the sound of tapping continues in evidence of the reconfiguration of the Fence.

The Fence is one of Barker's most strange and haunting plays, notwithstanding, or perhaps because of, its vivid scenes and strong narrative drive and detail, through which irreducibly complex events occur with relentless urgency. The final image exemplifies this: it is possible to interpret Istoria's release of her friends from their humiliations and encumbrances (symbolized by Photo's pram and Algeria's gown) as a release into the death which Istoria wished for them, in an aesthetic triumph over a demeaning world, which parallels Charmian's final adjustments of Cleopatra's ecstatic escape from diminishment and into death in Shakespeare's erotic tragedy *Antony and Cleopatra*. On the other hand, Istoria's actions may release her friends – in ways which are beyond them, unaided – from worldly attachments and impositions, in a way which is both singular and physically immediate:

'the duchess undressed by her loyal friend for the eyes of her boy lover –
eyes which cannot ever see her beauty'[10]. Here, Photo's blindness becomes
another 'fence' to challenge, and even inspire new intimacies, and a
poignant reminder to the theatre audience that they, however, can bear
full witness to this public manifestation of a (nevertheless) irreducible
privacy. The names of the characters hint at some unfathomable (even
fearful) symmetry, congruence, ironic function (Doorway) or even anti-
pathy (as when Kidney strangles Youterus) – and unusually, the characters
themselves comment overtly on the wry relationships between the chosen
names Photo, Camera and Film – as if the events of the play delineate or
allude to some process, for which the key to finite interpretation lies
provocatively just out of reach. This parallels the activity of the characters
(particularly Photo) who approach and draw back from fathoming the full
ramifications of a plot that combines Dickensian intricacy (redolent of the
various, resolutely elliptical yet interwoven seams of *Bleak House* or *Our
Mutual Friend*) with Oedipal transgression. The characters in *The Fence* are
also particularly *strangely engaging*: Lou degenerates into an authoritarian
ideologue, but the other forces and agents of repression are significantly
anonymized; other characters are, variously, persistently and surprisingly
insightful, wryly witty and emotionally expressive, and therefore theatrically attractive, even as their actions and discoveries challenge the most fundamental terms of given morality. The freedom demonstrated by these
actions and discoveries is compulsive: in Photo's words, 'freedom from
choice paradoxically', which makes the characters hard to censure. The passionate and painful pursuit and embrace of the forbidden, on which the
protagonists stake their lives, manifests a charge beyond the dutiful impositions of moral pedagogy: another example of how the ostensible "limit"
invites (even demands) its own rupture; in words from Barker's programme
notes, 'To *find out* is to admit to consciousness what was known but
fenced...'[11]; and *The Fence* involves the audience in the compulsion to *find
out*, even as it remains constantly surprising.

Barker's 2005 premiere production of *The Fence* was a *bravura* example of
The Wrestling School at its most bold and expansive, with the core of
eight actors augmented by a chorus of local amateurs and students, drawn
from each city to which the production toured, and drilled in advance by
an assistant director. Barker/Leipzig's setting made unconventionally full
use of height: in the exordium, for the slow ascent of a baby from the
pram of its negligent governess, whilst behind the steel Fence lowered in
gradual but inexorable counterpoint; for the "flying in" of railings and
tombstones for the cemetery (simultaneously descending in various
uneven shapes), of a chandelier which cast huge, predatory shadows, of
the zoo enclosures behind which actors huddled beneath drapes, promptly
assuming suggestive yet unsettlingly indefinite animal forms; in the slow
descent of the child Film towards his mother, striving in agonies of labour,

which Photo "conducted" with his rattle, the baby arriving at birth as his father was killed. In nocturnal scenes, characters braved and challenged roving searchlights, and the full depth of stages such as that at Birmingham Rep permitted the shadowy but seething emergence of the licensed thieves, in their sexual clamour, rapacious for the solitary elegant duchess, submitting herself to a bizarre passionate ritual, part formal dance, part ordeal. In a production which was Barker's most pictorially energetic, refined and memorable, the performances were also *differently effective*: Wrestling School stalwarts Jane Bertish and Séan O'Callaghan anchored the production, as the faithful, vehement and inventive Istoria and the sly, leering Youterus, a darkly ominous presence and weight which moved from the edge to the heart of the play; elsewhere, Nigel Hastings as Kidney developed the intentness of his low status into lethal loyalty; Alan Cox gave an engaging, sympathetic portrayal of Doorway in a style quite alien to more seasoned Barker actors, but which aptly caught Doorway's semi-comprehending efforts to manage his reactions to Algeria and made them honourable; Philip Cumbus brought a youthful luminosity to Photo in his confidence, pain and disturbing regression. Victoria Wicks, overcoming some initial reluctance to accept the part, provided a central and crucial performance of Algeria which equalled her *tour de force* as Gertrude; her imprisoned defiance of her captors was a particular high point of modern tragic acting in its unforgettable combination of heartbreak, anger and grandeur.

The Conditions of Death: *Heroica, Adorations Chapter 1, Dead, Dead and Very Dead, The Road, The House, The Road, Let Me, A Wounded Knife, Lot and his God*

'It is not enough to be dead one wishes to find death under the right conditions', claims a character in Barker's play *A Wounded Knife*. The search for what these conditions might constitute – and the question of on whose terms (and in whose interests) they are defined – reverberate through several Barker works of this period. The unproduced short filmscript *Heroica* (written 2005) is set '*circa 1920*' and begins with Sixt, a passenger '*in a high-collared uniform*', taking his seat in the rear of a limousine, his face '*drawn, fatigued, even bewildered*' as he reflects (in voice-over) how 'It's over' ('It had to be over and if it is over in this way and not in another still it is over that must surely be a cause for rejoicing much of me rejoices'). Three determined but unspeaking women, in elegant black costumes and hats, advance against the horizon of a high ploughed field to meet the car. They attend where Sixt is accommodated (held?), at '*an isolated, substantial house*', where he shaves and eats beneath the women's' '*unyielding*' gaze, reflecting 'We say of an ordeal thank God that is over but perhaps we are saying ... What will fill the silence?'.

The neighbouring field is filled with a row of differently tilted chairs, seating injured men whom Sixt insists he does not remember, including one youth whom one of the women describes as 'a child when you left'. Sixt's tunic is suspended from a solitary tree in a field and set aflame, whilst he tells himself 'I failed ... A beautiful word failure I must not become attached to it'. As various assertive and anxious labourers arraign Sixt as if in a tribunal, he deduces that failure is not just a single event, but its consequences: 'so one must say of failure it is not a moment like the closing of a door but rather it resembles a wound which never heals a wound which might have seemed innocuous but by never healing reveals itself to be fatal'. Whilst Sixt attempts to deny the maimed youth the right to hate him ('If I concede your right to hate me I will admit to being hateful and I am not hateful'), he realizes that he cannot elude the process ('I am going to die') over which the three women preside with a blend of eroticism (as when one presses her exposed breast to his lips) and formality. Whilst part of him protests that 'I have done nothing nothing but my duty', he tries to accept that, whilst the agitatedly human labourers continue the argumentative elaboration of value, he is moving inexorably to the realm and perspective of the dead, which are distinctively beyond such considerations of, and insistences upon, consequence : 'It does not matter what I did or did not do ... I have to die it is time to cease asking myself the question why to cease all questioning', even to cease 'to require conclusions'. Ultimately, Sixt is brought to a confrontation with an unmoving figure, *'leaning wearily'*, which causes him to exclaim (out loud for the first time) 'Is my father here?'. The women light cigarettes together, and the naked dead body of Sixt is passed by the labourers *'overhand and over shoulders, each reaching out with infinite care'* to support the body and bear it out of the house and over the furrows of the field towards its summit. The three women fling open the windows of the house (*'one on each floor, methodical, final'*) as the destroyed tunic twists on its hangar; they observe the procession, *'fidgeting slightly'*; the final direction is *'They are free. They know nothing'*.

Importantly, the audience never learn the defeat, disloyalty or war crime of which Sixt might be accused; rather they are associated with his perspective on how those around him demand that a process be purposefully and methodically fulfilled, imposing an order which renders his (self-) questionings redundant. The process and landscape, more dreamlike than nightmarish, provide a surrealistic visual "painting" of the process of historical judgment: those who are upheld, those it consigns, and the unpredictability and intransigence of both. The inscrutable formality of the images and events is reminiscent of Kafka and Bergman, and crucially ministered by the women who impel a resolution which remains beyond question, revelation or ultimate comprehension.

Adorations Chapter 1 (written 2005) is a second unproduced filmscript from that year which also addresses the theme of *conceding*, with due formality, to the demands of life and death. Ostin, a painter aged 60, works in a building which *'purport*[s] *to be a church'*, studying the figures in *'a pornographic/fashion magazine'*, *'absorbed but in a manner more academic than prurient'*, for the basis of sketch for composition on a wall. He is haunted by the distinctively audible gait, approaching presence and memories of the 45-year-old woman Gollancz (whose name carries the association of persistent, infuriating and inescapable allure from Barker's *The Ecstatic Bible*), in a climate of critical expectation; he even finds himself drawn into posing, in the act of destroying his draft sketches and drawings, for the documentation of a Young Woman photographer. In her distinctive hat and gloves, Gollancz conducts an assignation with a male Figure (which causes her to be *'pained by a decision she has not yet made'*). Ostin photographs them both with the Young Woman's camera and works their images into his nave painting, conscious meanwhile of the limitations of his achievements. He recalls how he sought to turn away decisively from Gollancz, maintaining 'I haven't the time … The time to suffer the time to make sense of the suffering to turn it into something else'; however, Ostin now seems brought to a personal *impasse*: when visitors expect him respectfully, *'He makes as if to speak, but doesn't'*; when surrounded in his workplace by an applauding crowd, he is *'unable or unwilling to proceed'*: whilst the visitors focus on *'the scale of what has been achieved'*, Ostin observes *'the expanse of unpainted wall, with an expression of grief and anxiety'*. The image of Ostin's painting coat, *'hanging pathetically from a hook'* on his unoccupied platform, crystallizes his apparent relinquishment[12], as he reflects 'It is magnificent to abscond magnificent and whereas to abscond from insignificance might be easy surely only those supremely wretched abscond from their … Apotheosis'. Ostin is ritualistically carried *'like a child'* by the Figure, along *'a wet road, empty, in a rural place'* to *'A field at dawn'*; Gollancz follows *'in impeccable black hat and coat'*. Ostin in placed in the earth, where he lies *'patient as if waiting'* while Gollancz and the Figure make love against a bordering fence (*'Her distant cry is mixed with the rook calling'*[13]). However, even as he impatiently wills himself to die, he hears gunshots: the bodies of Gollancz and the Figure lie by the fence. He collapses by her body in grief, *'as if axed'*; then, *'with a cruel gesture, tosses her hand away from him'*. The Young Woman finds him at the building where he has been working (but is now closed, bolted and abandoned), his head on the door, hands hanging *'like an animal'*. As she photographs him, he considers going away, but decides not to; *'His face … reveals his concession to being an object of veneration. The shutter closes again. His image is frozen'* in an apotheosis of sorts: unenviable, and not of his making.

Thus, *Adorations Chapter One* depicts in Ostin 'the sour satisfactions of abject surrender'[14]; even when his would-be fatal gesture of renunciation is undercut by the catastrophic violation of promised sequence, he discovers a further witness for whom '*To lie exquisitely thwarted*'[15], as a frozen image. The formality of the *danse macabre* which is the process of *Heroica* is paralleled, but disrupted, as Ostin is consigned to further life after Gollancz is gone. It remains arguable and ambivalent as to whether his consent to be frozen as an object of veneration constitutes a triumph or a corruption (recalling the concession of Galactia at the end of *Scenes from an Execution*).

The stately formality of the *danses macabre* in these two filmscripts, and their emphasis on an attempted dignity in conceding and surrender, is contrasted by the feverish activity of *Dead, Dead and Very Dead*, an opera libretto written by Barker in 2005 for music by F. M. Einheit and staged in Copenhagen, Denmark, in 2007. In this black comedy of frantic futility, the catastrophe of plague strikes, problematizing and heightening all issues of social value, reasoning and loss. The industrialist Loud buys a gun and six bullets with which to protect his advantages, pre-eminently his art collection; his wife has died of plague and his son, Houth, is marrying Loud's former mistress, Beza, a (literally) fading singer ('I have lost my voice why else would a singer marry?') who is twice Houth's age. The presiding clergyman resonantly remarks on the sudden demand for his services: 'Faith and Death they rise in parallel like lifts in great hotels', as plague becomes 'the only truth'. The catastrophe throws into relief what formerly has been, and now is, invested in the particularly frail terms of possession (the 'property', both the dying and infectious body of the bride/mistress, and the art object) and beauty (which Loud sentimentally asserts, 'oscillates on the perimeter of its own annihilation'). Loud's daughter Lacantia jeers at the hopes placed in his drugs laboratories and the 'ROTTING BRIDE' Beza: 'we are participating in a grotesque ritual a decadent parody of a social contract rendered meaningless by promiscuity and profit'. Loud himself oscillates between resignation ('we are dead all of us') and nauseous veneration (entreating Beza 'let me kiss your arse'). Indeed, the volatility and vehemence of all characters increases under the pressure of the social and personal crisis: time seems to both slow down (as energy flags and further decisive breakdowns are awaited) and accelerate (as they seek to do or say something particularly personally meaningful in the suddenly brief time available). Beza dismissively crumples Loud's most valued sixteenth-century drawing, to his horror; she locates her sense of the sacred in the bond of matrimony, whereas he had been transfixed (with a mixture of horror and ecstasy) by the sketch from 1505 which reflected back 'IN FIFTY-SEVEN LINES OF INK THE WHOLE OF MY CONSCIOUSNESS'. Compared to this sketch, he dismisses Beza as 'human rubbish' ('what you have destroyed is greater than the volume of your entire existence'[16]) who will not live another week, even as he '*squirms*' in susceptibility to her physical attractions. Lacantia reflects on the 'degenerate passion for understanding everything', now as thwarted and redundant as speculative possibility: 'I should stop declaring what I would do IF there is no IF'. She instigates the (ostensibly merciful but unmistakably vicious) killing of her dog, who has previously formed an ironic commentary on the activity and speeches of the humans, in both its unwitting dependent loyalty and its bathetic whimpering anxiety; Lacantia and Houth strangle the dog *'by slow degrees'* as the other characters watch *'in horror and disbelief'*. Beza infuses the form of erotic promise with the deathly, appearing for her wedding in a surgical mask: 'A proper bride is veiled isn't she? / So when her lips show they are for her husband's kiss and his only?'. Loud shrinks from this grotesque and macabre threat of his son's corresponding fatal infection; however, Houth insists that it is 'GOOD' that Beza is dying, 'It means we cannot lie / It means our marriage will be perfect', and he kisses her *'fully and passionately'* on the mouth, exhorting the aghast witnesses to applaud. Loud rejects Houth's plea for forgiveness as redundant, like all bids for consolation and atonement ('FORGIVE YOU WHAT FOR EVERYBODY'S DEAD'), as the coughing and vomiting Beza improvises faltering songs from the surrounding utterances. When Houth attempts to orchestrate his own death by shooting himself, the gun repeatedly misfires; his persistently undercut renunciation is accompanied by an improvised song by Beza, identifying the slippages of catastrophic time: 'When the roof caves in it waits a little bit', detectable to the whores and thieves who have the instinct to escape while others, about to die, 'only laugh and sip a glass of wine'.

Dead, Dead and Very Dead is a brief but concentrated, highly memorable choreography of anarchy in which pitiful and ludicrous effects alternate frenetically, then unpredictably pause to co-exist. The articulate heartlessness of its savage comedy mocks the drabness of decency and the hollowness of social promise, even as the freedom of the predatory drive to personal survival is also evacuated, as an alternative form of stability. The mockery of responsibility – by events which nevertheless demand a response – releases a desperate energy in characters whose prospects fundamentally question the validity of all social forms of duty and investment (familial and social commitment, altruism, dignity and transformation through suffering), but who can nevertheless lay no enduring claim to immunity or strength, either physical or emotional: the freedom in survival which comedy would normally extol. It is Barker's writing at its most profoundly – hilariously, appallingly – macabre.

Two radio plays by Barker, *The Road, The House, The Road* and *Let Me*, were broadcast by BBC Radio in 2006: both draw the listener into a discomfiting intimacy with a protagonist who is gradually but systematically stripped of the tokens of his identity and comes to perceive himself at a last ditch, faced with the challenge of undertaking an existential murder, which may prove decisive in his self-becoming, or undoing, or both (a defining hinge of willed action, crystallized by Copolla in *Let Me*: 'There is before the act and after the act who I was before the act will cease to be').

The Road, The House, The Road (directed by Richard Wortley for Radio Four) follows the ominous journey (something of a *via dolorosa/negativa*) of the scholar Johannes[17], published and known simply as 'J', seeking to return to his wife and his as yet unseen child, in 1519. J is self-consciously speculative, reflective and analytical, prone to articulating 'a preposterous or outlandish idea' and then being 'obliged to contemplate it to evaluate it painstakingly for its relevance and accuracy'. He is disconcerted when Biro, a fellow traveller on the road, harangues him and threatens him, claiming the road as his own. This abusive former soldier (whose very name carries the modern association of mass functionalism in inscription and sudden disposability) releases J, but directs him to turn left after a mile, straying from his intended direction, and instead stumbling across a house. Here, a woman, Barbara, offers him brief shelter but laughingly explains that it was her husband who intimidated and threatened J, at her instruction. Barbara claims to burn books: an unsettling prospect for the *'simultaneously complacent and apprehensive'* J, who locates his own 'estate' and 'property' in scholarship, such as that based on the documents he carries, which have been dangerously soddened by the journey. Barbara provides further challenges to J's self-satisfaction: her library, both selective and assiduous in its chosen holdings, contains not only all J's published work, but also an embarrassing extent of his unpublished writings, 'fragments of an earlier self', demonstrating an uncanny form of research, which makes him 'more disturbed than gratified'. Barbara claims this research necessary in order to strip the persona 'J' from the 'naked' boy Johannes and know him as a murderer – in the potential and prospective senses; she also tells him that his philosophical thinking and writing is 'fortification' and 'warfare', part of 'The lying that keeps us alive', even as he 'LONGS TO BE CORRUPTED'. Barbara mesmerizes J into burning his letter of introduction from his mentor and rival Erasmus, insists that he wish his wife and child dead, and – as part of his 'becoming' and discarding of 'the dance of civility' (its 'charm' and 'method') that he agrees has 'warped' his life – that he should do his 'greatest work' by stabbing Barbara to death, at her own insistence ('Think of it as payment for my hospitality'). J complies, even as he exclaims with *'a sudden plunge of horror'* 'I DON'T BELIEVE I'M DOING THIS', and Barbara accepts her death blow with determined composure and a strangely erotic intimacy[18]. As J escapes with his miraculously restored papers and dried coat, he encounters Biro on the road ahead, an 'impossible' overtaking by someone left behind him in the house. J anticipates a terminal conflict, the aesthetic resolution and 'BEAUTIFUL CONDITIONS' of his own death ('oh one could not design it better'), but instead Biro (like his wife) insists on his own killing, and even assists J with the achievement of this. J lurches onwards, with a new sense of freed malice and purpose: 'THEY ARE QUEUING TO DIE AND I MUST GRATIFY THEM / It's / Only / Manners'. J, fleeing in the freezing winter, pauses to discard his bloodstained shirt, berating himself in the terms of the soldier, (potentially fatally) stripping himself of another layer only to see the shirt caught on a bough, flapping like a red and white banner (manifesting J's initial sense of a speculation 'left hanging in the air or like an old coat on a peg the owner simply disappeared no need for an inquest', and recalling the terminal vestige of the coat in *Heroica*). He strives (like Shakespeare's Hamlet) to avoid misunderstanding, even in the knowledge that, as in a nightmare, every action he performs 'serves only to deepen the condition' he is struggling to emerge from: until he achieves a strange relaxation in sensing 'An inevitable and incontrovertible misinterpretation of the facts', not least with respect to his child, his wife and the world, and decides, with '*low laughter*', 'That's good…!'.

As well as the hallucinatory and inexorable quality of a nightmare, *The Road…* may remind the listener of Kafka (particularly *The Trial*, with Barker's J providing a fictional antecedent to Kafka's K) or Beckett (particularly the existential riddle and harrowing decline in the scarified landscape of *Molloy*). The precise sense of objective detail with which it delineates the dismantling and transformation of a subjective viewpoint may also recall one of Guy de Maupassant's *contes*, or William Golding's narratives of existential harrowings (such as *Pincher Martin*, *Free Fall* or *The Spire*) in which the protagonist, from whose perspective we see the world, is effectively turned inside out. Barker's play shares with these works a sense of formal perfection and aesthetic design, which makes the process it charts all the more unsettling. This power to disturb is heightened by something that distinguishes Barker's play from the works above: the protagonist J faces the problem of steeling his will to perform an appalling cruelty, rather than (like Kafka's K) eluding or suffering one. J's experiences may be interpreted as the final renunciatory raptures of a tragically disintegrating sensibility (like the Golding novels), and his death seems probable, but nevertheless it is not specified. Alternatively, J's final arrival at the strange relaxation in accepting an 'inevitable and incontrovertible misinterpretation of the facts' might propose and suggest something even more disturbing than his death, namely his continuing existence, discarding conventional moral significance and pursuing his newfound purpose and lethal 'civility', accepting and willing a thorough reversal of what's 'good', in which his bloodstained shirt serves merely as a token of passage (like the rag of sanity, discarded to permit access to knowledge of the next room, envisioned in *The Bite of the Night*). Richard Wortley's radio production achieves a stately and chilling *gravitas*, anchored by Michael Pennington's outstanding performance as J, which appropriately blends vocal effects of lightness, loftiness and surprise with an increasingly icy determination.

The profoundly unsettling effects of *The Road…* are, if anything, surpassed by the astonishingly remorseless black depths of *Let Me* (directed by Peter Kavanagh for Radio Three, to mark both the sixtieth years of both the station and the dramatist). Barker's draft script of *Let Me* features an epigraph from Pasolini's *La Realtà*: 'The Timid Whom Fear Makes Ruthless...', and is partly inspired by details in the historical writings of Sidonius, of taunting desecrations carried out by the encroaching Goths; desecrations the likes of which Aaron boasts, in Shakespeare's *Titus Andronicus* (V.1.): the digging up of 'dead men from their graves', to be set 'upright at their dear friends' doors'. Set (like *Titus*) at the twilight of the Roman empire, *Let Me* involves the audience in a relentlessly persistent inhabitation of the perspective of Copolla, an unambitious 70-year-old aristocrat widower, 'Roman in every way', who is increasingly and literally besieged by intruders; and who moves from the contraction and recoil of horror to the unenviable 'imaginative expansion of one's sense of self' which characterises terror[19]. The isolation and definite limitations which he formerly experienced as a 'privilege' is now recharacterized as an attrition, and a desertion by the police; even his son Euclid is too preoccupied with characteristically urban 'frantic and indispensable activity' to manage frequent visits, and moreover is prudishly appalled by how Copolla 'wrings wit from his despair' in his wry commentary on the corpses of his former friends and lovers which spring up, like bouquets of death, positioned in jeering adornment of his property, where he has lived for sixty years. Euclid and Copolla's slave Aphrodite maintain a faith in social promise, that the authorities will 'restore the situation in the outlying districts' and that 'Things can only get better'[20], to which Copolla wryly adds, 'or what's the use of cleverness'. Copolla possesses the bleaker intimation that 'we will neither of us know another decent hour' and tries to dismiss Aphrodite whilst begging her to contradict him, in reversal of their conventional terms of power (Copolla realizes 'now I sound like a slave while you regard me with a mistress's disdain').

Whereas Barker has frequently written from the point of view of the disruptive and reviled abject figure who compulsively re-presents the border of a society in recoil (Sleen, Dancer, Defilo, Lazar), *Let Me* inhabits the perspective of one who seeks identification with a social continuity (but not naïvely), and who is driven to extremity through outrage at the increasingly irrevocable damage and desecration of his reference points, which are re-presented to him in forms of vicious macabre parody. Copolla attempts to place himself beyond the emotional reach of his invading enemies, asserting 'I'm dead' and wishing himself reduced to ashes, mingled with those of his wife in her funeral urn. However, successive violations of order drag him, like Shakespeare's Titus and Lear, back to torturous and appalling life; indeed, Copolla achieves a lethal resolve and will, the loss of fear which accompanies the loss of hope ('EXPECT TO SUFFER that must be my motto from now on'), which is redolent of Titus's discovery of a terrible resourcefulness in response to incomprehensible atrocity: Copolla retracts all faith and determines to stand his ground as 'solitary absolute and utter me', in 'the eradication of every wretched and melancholy illusion' and 'the whole basket of discredited solidarities', whilst refusing the lure of suicide. However, even that ground is subject to a cravenly political external redesignation: a postman brings the news that Copolla should leave his family home, expressed in the merely theoretical platitudes of the Department of Justice (Region Five North Rhone), which claims it 'regrets to inform you that it is no longer able to extend protection civil or military to citizens living in estates or villages lying outside' the frequently shifting line 'described as A to B on the accompanying map', with the catch-all disclaimer 'the Department can assume no responsibility for personal loss or injury resulting from failure to act upon this advice'; this political use of language (here, in its deliberately and resonantly modern idiom) as the purposeful avoidance of human specifics has always been a target of Barker's moral analysis[21]. Copolla's anxieties that his house has been broached are realized, in terms of further sacrilege: a barbarian child has gained entrance and plays, uncomprehending, with the jar containing his wife's ashes, until it smashes. Ascribing a significant clumsiness and stupidity to the child, who laughs carelessly, he is driven to imagine a depth of specific and distinctive cruelty beyond the indiscriminate ugliness and insensitivity which characterizes the malice of the invaders: 'real cruelty beautiful cruelty cruelty refined that also is a culture LET A ROMAN SHOW HOW DARK AND LOUD IT IS'. Yet he recognizes that the child, either in blunder or as strategic sacrifice, has strayed and, like him, fallen onto 'THE WRONG SIDE OF THE LINE', the arbitrary and insentient 'thing called A to B', which decrees 'Rome is on one side of it and your lot's on the other'. He makes the child confront the postman's head, severed by the barbarian relatives of the child and positioned outside his door, with a characteristic 'STRICTLY ROMAN OBJECTIVITY', persisting with this confrontation and delineation through the child's reluctance (which may reflect the imaginative recoil of the audience), 'I ALSO FEEL SICK BUT SICKNESS IS A DOORWAY / COME THROUGH IT WITH ME'. Yet further intensification is demanded: Aphrodite unexpectedly returns and is moved to pity by the crying child, whom Copolla is using as a human shield when the barbarians try to smoke him out. Aphrodite encounters no such pity when she leaves; she is audibly tortured in an attempt to make Copolla release the child, straining his sense of strictly personal (and arguably regressive) process ('To stay can also be a journey ... To why I am / To what we were / To all that was'). When Aphrodite is killed, Copolla makes the child watch this, and the desecration of the corpse, guiding her educatively through the 'self-indulgence' of conventional limits of offence and nausea. In response, he considers blinding, hanging or stabbing the child as different ways to find a superior aesthetics in death (imagining the ways the barbarians might discover them, 'the actors beautiful and dead', perfectly posed in the pierced shell of the besieged house), surprising himself in the process with his own speculations and calculations ('these events so brilliantly orchestrated by your family send blood surging into the abandoned quarters of my brain rather as sudden storms flood buried cellars up come forgotten items INSTINCT CUNNING RAGE ... I am the archaeologist of my own personality'). Euclid is sent in by the barbarians to remonstrate and negotiate, but the events have widened the gap between father and son: Euclid affects hardly to recognize Copolla, who claims 'yet I am so much more myself', and sees on Euclid 'the pale and trembling mouth of the conciliator'. This is confirmed when Euclid adopts the speech of officialdom: 'a decision has been arrived at and most reluctantly to abandon the outlying districts in order to concentrate our resources on protecting areas with a higher area of population this policy will be reviewed in a month or two'. Copolla cuts through this semiotic miasma with visceral specificity, referring to the scattered ashes of his wife: 'You are standing in your mother'. More inventively, Euclid invokes a picturesque idealism based on a national creation myth: 'Let me like Aeneas put my father on my shoulders and leading the child by the hand walk out of Troy together'; however, Copolla rejects all promise and concession, 'I decline to play another man'; for all the seductiveness of the ideal, Copolla knows 'that was Rome's beginning this is the end an altogether harder metaphor' which he embodies: 'let the sculptors of the future understand my beauty my sword in one hand and the child's head in the other ONE MORE ABIDING IMAGE OF THE CRISIS OF OUR TIMES'. Like Titus with his obstructive son Mutius, Copolla draws his sword against Euclid, driving him out to torture and death at the hands of the barbarians. Copolla has gone beyond concession or conciliation, discovering within himself the will to choose to take the premise of his Roman nature to the hilt and beyond, even to the extent of becoming (to others) the catastrophic figure who incarnates the abject – the appalling incarnation of leprous vitality which unforgivably parodies the limits of identity endemic to social psychosis and diffusion – as when he chants wildly 'THE LINE THE LINE FROM A TO B RUNS EVERYWHERE BUT NOT THROUGH ME'. Copolla kills the child, preparing 'an enduring spectacle' for those who break in: 'I do have an eye for / Eloquent gestures', 'We live' – and, indeed, die – 'by such / Contrivances'; and, though the timing of the final fatal incursion is beyond his aesthetic control, it is surely beckoned and insistently invited.

Let Me constitutes Barker's most shockingly powerful command of the radio medium, drawing the listener into a sustained imaginative confrontation with the darkest depths of tragedy and atrocity which significantly occur at the very limits of social boundaries and human singularity. The form and events demand that the listener engage in an activity analogous to that of Copolla, pursuing imaginatively ('with a strictly Roman objectivity') the details and consequences of successive violations of a private space, so that (as when one engages with the narrative of *Macbeth*) to anticipate the next move, as the play demands, involves an admission of a capacity for extremity which would, in the "more rational" circumstances of ostensible social stability, be assiduously avoided; however the play reminds us of the extreme fragility and historical brevity of such circumstances, reminding us that, in the future as in the past, even the timid may be made ruthless by fear.

In terms of radio drama, only David Rudkin's *An Euripedes Hecuba: Queen into Darkness* (broadcast BBC Radio 1975) constitutes a comparable exploration of extremity in writing and production; and Rudkin's stage play *The Saxon Shore* (written 1983, staged 1986) also pursues the individual consequences of the Roman withdrawal of military forces, experienced as a breach in the continuity of the given world and the death of values, where the fragmentation of a social body and psyche is individualized through the image of one man, his isolation compounded, choosing to stand his ground alone, however mired, rather than submit to dispossessive conciliation. The final effect of a rejection of reconciliation is, of course, previously apparent in Barker's *The Europeans* and *Animals in Paradise*; and Rudkin's *The Saxon Shore* also repudiates conventional reflexes of appeasement, both moral and dramatic, when an unpunished killing dashes facile hopes for intercultural at-one-ments. *The Saxon Shore* and Rudkin's *Hecuba* depict how the boundaries of human definition can dissolve into a murderously territorial animality; in *Let Me*, a specifically human inventiveness emerges through broken forms, defiantly orchestrating the atrocious, and discerning a strange musicality in the varying forms of universal agony: as when Copolla observes, as his son and the child wail, 'Did any man hear such a music any man in history / She counterpoints him / The only missing instrument is me'. This disturbing sense of musicality is accordingly required in the vocal performances of *Let Me*, and Kavanagh's production was spearheaded by the remarkable performance of Edward Petherbridge: significantly a veteran, like Michael Pennington, of the vocal training and late 1970s/early 1980s productions of the RSC. In *Let Me*, Petherbridge demonstrates a vocal and emotional athleticism: his rendition of Copolla is sharply, even startlingly, brisk in its determinedly contained precision, then shifts gear to infuse a phrase such as 'comes flooding through the garden', or the designation of a blinding as 'EASY', with a precisely located resonance of emotional release which appropriately suggests a man whose limits are self-consciously being stretched and overwhelmed. The supporting vocal characterizations effectively suggest, even with subtle comedy, characters whose sensibilities are limited, before they are forcibly expanded through suffering into pitiful incoherence, wordless wailings. At the end of the play, one is left pondering Copolla's observation: 'it is an abyss the self you must equip yourself with long ropes very long'. *Let Me* avoids the successful strategies of Barker's other radio plays: the elegance of *Scenes from an Execution*, the hallucinatory cityscape of *The Early Hours*, or the transformative narrative drive of *Albertina*. Rather it achieves a concentrated, unflinching, severe focus which makes it Barker's fiercest exploration of the form, and one of his most awesome excavations of the human imagination consciously at the limits of language, suggesting how 'the yawning chasm of oblivion' at the point of death can lend 'some final' indefinable quality, such as the excavation of dignity 'even from the most corrupted soul'.

Barker's stage play *A Wounded Knife* (written for Odense Drama School and staged in Odense Theatre, Denmark, 2006) continues the theme of killing – as both incarceration of the self and a gateway to heretical new life. *A Wounded Knife* also unmistakably presents some variations on the themes, tropes and characters of *Gertrude – The Cry*[22] and *The Fence* (in particular, *Knife*'s first scene of a man, sexually fixated on his aunt, denouncing his uncle at the scene of his funeral, replays the opening gambit of *The Fence*), and finds new resonances for the familiar character names of Biro (here, a prince who might be considered incompletely Byronic) and Houth (on this occasion, an unsentimental and remorseless killer rather than an idealistic failed suicide). However, *Knife* develops into a compelling original narrative in its own right, driven by strong characters and surprising developments: the dominant sense of gamesmanship (apparent in the recurrent chess imagery) is established when the widow Sleev publicly calls her former husband 'filthy', as Biro commands, but turns the term from a denunciation to a collusion ('oh how I love you you filthy filthy man'), based on her excitement in proximity to a killer (even a dead one). Biro proposes that she come to him naked; Sleev, a consummate poetic self-dramatizer, responds to Biro's initiative by tossing one of her high-heeled shoes into the grave of her former husband and laughing at her own briefly uneven walk. She agrees to marry him. The royal tailor, Globe, is, however, preoccupied with something more sombre than wedding gowns. In an attempt at punitive therapy in a state prison, he cradles the corpse of his murdered daughter as he confronts her killer, Houth, who nevertheless inconveniently spurns contrition, denouncing pity as a 'prejudice' (this scene provides a memorable riposte to the iconic dramatic power of the last scene in Shakespeare's *King Lear*, perhaps the ultimate in pitifulness, in which Lear cradles the corpse of the daughter he was only just too late to save, notwithstanding Edmund's improbable conversion: Houth, conversely, demonstrates and expresses a staggering lack of pity).

One of the play's principal strengths is the relationship between Biro and his confidant Quittur, a poignant and irreducible friendship crystallized when '*they look at one another with a mixture of contempt and love*'. Biro frequently contemplates death, the deliberate stopping of his own heart; whereas Quittur maintains that the self is not always decisive; the heart also attends to the 'voices of those who love you', which sometimes prove louder. Quittur, appointed Minister of Justice, rescinds the death penalty on Houth, on an apparently and outrageously arbitrary directive from Biro, which Quittur believes is designed specifically to flout and enrage his own sense of propriety. In fact, this consignment to continued existence reduces Houth to '*mournful and bitter cries*'.

When Sleev attends a fitting for her wedding gown at Globe's *maison du haute couture*, her daughter Fashoda, a surgeon, recounts her loss – or inadvertant 'killing' – of a patient. The dead man was a 22-year-old poet, and whilst Fashoda is quick to dispel any romantic illusions about his profession ('a bad poet is worse than a bad lorry driver'), this poet was good, and his death triggers the unwelcome advances of his father towards her (confirming the eroticizing power of death's imminence, in hospitals and at funerals). When Biro visits to gaze at Sleev, Globe is enraged that the prince is accompanied by Houth, his hair and clothing '*exquisite*'.

Biro is accompanied by his entourage of Houth and Quittur, louche murderer and discreet moralist who attend him like Faustus's Bad and Good Angels. Biro relates his amusement at this to Tooshay, a palace maid with whom he has a dalliance even on the eve of his wedding. Quittur is agitated, not least by Houth's eyeing of Tooshay, whose sudden kiss charms Quittur's arms to '*rise, float in the air*', when she finds his uncharacteristic ferocity attractive. The play is built of such stark but engagingly disarming, persistently surprising counterpoints: the magnificently irresponsible compulsive seductress in a white wedding dress, and her black-clad lifesaving daughter, who is both frustrated and harassed to the point of contemplating suicide; the self-protectively laconic murderer and the man who pities women disproportionately, to the point of rage; the father who admits that his revenge would not provide an 'ecstasy' comparable to that of his daughter's murderer, and the sexually impulsive but persistently loyal servant: and, at the centre of them, the strangely indeterminate but constantly speculative figure of Biro, experimenting with those around him, testing their reactions and his own by self-consciously pushing loyalties and affections to their limit, as if to find a reason to live, himself. In a contest of wills with Sleev, he denies the promised intimacy of their wedding night (even as Houth, like the agent of an otherwise silent God or Devil, gravitates towards her and provocatively strips her before Biro's eyes), until she can speak the word 'darling' to Biro, with the surge of emotion she reserves for the self-conscious killer.

Fashoda contemplates suicide at the head of a cliff, but lingers in search of 'the right conditions'; Globe appears, and observes 'you remain discriminating a suicide has ceased to discriminate by definition her intellectual life has ceased', and adds notwithstanding or because of his distinct and unenviable perspective: 'Murder on the other hand I think is possibly life-enhancing ... even I am forced to recognize the possibility of it'. Fashoda offers to break her Hippocratic oath and assist Globe in the murder of Houth; she tells the bemused Quittur and Tooshay how her accidental killing of her patient has altered her for the 'better' ('It was a sacrifice you see he had to die to save me'), and darkly hints at the plans she is hatching with the royal tailor, Globe ('I cut he sews or it could be the other way round'). Sleev, formerly deemed sterile, is pregnant with Houth's child; Houth ruminates on Biro's strategy in letting him live, itself a figurative form of long-term but inscrutable seduction, and his own potential response ('If you don't make the move … that everything affirms is the proper and the necessary move … other moves follow').

However, Biro does not seem motivated by the charity, mercy or vanity Houth ascribes to him. Biro walks his garden by full moon, identifying the nocturnal cries of fox, child and woman (a servant with an assignation) which form a 'terrible orchestra'. He resists the blandishments of Sleev's infantilizing affection, encounters Fashoda and engages with her in a surprising reflective dialogue, explaining how he pretends 'not to know things', 'it is so much more … *human* … to be obliged to speculate'; his ambiguity towards Sleev and Houth affords a pleasure, and a Godlike knowledge. However, Fashoda points out that the 'blind energy' of Biro's heart is the sign of the limits of his control, 'A STRANGER IN YOUR CHEST … NOT LISTENING TO YOU'; a limitation which may also extend to God, whose 'agony is His immortality'. Biro embarks on his endgame when exhorted by Sleev to go for an excursion with Houth and Quittur, which Fashoda ominously terms 'Their last drive'[23].

In a scene both comic and terrifying, the three young mens' energies and mutual resentments charge their observation of, and commentary on, a rural landscape. Biro mysteriously provokes Houth into an account ('of infinite significance') of the distinction between a girl (such as he killed) and a woman, which Houth begins with the faltering statement that 'The body of the woman isn't what it seems to be'; he continues, on how it eludes and thwarts the limits of definition, and therefore control: 'You want this body you want this one and not another the universe stops there it seems the whole world's in her even when she has no idea of what's in her she looks at her body she says it's only me'.

Quittur is enraged by Houth's admission that Sleev urged Biro's murder, but Biro remains phlegmatic; the situation intensifies when Fashoda and Quittur converge in a mood of murderous practicality, having disabled the car. Houth pleads for mercy, as he did not when Biro previously pardoned him; then he did not wish to be saved but now he feels (or at least invokes) a paternal love: 'I have a little girl'. The surrounding ensuing '*silence of terrible pain*' gives way to disbelief and outrage; the moralist Quittur is '*inspired by rage*' and throttles Houth (a significant reversal of character which parallels Sonya's strangulation of Astrov in Barker's *(Uncle) Vanya*). This gives Quittur '*a certain dignity*'; but Biro gains Globe's complicity when he insists 'I killed him'. Indeed, Biro pushed Houth to his limits, of personal articulation and emotion, with fatal effects; but, moreover, this forces Sleev to identify Biro as 'Killer of the father of my child' when he returns to her. Knowing her susceptibility to killers (and concomitant fertility), Biro charismatically overwhelms her grief and orchestrates what will be their wedding night, in the deepest sense. He becomes aware of the effects of renewal, in others, in himself: '*He hears his heart. The sound of his heart fills the stage*'.

A Wounded Knife is a highly intriguing and engaging example of Barker's dramatic art. If Sleev, the Wicksian female lead, seems less developed than Gertrude and Algeria, there is instead a dynamic focus on the mingled intimacies and resentments that link the three young men, Biro, Quittur and Houth. As a startlingly unsentimental enquiry into the resonating shock of love, *A Wounded Knife* is a constantly surprising and *oddly* life-enhancing play, as was Barker's 1986 reanimation of Middleton's *Women Beware Women*. The play is properly wary of the transformative power and claims of love: significantly Biro manages to maintain a strategic distance and study of emotions (pursuing enquiries such as 'Is everything a deal I sometimes think so love for example what's that a transaction surely a transaction drenched in perfume draped in silk?') whereas the breathtakingly callous murderer Houth is forced to deduce 'I THINK LOVE IS DANGEROUS I THINK FOR SOME IT'S BETTER THEY PLAY CHESS' when he finds his personal boundaries redrawn by experience, and consequence. Formally, it might be most pertinent to consider *A Wounded Knife* as a startling dark (or 'problem') comedy, like Shakespeare's *All's Well That Ends Well* or *Measure for Measure*; indeed, *A Wounded Knife* has several contact points with the latter: questions of justice amidst a complicating atmosphere, rich with the tangles of eroticism and death; an inscrutable, deductive protagonist, with a relish of symmetry and contradiction, driven to test the limits of others and himself; indeed, Houth recalls both Barnardine and Angelo in Shakespeare's play, as a resolutely objective and unrepentant man, who is given a new lease of life by a self-consciously dramatizing ruler, until his strategy is undermined by the subjectivity which he discovers as part of his own instinctual *volte-face*. It is also simultaneously remarkably challenging and remarkably (unconventionally) comic, in the way that it insists on confrontations and reconsiderations of ethical boundaries surrounding death, and the way that it finally insists that we may prove surprisingly answerable to the emotions of others, with regeneration occurring precisely in the decay of former values.

Lot and his God (written 2007) approaches, from another perspective, the drawing and maintenance of a moral line. As in the story from Genesis, an angel visits Sodom, and declares its people 'filthy', in their acts, thoughts or permission; his report entitles him to be the agent of God's wrath, which should nevertheless spare the scholar Lot. The imposition of these clear moral divisions is tangled by the impact of Lot's wife, Sverdlosk, a hatted and gloved Barkerian *belle dame sans merci* who meets the angel, Drogheda, in a café, refuses to leave and proceeds to seduce him. It is appropriate that Barker finds himself drawn to this opaque story which commingles sexuality, the desire to look, remembrance, cities and catastrophe; and in which mass death is offset by individual catastrophe, when the disobedience of Lot's wife, whose transgressive backwards look is 'legible as the refusal of subjection'[24], draws divine wrath.

Whilst Drogheda denounces her 'obsessive and frankly pitiful self-regard', he also notes Sverdlosk is 'perhaps incapable of shame apology or even mild regret a thing oddly attractive to me and not me only many men', as he increasingly finds her infuriatingly fascinating. Lot joins them, but the café service is offensive and slovenly; enraged, Drogheda strikes the loutish waiters blind, then, when their distraught laments hinder conversation, dumb. Lot is appalled and pities them, but also realizes an angel 'is not inhibited by considerations of humanity': Drogheda dismissively proclaims them effectively 'already dead'. Nevertheless, Drogheda is captivated by the compulsively performative Sverdlosk, who crucially challenges the advantage of his supposed omniscience: Drogheda realizes the power of her insistencies: 'I cannot leave until I have enjoyed some intimacy with you'. Lot is not shocked by this, indeed Sverdlosk claims 'He's used to it'. Drogheda begins to wonder if he is not the subject of some sexual intrigue ('marriage is an unpredictable and frequently delinquent institution'), whilst acknowledging that, as surely as Eff in *Dead Hands*, he has succumbed ('the only question is … What happens next?'). Lot is indifferent to Drogheda's discourse, but pities the wretched appearance of the formerly insolent waiter, crawling over the floor and grasping the air in a way which makes Lot exclaim 'put him out of his misery why don't you'. Drogheda claims to represent an ultimate law and determinism: and that Lot knows God's will, 'the only question is when and if you are conforming to it'. Everything Lot says contributes to the misery of those around him, but he is spared punishment because God loves him; though as Drogheda intimates, this is not to say that Lot is 'immune from His resentment' given Lot's 'reluctance to obey Him'. Sverdlosk draws Drogheda into a lingering kiss; Lot wryly deduces that this 'lapse' Drogheda's 'own high standards' will only intensify his contempt for them; but Sverdlosk recognizes Drogheda's simultaneous experience of wonder and sadness ('you sense your passion for me will damage your life but at the same time you long to be damaged'), even as she insists on the addictive lure of transgression, in which she and Lot have built an equilibrium of complicity: 'often I am stolen Mr Drogheda but I remain Lot's wife this will frustrate you as in the end it frustrated all the others'.

Struggling with his own suddenly manifest contradictions, Drogheda realizes the excitement of the promise of sin, and Sverdlosk proposes 'in my own home let us violate all those things conventionally described as homely': Drogheda concedes to the initiative, which Lot knows will extend to intimacy in his library ('as if the books were dead men gazing on her'), the 'integrity' of which he fervently and passionately imagines from a 'cold but immaculate' distance. Whilst Lot awaits their ominously belated return, God suddenly speaks to him through one of the stricken waiters, knowing the combination of pleasure and dread which 'constitutes Lot's ecstasy', and probing this 'vast but mortal', even 'pitiful', love by enumerating her infidelities. Nevertheless Lot (like Photo with Algeria in *The Fence*) locates and discovers the revelatory in the sexual: 'How beautiful are the lies of Lot's wife how beautiful her truth'. Moreover, Lot's investment in the sexual is truly religious, in its sense of the mysterious and the paradoxical: 'how terrible it is to know all things I pity God I would not be Him'. This stirs God to an awareness, and tears of infinite loneliness: '*GOD moans in his solitude*'. Sverdlosk reappears, reporting that Drogheda, in his temper at his new sense of deepening 'servitude' to her, has burned Lot's library, the books Lot had intended to console him beyond the sensed 'destiny' of loss of his wife. Lot observes 'Churlish and arbitrary is God but my ordeal is obviously necessary to me', as a divine (and enviously punitive?) refinement of 'a man more sensitive than God thought men could be'. Sverdlosk divulges, that at the pitch of intimacy, Drogheda disclosed that God wants her dead, in order to save Lot; she refused Drogheda's plea that she abscond with him, in the sincere assurance that she loves Lot. They embrace with '*an insatiable possessiveness, at once sublime and pitiful*', near which '*the stricken WAITER, on his back and writhing, is a sculptural tribute to their passion*'; but he remains necessarily excluded from their prospect of escape.

This prospect is, of course, rendered all the more poignant by the audience's awareness that, traditionally, Lot's wife glanced back at Sodom and was also stricken dead (in what Martin Harries proposes as 'the model for the individual's self-destructive, even masochistic, retrospection'[25]), in what would be confirmation of Drogheda's prophecy, and part of his drive to avert it. *Lot and his God* is the most substantial of Barker's re-visionings of Biblical stories of 'wrestling with God' (*All This Joseph*, 'Sarah'); indeed, it offers a deeper and more theatrical proposition of the seduction of Almighty God than the earlier play of that title. Drogheda is a profoundly self-consciously riven character, a righteous scourge who finds his self-defining aloofness overwhelmingly compromised. Sverdlosk develops the voracious resilience of the *femme fatale* to points which are both comic and deeply poignant: initially she seems a deliberate reversal of the vulnerable woman in 'Cruel Cup Kind Saucer'. Her splendid egotism and devotion to shoes help her to overcome Drogheda's vengeful and demeaning (possibly fetishistic?) demand for removal of her shoe ('Hobble your absurdity will mock my idiot's infatuation'), when, on his departure, she unzips her bag and removes an identical shoe '*necessary to restore her equilibrium*'. Her command of such artifice, 'most precious' to Lot, is a further dimension of her continuously and surprisingly resurgent love for him; though the self-consciously social terms of her seductive personality make her ill-suited to the demands of exile, as Lot observes (her fragile shoulders are not designed to 'haul buckets out of wells'). Barker's dramatization of the waiter, as an index to the surrounding social mayhem, is also a small but remarkable *tour de force* of theatrical complexity: the character progresses from louche insolence to being the subject of savage comedy when stricken with successive debilitations (not conventionally grounds for humour, but, in the context of the story, divine and intrinsically unquestionable justice) which progressively diminish him to '*silently turning on the ground like a dying fly*' before briefly becoming the mouthpiece of God, who then admits an infinite sorrow. The narrative context makes the character laughably one-dimensional, first in his deliberate offensiveness and then in his punishment; then pitiful, in both his divine and abjectly human testimonies of self-conscious limitation. Lot initially seems a relatively conventional character, who discloses surprising sexual and imaginative subtleties[26] as part of the increasingly suspenseful narrative, in which the route to the apparently foreknown conclusion, 'what will happen', is increasingly defamiliarized. In its comitragic speculation on the terms of human and divine perfection, the play finally suggests that, to an infinitely and enviously lonely God, Lot is briefly but unendurably fortunate in his knowledge of a woman who is a 'perfect liar'; and that only the loss of her can complete and refine the 'necessary ordeal' of his sensitivity, which God both loves and resents.

Contesting the world: *The Forty*

Barker's explorations of both the singular moment and the limits of language achieve a new formal concentration and beauty in *The Forty (Few Words)* (written 2006): a compendium which stretches beyond even the ambition of collections such as *The Possibilities* and *13 Objects* to present forty short plays, each concentrating on a moment of extreme emotional tension, in which a line or a few words of speech or even a series of silent actions offer an excavation, through a gestural choreography of physicalized imbalance, of the poetic crystallization of an agony.

The Forty also foregrounds the ways in which words and gestures provide currency for *negotiation*, as in the first play, in which a man and a woman barter the hope invested even in a verbal formulation of ostensible contraction and avoidance, 'I / You do not wish to be hurt again'. Their actions manifest and query the terms of these statements, what one might mean in relation to another. This is a logical development of a feature in Barker's work noted by John O'Brien and myself in our 1991 production of *The Europeans*: what we called the 'Love/Siege' dynamic, in which the meanings of each word (or action) are defined and manifested differently in each context of its use; it can even be played/worked for (and into) the opposite of what was first suggested as its meaning, thus manifesting a dangerously promising dance of individual wants and extreme provocations, a (possibly enticing) declaration of war for the terms of life. Many Barker plays are rhythmically, almost musically, driven by a series of linguistic *leitmotifs* (or, as I prefer, *leidmotifs*) which demand that the performer hit certain patterns of stress (both linguistically and physically) to build tension and/or generate emotional release, so that the development from one scene to another operates through combination and collision, in metaphoric and emotional, rather than literal or linear, connections[27]. *The Forty* develops this poetic analysis of interpersonal negotiations into a principle of its focus. Play 'Two' mines the different registers of sorrow in the constituent words spoken by a conscripted soldier to his parents – 'If I go I shan't come back I know' – when individual words and combinations are divided into separate sonorous knells. 'Five' discovers an elegant woman at the bedside of a mortally ill man: her words 'All said and nothing more to say' might provide a (Goyaesque) title for many of *The Forty*, as the characters find themselves at the limits of language, yet required to persist; in this instance, the woman discovers a new resource, exposing herself to provoke in the patient '*a long, slow cry of love*' which the nurse cannot censure: her hand goes to signal disapproval '*but the fingers curl in sympathy, and are restless…*'.

Many of the short plays depict an existential process, of remaking the self, through re-presenting the self – or else the unsuccessful attempt to fracture previous definition. In 'Seven' a young woman mesmerically mines the repeated utterance of her words 'Apparently I'm ill' as a seductively enticing incantation, hinting at her catastrophic power – as does the precarious balance of her tilted pose; together, they drive men to desperation. In 'Nine' we observe the undermining of self-definition through words ('Oh, you were very different once') and deed; this in turn contrasts, in 'Sixteen', with the self-presentation as pitiful, but unsuccessful (a man unsuccessfully pleads to a woman 'I'm ill / So can I come back?'); and with the fortification of self-definition in 'Eighteen', which a writer distils from his brush with totalitarianism (glamorizing his action, 'In mid-wickedness…') . In 'Ten' we are invited to contemplate the physicalized dynamics of an isolated breakthrough: a man articulates his unwelcome conviction 'I realized something has changed', and reconstructs himself: '*the tension falls away*' as '*He no longer fears his knowledge*', and consequently '*does not feel the need to communicate*'. By contrast, in 'Eleven', an unfaithful wife maintains that her liaison is 'nothing or nothing much'; yet it is still 'Not nothing' as her husband's consternation and her tears testify, even as she '*resents the innocence*' of his gesture of '*pitiful appeal for reconciliation*': the emotional lines of tautness between them increase their torsion, rather than relax.

Many of these characters are glimpsed at crisis points, as in that resonant phrase 'at the end of their tethers': such as the man in 'Thirteen' who insists 'I will say it and go', until he and his female addressee discover a limited reassurance, or perhaps a tensile eroticism, in his not-saying. 'Fourteen' depicts the urgent promise and possibility of an assignation at an art gallery, when one party insists 'I am here today but only today' but circumstances conspire to prevent consummation, so that (in one of Barker's most eloquently precise, yet infinite, stage directions) *'Time hurts the unfulfilled lovers'*. Even this drives the woman into a *'contest with the world'*, as she approaches another man, but her *'terrible tenacity'* makes him think her mad; *'with infinite sadness ... She resumes her relations with reality...'*, predicated on (self-) limitation. A similar, pained resumption occurs in 'Nineteen': a model, *'inebriated with her own divinity'*, cavorts next to an old man on a park bench for a photographer, who is enthusiastic until her self-control gives way to an *'ecstasy'* of sexual arousal for the old man, which fills the photographer with unforgiving dismay; however, the despair briefly expressed by both photographer and model lacks the profundity of that of the old man, who is, once deserted, briefly blessed and forever cursed with the memory of experiencing a promise otherwise beyond his imagination. The model is briefly seduced by her own performance, destabilizing not only herself but all those around her. This contrasts with the official in 'Twenty' who eventually subdues a defiant fugitive by repeating the phrase, 'can't you see it's all over', and its component words: the official maintains his confidence, and in his persistence discovers an exhaustion in the fugitive, who submits to the emotion with tears of relief (the sound of weeping rises and falls *'orchestrally'* around him). When the fugitive is led away, the official relaxes his efficacious performance, and submits to his own tears of relief, exhaustion and perhaps even pity.

The prospect of death is dramatized differently by two subsequent plays. 'Twenty-Four' shows an old man deliberately courting death through exposure to cold night air, but next morning he finds himself still moving, in *'half-irritation, half-concession to his continuing life'*. In 'Twenty-Five', two women pursue a suicide pact on a clifftop, until a servant grasps the waist of one who proves crucially reluctant, and he is physically overwhelmed by her passionate gratitude.

In fact, there is generally less speech as *The Forty* progresses; rather the plays flow into a cumulative sense of transient human traffic in brief surprising contact, testifying to the seizing and/or dying of an impulse. In some ways *The Forty* recalls and develops theatrically the premises of Barker's 1986 poem sequence *The Breath of the Crowd*: the tides of witness, movement and unexpressed potential which can briefly disintegrate familiarity, in emotional defiance of a social climate where physical presence and proximity is usually only permitted and excused through a rigorous "privatization" of the self, where avoidance and indifference are elevated to the systematic. Theatre and dance offer specifically singular exceptions to these rules, by insisting on the exercising of physicality, proximity and the gaze – all of which can prove paradoxical, reversible and mutinous in their surprising acknowledgements of separateness and independence, lost and perfectly exhausted choice. 'Thirty-Three' offers a particularly poignant distillation of such elements: a man meets with a woman who is *'drained of any pride but fixed in the remnants of her fascination'*; they smile wanly, *'The desire to run to one another lives in their limbs, but is gone from their minds, and in vital seconds, dies'*; and the woman, *'as if overwhelmed by a deluge'*, turns and strides to a second man, who catches her in his arms. The first man works through his raging emotions (expressed as deep groans) before discovering an acceptance, and decision to walk on in his own separate direction. 'Thirty-Nine' shows how the promise 'I said I would always love you' can turn into a belittlement, even a source of resentment to the recipient: an elegant woman accommodates herself to the slow gait of her decayed lover, who humiliatingly drops one of his walking sticks: when a younger man performs assistance, the old man senses condescension, and the woman recognizes his purpose as not *'altruistic'*; the old man in turn humiliates the younger man, and then the woman, by refusing to accept the stick which they have retrieved for him, until they concede to him; the woman finally repeats her litany, part-assurance, part accusation.

The Forty concludes memorably with a tidal ballet of presence, avoidance and instinct: a throng of people, covering their eyes with a single hand, jostle across the stage and flock like birds: a naked woman crosses what might have been their line of sight; *'She makes no reference to them, nor do they seem to register her, but when she has gone, one of the throng drops his hand and stares, haunted, tortured by imagination'*. She returns twice, more fully clad and studying a map, yet invisible to the man who remains stricken, *'pained by the brief image that crossed his fixed stare'*, and exclaims 'My dear'; she does not hear his appeal, responds to an insolent summoning whistle, and hurries to an encounter with brusque familiarity; the man replaces his hand over his eyes. This offers a poignant, rhythmic image of proximity and imaginative insight: the plenitude, and forsaking, of possibilities in both.

A Wrestling School production of *The Forty* was scheduled for touring in Autumn 2007 and (like *The Fence*) planned to incorporate local community performers and students into an orchestrated silent chorus of movement; and also to incorporate a soprano singer (building on Barker's discoveries in opera through his libretti for *Terrible Mouth* and *Dead, Dead and Very Dead*) who would add an elements of sung utterance and wordless song as an enhancement of, or substitute for, speech. However, in 2007 the Labour government diverted 112.5 million pounds from the Arts Council of England to pay for preparation for London's 2012 hosting of the Olympic Games. Rather than oppose this with any public demonstration of eloquence or conviction, the Arts Council of England responded by systematically excising the criterion of

'Artistic Development' (the principal ground for The Wrestling School's funding) from consideration in its awards, and rejected the application for *The Forty* as 'insufficient priority'. This decision not to fund this particularly innovative project was a politically ominous and significant outcome of an arts funding policy which, in its documented 'assessment criteria and priorities' of May 2007, privileged an ostensibly quantifiable social utility (such as 'increased participation' by 'groups who may be at risk from social exclusion' and 'in places experiencing significant growth in new housing') over any commitment to, or mention of, artistic development; thus, as Barker noted, the Arts Council 'yet further diminished the range of theatre practice when its very purpose was to extend this range'[28]. In consequence, there was no Wrestling School production in 2007: one year short of its potential celebration of twenty years of highly acclaimed international work as an independent theatre company.

'Beyond shame there is ecstasy': *I Saw Myself*

The situation of the *singular* in relation to the borders or boundaries, and the consequent effects of self-consciousness, is further investigated in Barker's stage play *I Saw Myself* (written late 2006)[29]. The compulsively transgressive widow Sleev[30] is reanimated from *A Wounded Knife* and placed at the centre of a play set in 'Europe in the Thirteenth Century'. As a war draws near, Sleev presides over the creation of a heroic tapestry of the battle in which her husband fell; she is assisted by two weavers, Ladder and Keshkemmity, and a maid, Hawelka. Behind the mirror door of her wardrobe, Sleev secretly keeps a naked man, Modicum[31], to whom she confides her secular and sexual confessions, even as she conducts a (literally) glancing affair with her son-in-law, Guardaloop (a criminal intensification based on heightened awareness of time, predicated on the subtlety that 'too little is enough'). Sleev delights in the seductive effect of the mirror, an improvement on the situation of Paradise, which had no mirrors, ensuring 'when Adam gazed on Eve what Adam saw was never seen by Eve'. Sleev deliberately exhorts Guardaloop to take her before the mirror, amplifying the aesthetic and transgressive power of the act through the (seductive) image of the self, seduced: she even personifies the mirror, as her 'eunuch'. Again, Baudrillard's ideas are pertinent here: he proposes that the mirror's power is its 'absence of depth', a 'superficial abyss' which others find 'seductive and vertiginous' only because they each think themselves are each the first to be swallowed up in it:

> All seduction in this sense is narcissistic, and its secret lies within this mortal absorption. Thus women, being closer to this other, hidden mirror (with which they shroud their image and their body), are also

closer to the effects of seduction. Men, by contrast, have depth, but no secrets; hence their power and fragility ... *'I'll be your mirror'* does not signify 'I'll be your reflection' but 'I'll be your deception'.[32]

But whilst Sleev implicates others in the ecstasy of deception, she also recognizes the mirror's capacity for unwelcome reflection: 'With desire came the mirror but the mirror only served to deepen the anxiety desire inevitably creates / *(She looks at herself)* / Eve now suffered the dread that she was insufficiently beautiful'. Sleev works at the tapestry with an intent scrutiny comparable to that of 'the young monk in the Abbey of Calcetto' (a reference to Hoik in *The Moving and The Still*) and an urgent concentration which may demand her sight, but her labour is not for the glory of Almighty God or martial fervour. Keshkemmity recognizes that 'when soldiers die in foreign places they die for women'; despite exhortations of 'Christ ... the people and the King', 'really it is women'. The tapestry's three streams of simultaneous imagery permit a radically contextual image of the battle: this is initially and conventionally conceived as a central depiction of the soldiers and their agony, bordered by the agriculture which is the source of national prosperity, and by the fidelity of women. However, Sleev was not faithful (her sexual adventuring meant that even her pregnancy was 'wild with desire and anxiety', 'ecstasy and dread': she was not carrying her husband's child); and she realizes that the tapestry will be greater if it includes her, depicting her passionate infidelity alongside the warriors' futile sacrifice. She charismatically commands her weavers and maid with the disarming power of her own compulsive volatility, inconsistencies and dependencies; as when Ladder is driven to strike her, but is not dismissed, with Sleev concluding 'to rage at someone is to love them I felt love in that blow'. However, Sleev's brief consternation at the blow destabilizes the rules of her agreed equilibrium with Modicum, who is fleetingly revealed to Ladder when he protectively surveys the scene. Sleev claims this 'kind act' to be 'altogether detrimental' to the 'divine constraint' which maintained the 'precious lie' of his (mute, inactive) perfection. She anxiously seeks a 'fat theologian', whose ministrations might subdue her 'by volume or by argument', to be a husband and settling force once she has completed the tapestry. The mirror's concealment of Modicum is also associated with death: the superficial Guardaloop sees a poignant contrast between the front of the mirror and the back, which suggests the 'nothing' of death; Sleev plays (to) the mirror in terms of Baudrillardian *trompe l'oeil*, an enchanted simulation which adds a new dimension to the apparent limits of the "real" world: it lends her the 'sensual, incestuous transaction' with image, double and death which Baudrillard associates with the power of seduction'[33], and permits her 'to reveal that this "reality" is naught but a staged world, objectified in accord with rules of perspective'[34]. However, death may be courted, thrillingly, but only imperfectly subdued. Modicum's unpredictability is manifested by his ability to escape briefly, by a secret recess in the wardrobe, at times unknown even to Sleev, which heightens both tension and relief. Thus the image of the naked man in the wardrobe, both surreal and farcical (in the classical trope of the concealed lover), becomes amplified to the status of an unreliable God, who may or may not be present or responsive. Sleev meanwhile reinterprets the story and lessons of the traditional marginalization of Eve, even as she (re)locates the singularity of her own ungovernable feminine sexuality in the tapestry of battle, so as to subvert traditional perspectives and create something greater in its complexity (she remarks, 'to be true you have to unpick the convention'): the principal stream of the tapestry becomes instead 'the history of a lady's secret life'. But her bruising presages further distortion of her relationship with the denizen of the mirror, as Modicum resists her sexual invitations and commands, torn between love and resentment. Ladder, now aware of his presence, castigates him as a 'USELESS MALE' who should be 'in the war' – that is, an instrumental functionary of traditional gender roles, social utilitarianism and martial imperialism – which reduces Modicum to an audible '*low sobbing*'.

Sleev purposefully insists on the incorporation into the tapestry of that which might conventionally be received as rebuke: Ladder's identification of 'high temper, arrogance and madness', through the proposed depiction of white around the eye of the central female figure, is approved by Sleev, who rather seeks to elaborate the degree of sexual provocation. However she admits that she is literally blinding herself through her compulsive investment in this (ironically visual) work[35] (whilst disarmingly acknowledging that even this briefly incandescent self-absorption mirrors the tapestry as further wilful strategy 'designed to draw attention to the woman known as Sleev'). Her efforts divert conventional perspectives and values into a different set of stakes, where 'Nothing is self-evident' and 'the war is … less important than my life': the Barkerian project of anti-history which undermines discourses which claim authoritative (and authoritarian) truth. The doubting and potentially mutinous disciple Ladder disputes the terms of scale, maintaining 'I want to tell the story of my people it is solemn to me and the story of your body is not'; however she recognizes the absence of real choice as the invaders draw nearer, and admits to Keshkemmity 'my lady is a great weaver of tapestry and her life is killing her as surely as her lord's killed him'; 'there is a war in my lady also'. Moreover, Ladder is drawn into loyalty by Sleev's self-overcoming in initiating and refining a consciously contradictory project which involves unreasonable investment by all involved: 'if your heart does not hurt then you have not made a tapestry'. Sleev, in her discomfiting passion, at times calls to mind a feminized version of Christ as figured in Starhemberg's vision in Barker's *The Europeans* ('you follow him who triumphs over himself, who boils within and in whose eyes all struggle rages' *OP*1, 140); elsewhere, Sleev repudiates the figure of Penelope ('the flaccid packet of fidelity') in Homer's *Odyssey* and identifies with Eve, and works determinedly to read out of the scriptures a new gospel of heretical feminized hermeneutics, which posits Eve as a Nietzschean/Barkerian superwoman:

> Eve lied because she found the world unsatisfactory knowledge far from being the prize it is made out to be only revealed the poverty of things the poverty even of paradise so Eve because she had imagination was obliged to lie how else could she spare herself the agony of her disappointment Adam was poor like all creation no wonder God was enraged with her He knew that everything he made Eve despised

Sleev's husband-to-be Club is perplexed both by her scriptural exposition and by the tapestry ('This naked bitch why is her mouth turned down ... if she isn't happy why does she go about with nothing on'). Contrastingly and proportionately, the women's loyalty grows fiercer, both with the imminence of catastrophe which ordains everything will happen 'unexpectedly' in terms memorably identified by Sleev: 'We live in crushed time' – a time which will (at least) overthrow the imposed terms of "reality", where 'to even contemplate the making of a tapestry', especially one in which 'loss is necessary', 'is arguably nostalgic a grotesque and comical redundancy'. The scene in which the women complete and seal the tapestry with their own blood constitutes a dramatic ritual with a passionate intensity beyond conventional or rational obligations (perhaps suggested by a detail in *The Possibilities*, specifically 'The Weaver's Ecstasy'): this is what Baudrillard would identify as The Rule, a passion which binds the players to the game which ties them together and permits them participation in an 'endless, reversible cycle' which can dissipate and abolish the 'terrorism' of externally imposed authoritarian meaning[36]. Club seems intent on imposing the sort of infantilizing love which Sleev associated with husbands in an earlier rebuke to Modicum; in fact Club is prevented by Modicum, who had been 'taken' by the army, now reappearing in the cupboard: *'an image of immobility and patience as before, but altered'*, speaking the catastrophic ravages of war. But Modicum is no benign *deus ex machina*, rather a reassertion of *das Unheimliche*, that which is usually kept out of sight, and now defamiliarized, no longer 'at home' in the cupboard; he comments wryly on the transformations wrought on himself ('I am if anything more me') and Sleev ('I have discovered speech whereas she can no longer see'). Identifying himself with the force of war into which he was impelled, Modicum has returned, unrestrained and vengeful, to wreak his own havoc[37], forcing Club to attempt a valedictory consummation with Sleev, but the pressure reduces Club to *'a spectacle of castrated misery'*; then Modicum coerces Sleev to fellate him ('if you cannot do it loving do it hatefully') under threat of death. The boundary which separates war from peace is broached.

Modicum surveys the tapestry, instituting his own reign of vicious instrumentalism by contemplating its destruction and use as a site for sexual desecration. When Sleev denounces him ('Vile is the king who came out of a cupboard'), Modicum mocks her own bids to compartmentalize and license 'chaos on her own terms'; in a particularly shocking repudiation of any possible reconciliatory impulse, he responds to her provocations by beating her savagely. The remorseless whirlwind of destruction involves the rape of Hawelka and the scattering of Sleev's family, though Sleev remains unsentimentally phlegmatic about her granddaughter's power and chances of survival: 'The girl is too much me to die of men ... if she survives a dozen miles she'll govern' her abductors. Spurned by her own daughter for the 'FILTH' and 'ABSURDITY' of her life, Sleev acknowledges that she has 'lived like someone ill' but that this illness has been 'experienced as ecstasy', parallel to that of Eve, forced by God to suffer: 'BUT PARADISE HAD TO BE USURPED beyond shame there is ecstasy forgive Eve forgive me our knowledge and we will forgive God accordingly'. Even Sleev's daughter joins the weavers in ripping and dragging away the tapestry, in what Ladder ordains an act of secular communion: 'The tapestry is my lady and my lady she is now ... the tapestry' (a parallel to the disciples' binding ingestion of Lvov in *The Last Supper*). Sleev proclaims her body and self 'A VOID'; however, this is repudiated when the mirror door moves on its hinge and the blind Sleev is *'paralyzed by the revelation'* of glimpsing her own reflection, exclaiming 'I saw myself'; her fingers trace the shape of her reflection (the familiar estranged and transformed into the unfamiliar) as *'The light dies in the room'* to return the audience to darkness with this ember-image of uncannily confirmed (transfigured?) selfhood on their retinas: suggesting, perhaps, that, in the realm of appearances, what usually passes for "reality", functionality and the locus of power may be merely the brief manifestations of a principle of perspective.

Along with *Gertrude – The Cry*, *I Saw Myself* is Barker's second masterpiece of the second millennium. As Séan O'Callaghan has observed to me in conversation (19/4/2008), Barker's major works in the period under review are dominated by a triptych of progressive dramatic portraits of female sexuality: *Ursula* depicts both the wilful maintenance and the wilful discarding of virginity; *Gertrude* shows a woman living at a ruthless peak and relentless pitch of transgressive sexuality; *I Saw Myself* dramatizes a movement into a (defiantly unrepentant) reflective awareness. *I Saw Myself* also self-consciously revises the premise of *Scenes from an Execution*, and the figure of the female artist (much more successfully than *A Rich Woman's Poetry*), to centralize the testimony of exploratory transgressive sexuality against Barker's other great themes of death, war, art and loss. One might go so far as to say that Sleev's personal *coup* reverses that of Galactia, who places an anti-historical 'people's narrative' at the centre of her art; Sleev subordinates that narrative to her own, elevating personal history to revolutionary status, and diminishing the collective. The crucial moment of the tapestry – the death of Sleev's husband in battle – is counterpointed precisely by betrayal, illustrated in ever more transgressive ways. Sleev suffers for her persistence in self-analysis, through the cruel return of the political, Modicum transformed into a vengeful war casualty; but her will and capacity for insight remains undimmed, surprising even to herself. *I Saw Myself* is a (perhaps the) crowning example of Barker's characteristic imaginative pursuit, extending the matter of style (and space) beyond art (and art history) to the political then metaphysical dimensions. Its disclosures subvert the sense of factuality/"reality" which seeks to determine consciousness – in parallel to the ways in which Baudrillard asserts that surrealism and *trompe l'oeuil* destabilize the principles of functionality, by 'exceeding the effects of the real'[38]. *I Saw Myself* opens up possibilities through the manifest play and artifice of drama and theatre: a sphere self-consciously limited and ephemeral but capable of 'a radical surprise borne of appearances, from a life prior to the mode of production' of our received terms of 'the real world'[39].

11
Inconclusion: Consolations in Extremity

A Style and its Origins

> Strange where one finds one's consolations in extremity
>
> – *Let Me*

A Style and its Origins, written in 2005 and published in 2007, is an (auto)biographical reflection on practice co-credited to Barker and his photographer pseudonym Eduardo Houth. Barker had originally wanted to publish it under Houth's name only; I and others argued that its divesting flourish, which exposed the scenographer *personae* Leipzig, Kaiser and Shentang as strategic masks, was mitigated by a maintaining the existence of the fictional Houth. However, the third-person *perspective* of Houth was and remains a powerful strength of the volume. Barker dramatized *himself* against the crowd, as if reversing the telescope through which he usually scrutinized the outside world, training it inwards. The result was a form of dramatic self-exposition, in which Barker nevertheless saw himself externally, writing as if he were dead.

A Style and its Origins should ideally be read in succession to (or else before) the book the reader now holds, in order to gain an additional internal (but externalized) perspective on Barker's writing and production work during the period 1988 to the present, which both volumes trace. After some initial locations of images resonant from his childhood and beyond, Barker/Houth's narrative principally follows the formation and refinements of The Wrestling School. Some claim the revelation of experience is the best argument of all; here, Barker/Houth moves beyond *arguments* for a theatre to an orchestration of experience, both confessed and dramatized, which have informed, opposed, encouraged and shaped the development of a unique theatrical (and personal) aesthetic.

Brendan Kennelly asserts: 'All that's left of anyone is a story / and all that's left of a story / is how you tell it'[1]. In *A Style and its Origins*, Barker finds a style for his story: a story of discovering a style ('something arrived at by painful study, a distillation of thought and practice, and essentially a moral decision': *SAO*, 32), in which dramatic poetry, physical choreography, *mise-en-scène*, philosophy metaphysical and hence political, sexuality as experienced and imagined, all constitute unstable elements in the alchemical composition of metaphorical experiences. This involves consciously operating at the boundaries of a culture's knowledge, like the conjuror[2] who recurs as a figure in Barker's imagery: he who demonstrates a 'passion to master the world of facts', evading the actual, defying gravity; in Barker's case, a conjuror of death, through whose ministrations 'we are privileged to know the supreme law of limitations but simultaneously to be *ecstatically uneducated by it...*' (*DOAT*, 40, 93).

Barker/Houth observes how Barker 'had a vision of acting as a form of religious practice, which in its most spiritual manifestation became an ecstasy, an ecstasy in which the actor would not know himself' (*SAO*, 28). This in turn draws the audience into 'an ecstasy by the spectacle of excess even to the *contemplation of death* which is the essence of tragedy' (*SAO*, 64), its fundamental questioning of life's conventional values and habitual priorities. The artful duality of the '*theatre poet*' is at the heart of Barker's work, his painstaking animations of which he finds both 'ecstasy and ordeal' to watch on stage: on the one hand, clamorous social gatherings 'constricted the flow of his thoughts, which under more intimate circumstances flourished in profusion' (*SAO*, 96–7); on the other, he seeks 'the public place to be unpublic in' (*GTGU*, 34) in the conviction that the theatre remains a place where (possibly imperceptible) redefinitions may occur which at a later date may seem significant (where 'the effects of the work of art are characterized by *delay...*', *DOAT*, 103). This paradox is often the theme and subject of his theatre, which 'ecstatically affirmed the destructive effects of passion on *values*, the way in which desire eroded responsibility and unpicked *laws*', yet nevertheless also affirms 'ecstasy as the only riposte to life's laws, but ecstasy with another, a defiant *duality...a* perfection of the 'we' outside the hounding conformity of the collective'(*SAO*, 90, 99).

On 5 October 2007, Karoline Gritzner and I travelled to the New End Theatre in London to be part of a small invited audience for a first reading of *I Saw Myself*, directed by Barker, organized by Colchester Mercury Theatre in their bid to drum up financial support to enable their commitment to a possible production of the play. Working quickly, Barker had assembled an impeccable nucleus of Wrestling School artists: Victoria Wicks (Sleev), Jules Melvin (Ladder), Julia Tarnoky (Hawelka), Justin Avoth (Guardaloop), Claire Price (Sheeth) and Nigel Hastings (Club); in the raw absence of the full production associated with this time of year, other Wrestling School Associates (Gerrard McArthur) and supporters gathered in pained anticipation. Sleev's passionate account of, and response to, living in 'crushed time' generated profound resonances: even the boiling heat of the theatre contributed to the feverish yet defiant quality of the play and performances, which told the story of a transgressive artist and her variously devoted servants fighting for imaginative breath, in the path of an advancing and uncontrollable war, by mustering and focussing impossible energies, to overturn even the imposed terms and perspectives of possibility and contingency. Wicks spearheaded the event with a *bravura* performance equalling her Gertrude; others responded with fierce, purposeful precision.

Afterwards, the Mercury administation and other invitees were wary or unwilling to approach Wicks and Barker, as if shrinking in disbelief from the dissolvent electricity of the play and production, or in emotional exhaustion. Only I walked towards them, to bring drinks, in token appreciation of their efforts in the theatre's parching heat. Nevertheless, that electricity lingered, in the hushed, semi-coherent, then increasingly exhilarated exchanges of those who had witnessed (not just a narration, but) a demonstration of the proposition, that the only responsibility we have, in engaging with the would-be imposed terms and perspectives of "reality", is to remake them, *on our own terms...*

The event drew forth an astonishing, decisive, splendidly unpredictable development: a member of the public who was in attendance was inspired to replace and enhance the denied Arts Council subsidy for the next three years, rendering The Wrestling School a truly independent theatre company which could expand its ambitions to mount two productions (a major production and a chamber work) per year.

In April 2008 The Wrestling School celebrated its twentieth anniversary, and this new phase in their creative freedom, with Barker's production of *I Saw Myself* at London's Jerwood Vanburgh Theatre (with plans extending to productions of *The Dying of Today*, and *Found in the Ground* and *The Forty* 2009–10). Barker/Leipzig's pale ash-coloured wood set assembled empty picture frames, steps to a wardrobe and suspended balls of coloured wool which hung like internal organs, then were hoist away and replaced by weaving desk-frames, at which the seamstresses took up their positions, seated on stools. Jules Melvin and Julia Tarnoky reprised and developed their roles of Ladder and Hawelka from the 2007 reading, with fiercely precise and delicate core performances; Séan O'Callaghan brought to Club a leonine vigour, poignantly appalling at its point of disintegration. All other actors were company debutants: Geraldine Alexander played Sleev with more humour and vulnerability than Wicks, her final lines more a deduction than an assertion; Nick Barber finally emerged from the wardrobe as Modicum, his former nakedness clad in military boots, trousers and braces, holding a knife. The production achieved a remarkably orchestrated vocal musicality and tension, their precision unique in contemporary British theatre: the sonic arcs of horror, wit and passion in the trajectories of the characters/performers were acutely counterpointed by electronically extended creaks and hums which accompanied the lowerings of the desks and the final tearing down of the tapestry's sections. It was a salutary offering at this hinge point in the company's achievements, reflecting a twenty year testimony of exhilarations and rage, yet seeming to glimpse prophetically future irruptions of 'crushed time' – a temporal dystopia, triggered by a crucial action, which re-presents the end of what you have known, and presages a new state of being which will not listen to former appeals (like Guardaloop's pathetically redundant phrase 'Excuse me').

Barker's work dramatizes and demonstrates life's brevity and intensity[3], and the mercilessness to self and others (theatrical, even when occurring outside the theatre) which is often required to create a moment of life at its fullest pitch, through reorganization of time, in which people are brought together, to make something happen, through the precise manipulation of expectation, interest, unusually heightened inter-/personal awareness, pressure, friction, which generates ecstasy: a momentary revelation of self and others which constitutes an intensely personal pitch of experience and knowledge ('in the midst of' but) beyond the received terms of history...

Appendix One: Testimonies by Barker Actors

**Julia Tarnoky: Cynthia in *Ursula*, Henderson/Plevna/Berezina/
Season in *The Ecstatic Bible*, Suede in *He Stumbled*, Lindsay in
A House of Correction, Hawelka in *I Saw Myself***

Barker's work presents vital invitations and challenges to respond to with all faculties: vocal, physical, intellectual, instinctive, and imaginative – you need to speak clearly and think fast. All the text – different in my experience from play to play and character to character – is muscular, sculptural and visceral. Strongly rhythmic and musical, it is best thought and spoken on the word, on the line – engaging with the vitality inherent in every word and each syllable. It may be complex and contradictory, inventing and reversing, so the way to understand the character, and be clear for an audience, is to commit moment to moment to connecting directly to the words – which constitute actions. Any speech or utterance for a Barker character has vital significance; even when they seem or feel lost for words, this is given eloquent expression. Every word is a key – which spoken, unlocks meaning. As distinct from other forms, in which speech is like the crest of a wave, the culmination of a process, in Barker the word has the force of the wave itself.

Ursula was the first time I worked with The Wrestling School so the pleasures of engaging with this works' sheer outstandingly original acuity of perception and expression were new – indeed, revelatory – to me. Amidst the panoramic landscape of *The Ecstatic Bible*, I had a range of characters to encompass. My aim was to make them all distinctly different which I was able to do by focusing on the shapes in the text, my breathing, and where I placed my voice to discover how each might best be inhabited. After the huge scale of *The Ecstatic Bible*, which felt like a great wasteland and an open excavation, *He Stumbled* seemed possessed of a darker narrower atmosphere, set, as it were, at great depth. In production it proved to be no less epic, showing that a close examination of the particular casts its own light and throws its own shadows to different but equally universal effect.

What I have found distinctive about performing Barker with TWS to any audience is a particular confidence in the text, the spectacle and the production as a whole. I think, in general, people are happy to be honoured and respected with the genuine complexity of something given and inhabited with vitality; true intensity is repaid by true attention. Barker is distinctive and unfussy as a director; he never fails to treat enquiry with courtesy, interest, patience, and an honest answer. It is like being in an orchestra: he expects that everyone can play their instruments and read the music, so we begin.

**Justin Avoth: Albert in *Gertrude – The Cry, 13 Objects*, Eff in
Dead Hands, Guardaloop in *I Saw Myself* (reading)**

Rehearsing with Barker is rigorous and liberating. There are no long drawn-out discussions of themes, meaning or what – in any given scene – we might aim for. The scene is read once for understanding and a second time if rhythmical inaccuracies have identified moments where there has been a lack of clarity; then the scene is put on the floor, and one looks to find the appropriate physical expression of thought and emotion. In the best instances the process allows the actor to work on a deeper, perhaps more sub-conscious, level, since it denies him the possibility of integrating pre-prepared ideas and physicalities.

The great pleasure of Barker's work resides in the complete structure of atmosphere, incorporating sound and design. These effects are not separate from the performer but rather inform the interpretation and act with the actor. They penetrate within and produce a response in the actor which informs the moment to moment drama of the play. Plasticity or dexterity of voice is essential, but the language of the play does not only reside in the voice but in the physiological response to the choice of words and the rhythmical structure as well as the movement of the argument.

The cast of *Gertrude – The Cry* comprised of a mixture of experienced Wrestling School actors and enthusiastic new arrivals. I was delighted to be a part of a theatre which was neither polite nor complacent and which demanded a participation from its audience rather than a passivity. I felt as close as one could be to working in European Theatre in Britain.

I developed an idea about Albert in rehearsals and performances, an idea which may not have been what Barker had in mind but was helpful in structuring him. I felt that there were three distinct phases in his development: in the first he is (in the framework of Shakespeare's play) Horatio – friend and fellow student; then he becomes a sort of composite Rosencrantz and Guildenstern, not insofar as he is charged by Gertrude and Claudius with observing Hamlet but that he develops his own agenda which is not one residing in friendship (but rather a lust for Gertrude); from this arises his critical view of Hamlet, his distancing from him and final return to marry Gertrude as a ruthless Fortinbras – both martial and marital at the same time. This is something I never discussed with Barker – perhaps it seemed too pat or reductive – but it gave me a shape for Albert's extraordinary development.

13 Objects presented many creative and technical challenges by constantly casting the actor afresh upon the stage. Playing a (small) multitude of roles and scenes meant one was always entering for the first time, as it were, so hadn't the luxury of establishing one's character and developing it towards its conclusion in the accepted way. One had to be both rigorous and flexible, able to excavate the individual scene immediately and inhabit its different linguistic and physical rhythms and atmospheres. *13 Objects* represented an almost perfect paradigm for the actor's training and art: the technical necessities of voice and body; sound diction and musicality; fluidity of expression bodily; swiftness of thought and emotion; great reserves of concentration and relaxation; variety in all these means.

Eff in *Dead Hands* took these challenges, albeit in a form which did allow for the development of a single character and situation, and drew them out to huge lengths. Barker described Eff as *'le battleur'* – the juggler, in tarot cards. Another appropriate image for him might be Sisyphus. There is a vertiginous inevitability to Eff's 'progression' in which resolution – such as it is – amounts to understanding and accepting his circumstances, but not refusing it. Barker delights in his character's dextrous examination and articulation of their immediate circumstances, particularly as their passions and desires are enacted and embodied: ever more layers of language accrete in their desire. The language of Elizabethan theatre requires the actor to pursue the meaning of the phrase until one has, so to speak, beached on the shore. Barker requires his actors to tack and tack and tack again before beaching; enacting an emotional autopsy while the passion is still warm; refusing a conclusion. Eff represented this challenge at its most thrillingly etiolated. Such challenges inevitably reside in the pursuit of variety of tone and atmosphere. Where the structure of conventional drama allows the scenic division to build a progression, here those changes in key could only be found within the language. Barker's ability to continuously find variety within the schema of a neurotic monomaniac amply demonstrates his dramatic and philosophical brilliance.

The extremity of the language and the situations Barker formulates can sometimes alienate an audience. There is furthermore a suggestion in the choices of costume of setting and sound of exclusivity and complexity which is out of step (sadly, in my view) with the prevailing cultural appetite, which demands either the logic and justice of a fairy tale or the spurious authenticity of 'verbatim theatre', as if to highlight or reveal what makes theatre theatre in a cinema-shaped culture. One is therefore aware that there are several strands of audience. Some are *au fait* with the work and its style through study or other familiarity. Many people – whether they have encountered the work before or not – respond to its mordant humour. Many people respond to the beauty of the spectacle and its completeness with the commitment of the actors. Some people are unnerved by what is demanded of them by the plays. On one occasion during the *13 Objects* tour an audience composed of sixth form students were totally unable to countenance certain circumstances within the various scenarios. Commonly these were moments when love was addressed. The extremity of the language was well within their scope. The idea of love was transgressive. Whatever the temper or response of the audience the performer can only be fully committed to the play; to its extremity of moral speculation; to the extremity of the language and the physicality. One is supported in this by the completeness of the form and by the strength of the ensemble. The actor can embrace the extremities of his character and even the extremity of the audiences' responses.

The challenges and pleasure of Barker's work reside in large part in the quality of his superabundant language and ideas. Technically one has to be able to achieve musicality of rhythm and pitch in the voice; one quickly finds that rhythm and meaning are indivisible, so that to be incorrect in the former obfuscates the latter. The language also releases and dictates the temper and tempo of the character, and what I think of as their emotional choreography. One of the great joys in performing Barker's work is how one can surprise oneself. He values his actors' courage: a particular sort of courage which insists upon the ensemble, not the individual – insisting upon an immersion in the *mise-en-scene*, whilst demanding the personal courage required to perform an act of revelation either emotional or physical (often both). Barker knows just what he is asking of his actors. The respect we have for him as a continuously developing and reaching creator enables our animation of such moments.

Edward Petherbridge: Copolla in *Let Me*

The radio play *Let Me* involves ninety minutes of talk, mostly by my character; it features rhetoric, hyperbole, weeping, laughter – from ironic, through tender, to wild and hysterical – elegiac nostalgia, madness and brutality, the occasional comic throwaway line, and much more. Myriads of the most finely set traps for an actor: I say traps because of the current climate, in which the worst thing an actor can be caught doing is acting. A distinguished classical director in a recent discussion of Shakespearian acting style was coldly dismissive of what he deemed to be 'rhetorical', according the term a status beneath discussion. However, to cite one Dictionary definition of rhetoric – 'exaggerated oratory or declamation' – that is something which certain well-known Shakespeare characters are not above using, even to unconscious comic effect (think of Master Ford), and that is only one definition of rhetoric: 'the art of effective speaking or writing' and 'the power of persuading by looks or acts' are others. Only caution of the narrowest kind proscribes the full exploration of how an actor is to manage the richness of Shakespeare's language, or Barker's for that matter: it is a caution exercised by people who, however intelligent and sensitive, lack the technique to investigate (let alone demonstrate, or persuade by looks or acts) what it is they mean, and what range of utterance they deem appropriate. Barker's elderly hero is no neurotic, his demons are real live Barbarians engaging him in an endgame he knows he cannot win. There can be something dangerously thrilling about playing a character in extremis: I woke dizzy on the fourth and last morning of recording, luxuriantly punch-drunk but eager for the onslaught on the remaining climactic scenes. I began to wish I were doing the play eight times a week in a theatre, so that I could fill my lungs to capacity and call to the Barbarians at the top of my voice out of the windows, weep, laugh, threaten and run from the cellar up the steps to the gallery overlooking the statues and fountains on the terraces, imagine being in Barker's great doomed villa, listening. And treading the fine line that allows the character to use rhetoric, without falling into the trap of giving a rhetorical performance (a crucial difference which I hoped I had the instinct to discover in time). The play's terrain allows only perilous footholds for the audience's empathy: will they manage to empathize with a man behaving badly in the face of the destruction of his home life, and of civilization as he knows it? The audience is in extremis. The title *'thespis in extremis'* seems to capture where the crags, the plains, the peaks of Barker's play had me: by no means an uncomfortable place; on the contrary, it seemed the best place for an actor to be.

Melanie Jessop: Gwynn in *Victory*, title roles in *Judith* and *Und*, Galactia in *Scenes from an Execution*

Barker is a poetic dramatist. But he is not just a poetic dramatist. The relationship his characters have to their language is complex and difficult. It's easy to be seduced by the beauty of the language and to forget the primacy of its function, which is to effect change. Barker goes beyond the naturalistic – his characters are changing themselves as much as the world around them; and language is the means by which they do this.

When I was playing Judith, directed by Barker, I remember the excitement of making the discovery of the extent to which Barker's poetic ear supports the actor in their intention. Judith is engaged in a struggle with Holofernes and is losing confidence in her ability to seduce him. At one point she responds with the confessional utterance: 'I also am unhappy'. I found this line difficult to say. The vowel sounds, particularly moving from 'am' to 'un' require dexterity. I kept practicing speaking this line, aware that the absence of any hard consonants was the problem. I began to understand that Judith is using sound to pursue and support her objective. There are no hard sounds in that line because her intention is to smooth, to insinuate, to placate. This realisation/revelation was important because it made me aware of the rewards of forensically examining such a rich poetic text and that often it is in the unconscious and intuitive part of his poetic imagination that Barker most aids his actors. Judith is not conscious of the specifics of her use of language in that moment. She intuits – like Barker – what is required, and the profound emotional clarity of that choice is demonstrated by its expression.

Und provides a paradigm for understanding the unique challenges and rewards of playing Barker. The significant development in his work is the increasing complexity of his characters' self-consciousness. This is also one of the engines of his pursuit of the non-naturalistic and presents very particular issues for the actor. In *Und* we are presented with a woman who is alone, throughout a drama that plays in real time for 1 hour and 20 minutes. The first question for the performer is 'Who is she talking to?'. Barker doesn't tend to write soliloquies, preferring to place his figures in a physically tense but connected relationship. But *Und* appears at first to present all the difficulties of soliloquy. Is the fourth wall broken? Is the character talking to herself? The first is untenable in Barker's aesthetic, the second reductive and energy depleting for the actor, because energy must always have a direction.

I imagined the fourth wall to be a huge mirror in which Und could see herself all the time. She is therefore the apotheosis of self-consciousness. She is her own audience. Her self-consciousness and self-awareness combine with her isolation to produce a piece of highly sophisticated and painful drama in which language becomes literally the engine of survival. The dexterity, vitality and intelligence of her command of language enable her to put up an impressive resistance to the inevitability of her own death. Language is life in this piece; when she stops speaking, her life is over. The ability to express thought, to command and structure response – and to will a 'choice in reaction', where there would appear to be none, through an engagement with language of such emotional, intellectual and physical tension – are the actions at the heart of Barker's dramatic world. His language inspires the energy for performance.

Gerrard McArthur: The Priest in *The Ecstatic Bible*, Vanya in (*Uncle*) *Vanya*, The Admiral in *Scenes from an Execution*, The King in *Knowledge and a Girl*, director of *The Dying of Today*

Barker, the Actor and Indeterminacy

Questions of identity are clearly touchstones for the actor, in himself, and in his giving over of himself to the suggested persona of a character, or 'the other', which is also himself. Nothing about this can be fixed, the currency of it is indeterminate. Indeterminacy is also central to the inquiry and development of twentieth and early twenty-first century art.

Indeterminacy is an obvious paradigm for the 'chemical' activity innate to the actor and his processes, and indeterminacy is the motor of compulsion in Barker. In this sense, the modernity of Barker's texts is that the surface is the depth. Great texts are concerned with the poetic exploration of psyche, and the explication of such in the body of the language. Like Shakespeare, Barker brings this explication to the open surface, where the characters are both the experience of the language and the objects of that experience, consciously, at one and the same time. This is the drama of the psyche defining itself in language, in a recurring sequence of continuous self-discovery and self-immolation, under extremes of circumstance and pressure – like a verbal and dramatic exposition of continual 'Big Bang' theory in the self-defining psyche.

This is not an attempted description of theory; this is how it feels to play it. Indeterminacy is a concrete experience, where with great discipline the actor needs to inhabit the unexpected in a continually transforming identity. In Barker, words aren't a way of emoting, or of demonstration of feeling, they are the exposition of unsettled possibilities, where even the deepest of convinced feeling is riven with the ambiguity, and then increasingly, the expectation that nothing is a settlement, and that to seek settlement is itself false, in fact, not pure.

The actor needs to contain and express this sense, at the core of his sounding out Barker's language, of this duality; that as certain as he is of the necessity to speak in seeking definition, that very definition is fundamentally insecure. This has a direct effect on the quality of the actor's speaking – a consciousness of sound, in theatrical space, as an experiment, as well as a need, or a function. The sound of seeking the indeterminate, the intangible, should be a concrete drama between the actors, and between the actor and the audience. It is a continuous, energized struggle to sound out propositions – a wrestling school in deed – engaged in a drama of psyche and persona in a will to identity out of an indeterminacy of some kind, enacted by an extremity of event arising from the catastrophic landscape.

If an understanding of this kind is the starting point for the actor, it clearly should lead to what might be understood as the actor's sensitization to a refined degree of being, a degree of precisely explored and resonated being, a super-conscious ' I am being like this'. This is a promoted degree of self-consciousness that is emphatically not 'stagey', where some might follow that false trail. In fact, the trail is one of hyper-conscious, liquid, mercurial trace-following of indeterminacy and discovery. This requires speeds of thought, and capacity for change in a spread of emotionally-justified ranges of sound and articulacy. When the persona is in a crisis of indeterminate and constantly refounded identity, the actor has to pursue a search for the precise, the necessary: the selected word, and word cadence, the apposite verbal and physical determination: the requirement is one and the same for both actor and persona under pressure. To descend into a gush, a slew, an emoting, is to fail completely to understand the disciplined opportunity and practice of the text.

Out of precision and consciousness comes, very naturally, humour and knowing: where knowing and describing how black the experience is, while experiencing it, is to be in possession, quite naturally, of the apprehension of folly, or futility, as felt and serious: how comic, how tricked you are. The actor has to apprehend and communicate the aesthetic process here – that of the masks of Tragedy and Comedy crushed forcibly together, grinding one on another in baleful, true excitation.

This aesthetic is embedded in the flow of Barker's cadences, the generative coils and spasms of line after (often) long, self-discovering, self-identifying line. The sound energy of these lines is like the loops of an accelerating, elliptical planet being flung out into further speed by close contact with the energizing sun of an idea or an experience. This is how it feels saying these lines, allowing the energy of them – crackling, dark, pained, funny – to mould you, as much as you are moulding them. This is the implication of indeterminacy: that you must actively demonstrate that you are as much subject to the lines and the way they form themselves in your mouth and your psyche, as they may be subject to you. This is drama in itself – a flux – and is required in the way Barker's imagination and intention is structured.

So Barker's text on the page needs to be seen more consciously than in other texts as an architecture of experiencing sound, and being energized by that drama. And energised in the truth that this is a drama in itself, and that it is a precise searching, and a precise, but liquid practice; it is not arch process, or one seeking abstraction. The dark, true, comic and serious fear is that living is a rootless, indeterminate abstraction, and that has to be played with consciousness, which is not self-consciousness. Abstraction is in the aesthetic, in the fabric of the landscape and the text, and it does not need to be semaphored.

Within such a context, being directed by Barker is giving yourself over on this basis to a consciousness of this architecture, and that the requirements of it will necessarily be expansive and elastic, as well as always absolutely precise. Nothing is 'something like'; it is exact and clear; it isn't non-rhetorical, it makes a natural aesthetic of trying out the sounds of the language, and such is the strength of the construction, the actor can hear pretty readily the true or the untrue sound. We're here not looking for the method of the actor, we're trying to hear the method of the dramatist. A ready acknowledgement to fulfill the function, the requirement, and not 'cost' it in Stanislavskian terms, is a straightforward necessity.

If the function is fulfilled, it will be true, because it is written true, and deeply, at the surface.

Appendix Two: Howard Barker: A Chronology of His Work

Dates are of first stagings, broadcasts or publications (not subsequent productions in the same medium). All places of publication London, unless otherwise noted. 'JC' = John Calder; 'OB' = Oberon Books.
All first stagings occur in Britain except where noted; 'HB' = production directed by Barker.
*Indicates that this unpublished text is summarised and considered in Rabey (1989)
# Indicates a non-professional university production

1970: *One Afternoon on the 63rd level of the North Face of the Pyramid of Cheops The Great*: broadcast BBC Radio 1970; unpublished*
Cheek: staged 1970; published in *New Short Plays: 3* (Eyre Methuen Playscripts 1972, volume credited to Barker, Grillo, Haworth and Simmons)
No One Was Saved: staged 1970; unpublished*

1971: *Henry V in Two Parts*: broadcast BBC Radio 1971; unpublished*

1972: *Herman, with Millie and Mick*: broadcast BBC Radio 1972; unpublished*
Edward – The Final Days: staged 1972; unpublished*
Alpha Alpha: staged 1972; unpublished*
Faceache: staged 1972; unpublished

1973: *Skipper*: staged 1973; unpublished*
My Sister and I: staged 1973; unpublished*
Rule Britannia: staged 1973; unpublished
Bang: staged 1973; unpublished

1975: *Claw*: staged 1975; *Stripwell*, staged 1975; published together (JC, 1977)

1976: *Wax*: staged 1976; unpublished*
Heroes of Labour: unproduced television play; published in *Gambit* 29 (JC, 1976)

1977: *Fair Slaughter*: staged and published (JC)
That Good Between Us: staged; published 1980 with *Credentials of a Sympathiser*, unproduced television play (JC)

1978: *The Love of a Good Man*: staged; published 1980 with *All Bleeding*, unproduced television play (JC)
The Hang of the Gaol: staged; published 1982 with *Heaven*, unproduced television play (JC)

1980: *The Loud Boy's Life*: staged; published 1982 in *Two Plays for the Right* (JC)
Birth on a Hard Shoulder: staged (Stockholm, Sweden); published 1982 in *Two Plays for the Right* (JC)

1981: *No End of Blame*: staged and published (JC)
The Poor Man's Friend staged*

1983: *Victory*: staged and published (JC)
A Passion in Six Days: staged; published 1985 with *Downchild* (JC)
Crimes in Hot Countries: staged#; published 1984 (JC)

1984: *Pity in History*: published in *Gambit* 41 (JC); staged 1984# (Dublin, Ireland), and broadcast BBCTV 1985
The Power of the Dog: staged and published (JC)

'Art Matters' (sketch) staged (published in *The Big One*, eds. Bill Bachle and Susannah York, Methuen, 1984)
Don't Exaggerate (performance poem): staged, published with other poems 1985 (JC)

1985: *The Castle*: staged and published with *Scenes from an Execution* (JC)
Scenes from an Execution: broadcast by BBC Radio
Downchild staged
The Blow: unproduced filmscript*

1986: *Women Beware Women*: staged and published (JC)
The Breath of the Crowd: (performance poem): staged and published with other poems (JC)

1987: *Gary the Thief/Gary Upright*: (performance poems): published with other poems (JC)

1988: *The Possibilities*: staged and published (JC)
The Last Supper: staged and published (JC)
The Bite of the Night: (written 1985) staged and published (JC)
Lullabies for the Impatient: poems, published (JC)
The Smile: published in *New Plays: 2*, ed. Peter Terson (Oxford UP)

1989: *Seven Lears*: staged and published with *Golgo* (JC)
Golgo: staged and published with *Seven Lears* (JC)
The Europeans (written 1987) staged (Toronto, Canada) and published with *Judith* (JC)
Arguments for a Theatre (essays): first edition (JC)

1990: *Scenes from an Execution*: stage production
The Early Hours of a Reviled Man: broadcast by BBC Radio; stage production#
Collected Plays Vol. 1 (*Claw, No End of Blame, Victory, The Castle, Scenes from an Execution*) published (JC)

1991: *The Europeans*: staged#
The Ascent of Monte Grappa (poems): published (JC)

1992: *A Hard Heart*: staged, broadcast by BBC Radio, and published with *The Early Hours of a Reviled Man* (JC)
Ego in Arcadia: staged (Sienna, Italy)(HB)
Terrible Mouth (opera libretto): staged and published (Universal Edition)

1993: *Collected Plays Vol. 2* (*The Love of a Good Man, The Possibilities, Brutopia, Rome, (Uncle) Vanya, Ten Dilemmas*) published (JC)
Arguments for a Theatre (essays): second edition, Manchester University Press
All He Fears (marionette play): staged and published (JC)

1994: *Hated Nightfall*: staged (HB) and published with *Wounds to the Face* (JC)
Minna: staged (Vienna, Austria) and published (Alumnus, Leeds)

1995: *Judith* staged (HB)
(Uncle) Vanya (written 1992) staged#

1996: *(Uncle) Vanya* staged (HB)
Collected Plays Vol. 3 (*The Power of the Dog, The Europeans, Women Beware Women, Minna, Judith, Ego in Arcadia*) published (JC)
The Tortmann Diaries (poems): published (JC)
Defilo (Failed Greeks) written

1997: *Arguments for a Theatre* (essays): third edition, Manchester University Press
Wounds to the Face staged
An Eloquence (film) written
The Blood of a Wife (film) written

The Seduction of Almighty God written
1998: *Ursula* staged (HB)
Collected Plays Vol. 4 (*The Bite of the Night, Seven Lears, The Gaoler's Ache, He Stumbled, A House of Correction*) published (JC)
Ten Dilemmas staged#
1999: *Und* staged (HB)
Scenes from an Exexcution staged (HB)
A House of Correction broadcast, BBC Radio
Albertina broadcast, BBC Radio
The Swing at Night (marionette play) written
A Rich Woman's Poetry written; *All This Joseph* written
2000: *The Ecstatic Bible* staged (Adelaide, Australia)(HB)
He Stumbled staged (HB)
Animals in Paradise staged (Malmo, Sweden)
The Twelfth Battle of Isonzo staged (Saint-Brieuc, France, in French)
The Swing at Night (marionette play) staged and published (JC)
Stalingrad (opera libretto) written; *All This Joseph* written
2001: *The Twelfth Battle of Isonzo* staged (Dublin, Ireland, in English)(HB)
A House of Correction staged (HB)
Collected Plays Vol. 5 (*Ursula, The Brilliance of the Servant, 12 Encounters with a Prodigy, Und, The Twelfth Battle of Isonzo, Found in the Ground*) published (JC)
Two Skulls written
2002: *Gertrude* staged (HB); *Knowledge and a Girl* broadcast (HB), BBC Radio; published together (JC)
Stalingrad (opera libretto) staged (Denmark)
Brutopia staged (Besançon, France, in French)
N/A (Sad Kissing) written
Five Names written
2003: *13 Objects* staged (HB)
The Fence in its Thousandth Year written
The Moving and the Still written
2004: *Dead Hands* staged (HB) and published (OB)
N/A (Sad Kissing) staged (Vienna, Austria)
The Moving and the Still broadcast, BBC Radio
Christ's Dog written
The Dying of Today written
Acts (Chapter One) written
2005: *Death, The One and The Art of Theatre* (essays) published (Routledge)
Animals in Paradise staged (Rouen, France, in French)(HB)
The Ecstatic Bible published (OB)
The Fence in its Thousandth Year staged (HB) and published (OB)
Christ's Dog staged (Vienna, Austria)
Two Skulls broadcast (Danish Radio)
Dead, Dead and Very Dead (libretto) written
Heroica (film) written
Adorations Chapter 1 (film) written
Let Me written; Howard Barker/Eduardo Houth: *A Style and its Origins* written
2006: *The Seduction of Almighty God* staged and published (OB)
The Road, The House, The Road broadcast, BBC Radio
Let Me broadcast, BBC Radio

    A Wounded Knife (formerly titled *A Living Dog*) staged (Odense, Denmark)
    Plays: One (Victory, Scenes from an Execution, The Possibilities, The Europeans) published (OB)
    Plays: Two (The Castle, Gertrude – The Cry, 13 Objects, Animals in Paradise) published (OB)
    The Forty (Few Words) written
    I Saw Myself written
2007: *Lot and his God* written
    Howard Barker/Eduardo Houth: *A Style and its Origins* published (OB)
    The Dying of Today staged (Caen, France)
    Actress With an Unloved Child written
2008: *Twelve Encounters with a Prodigy* staged (Odense, Denmark)
    I Saw Myself staged (HB)
    The Dying of Today staged (London)

Notes

Chapter 1

1 A. Bogart, *And Then, You Act* (London: Routledge, 2007), p. 115.
2 Barker, in the radio documentary 'Departures from a Position', BBC Radio 3, 14 February 1999.
3 Barker, *ibid.*
4 Roger Owen, in *Theatre of Catastrophe*, eds. K. Gritzner and D. I. Rabey (London: Oberon, 2006), p. 192, 193.
5 Quoted by Aleks Sierz in *The Theatre of Martin Crimp* (London: Methuen, 2006), pp. 66–7. Crimp has, also like Barker, found more receptive audiences in France, where writers enduringly influenced by Beckett, Genet, Duras and Sarraute tend to reject linear narrative for more experimental treatments of time and viewpoint, and characters who are existentially constituted as 'present to themselves' through the language they speak (Noëlle Renaude quoted in *ibid.*, p. 74).
6 Michael Mangan, *Performing Dark Arts* (Bristol: Intellect, 2007), p. 93.
7 *Ibid.*, p. 94.
8 Barker, on the BBC Radio 3 programme *Private Passions*, 11 June 2006.
9 Rabey, in Gritzner and Rabey, p. 18.
10 The term 'anti-history' first appears in the subtitle to Barker's play *The Power of the Dog* (staged 1984).
11 For example: *Downchild, Crimes in Hot Countries, The Europeans, Brutopia, Hated Nightfall, (Uncle) Vanya, The Early Hours of a Reviled Man, Golgo, The Fence in its Thousandth Year.*
12 Julia Kristeva, *Powers of Horror*, trans. L. S. Roudiez (New York: Columbia University Press, 1982) p. 5.
13 Murgatroyd, in Barker's 1984 play *Pity in History*, is the most savagely comic example of this.
14 Interview with the author, 19 August 2007.
15 Savill, quoted in Cornforth, Andy, and Rabey, D. I., 'Kissing Holes for the Bullets: Consciousness in Directing and Playing Barker's *(Uncle) Vanya*', in *Performing Arts International*, Vol. 1, Part 4, (1999) 25–45 (43–4).
16 Quoted as an epigraph to Barker's 1988 poetry collection, *Lullabies for the Impatient.*
17 Oliver Ford Davies, *Performing Shakespeare* (London: Nick Hern Books, 2007), p. 117.
18 Interview with the author, 19 August 2007. Bertish adds: 'It is even more crucial to be rigorously specific with Barker's texts. I know that sounds obvious but I feel that sometimes when audience members say they don't understand what they are watching it is because an actor has been generalised or indulgent – traps which are essential to avoid. Indeed, Howard's writing exposes actors if they in are in any way non-specific in their performance'.
19 M. Mangan, *A Preface to Shakespeare's Tragedies* (London: Longman, 1991), p. 37.
20 *Ibid.*, p. 40.
21 *Ibid.*, p. 42. Consider, in this context, George Steiner's observation: 'Man acts as if he were the shaper and master of language, while it is language which remains

mistress of man. When this relation of dominance is inverted, man succumbs to strange contrivances'; *After Babel* (Oxford: Oxford UP, 1975, p. xi.

22   'Wood savagely and eloquently depicts war as a populist sideshow, rigged by those in power, in which soldiers self-consciously perform "turns" of momentary glamour in a system which promptly consumes them; and entertainment as a reconciliation which barely masks mutual exploitation' – Rabey (2003), p. 243.

23   *Ibid.*, p. 18.

24   Wood, Introduction to J. Whiting's *Saint's Day* (Heinemann: London, 1963), pp. vi–vii.

25   See Rabey (2003), p. 23.

26   *Ibid.*, p. 27.

27   *Ibid.*, p. 26.

28   Whiting's theoretical concerns and objectives, as formulated through his collection of essays, *At Ease in a Bright Red Tie* (London: Oberon, 1999), also reveal several contact points with those of Barker: 'The theatre is the only art which lacks an articulate form of scholarship. So it is always being mistaken by its audience for something it is not' (p. 95); 'A work of art ... does not necessarily entertain, instruct or enlighten ... The thing is there: an audience taking from it what it can. It is not the artist's job to simplify the means of communication' (p. 17); 'I want to achieve something very raw: not coarse in texture, no; raw in the sense of the agony of an exposed nerve. As such it must carry at its beginning the sob of pain, the half-laugh, and then, in progress, rise through the crescendo scream to a finale of realisation and awe' (p. 91).

29   Saunders, *'Love Me or Kill Me': Sarah Kane and the Theatre of Extremes* (Manchester: Manchester University Press, 2002), p. 9.

30   James Macdonald, quoted in *ibid.*, p.8.

31   One may perceive the influence of Barker (and, most specifically, of his 1983 play *Victory*) in the subsequent plays of Nick Dear (*The Art of Success*, 1986) and Stephen Jeffreys (*The Clink*, 1990; *The Libertine*, 1994); and (more widely, unsurprisingly) in my own drama.

32   *Francis Bacon in Dublin* (Dublin: Hugh Lane Gallery and Thames and Hudson, 2000), p.20.

33   *Ibid.*, p. 15.

34   *Ibid.*, p. 23.

35   Barker, Introduction to *Landscape with Cries: Howard Barker Painting 1996–2006* (The Wrestling School: London, 2006), brochure accompanying the exhibition at the gallery Shillam Smith 3, 12 June–1 July 2006. Barker's drawing and painting deserves further exposition in its own right, which I have neither the space nor the expertise to undertake here.

36   *Ibid.*

37   Barker, on the BBC Radio 3 programme *Private Passions*, 11 June 2006.

38   Compare Barker's assertion that, in his theatre, death is 'the condition of beauty' and anxiety 'the price of its revelation': *DTOAT*, p. 26.

39   Juliet Stevenson in the radio documentary 'Departures from a Position', BBC Radio 3, 14 February 1999.

40   Jean-François Lyotard, *The Lyotard Reader*, ed. A. Benjamin (Oxford: Blackwell, 1989), pp. 190, 191. In an unpublished examination essay at Aberystwyth University in 2004, Thomas J. Barnes observes how Barker's theatre gives the audience a great deal (of physical, linguistic and scenographic detail) , yet 'threatens to take away so much more'. The characters may appear to have symbolic aspects (not least in their names) and their actions ritual qualities which may

appear to give the audience 'clues to unlocking their intentions', only for the language and/or actions to 'shift and mutate into something new', eluding this grasp. Barnes claims that in a play such as *The Twelfth Battle of Isonzo* Barker uses language as a 'dark magnet' in which the characters' performative utterances alternately belittle and excite each other, and provide 'a strange and unnerving puzzle' which gives the audiences clues as to the impulses informing the lines rather than a single or definitive sense of their emotional meaning for the speaker. Barnes also likens Barker's language (particularly in *Found in the Ground*) to 'a map from which the signs and figures are gradually removed', involving 'a laying open of our social selves' in a terrain which the audience are nevertheless 'left to negotiate blind, feeling our way through a landscape which has the potential to lead us somewhere we have never been before' (quoted by permission).

41 Lingis, *Dangerous Emotions* (University of California: Berkeley, 2000), p. 165.

42 Lyotard, *op. cit.*, p. 206.

43 *Ibid.*, p. 246.

44 *Ibid.*, p. 393. This is a major disjunction between Barker and Edward Bond, who increasingly places his faith in an ideology of "rational" political analysis, the very foundation of which may seem socially deterministic and inscriptive.

45 Russell, e-mail to the author 31 July 2007.

46 Pertinently, Alphonso Lingis notes how Nietzsche, one of Barker's favourite philosophers, attempts the reversal of principal norms and values to 'evaluate positively the bodily, the instinctual, the earthly and the transitory': *Body Transformations* (Abingdon: Routledge, 2005), p. 89

47 Lyotard, *op. cit.*, p. 211. He further notes how 'This failure of imagination gives rise to a pain, a kind of cleavage within the subject between what can be conceived and what can be imagined or presented. But this pain in turn engenders a pleasure, in fact a double pleasure ... [the imagination] striving to figure even that which cannot be figured ... This dislocation of the faculties among themselves gives rise to the extreme tension (Kant calls it agitation) that characterizes the pathos of the sublime, as opposed to the calm feeling of beauty', *ibid.*, pp. 203–4. For development of these ideas, see Gritzner, 'Towards an Aesthetic of the Sublime in Howard Barker's Theatre', in Gritzner and Rabey (2006), pp. 83–94.

48 Gritzner, 'Catastrophic Sexualities in Howard Barker's Theatre of Transgression' in M. Sönser Breen and F. Peters (eds) *Genealogies of Identity* (Amsterdam and New York: Rodopi, 2005), pp. 96–106 (95, 96, 104, 97). In several Barker plays, particularly *Lot and his God* and *Ursula*, love is characterised as the sublime loss of self.

49 J. Stevenson in the radio documentary 'Departures from a Position', BBC Radio 3, 14 February 1999.

50 Barker, in the radio documentary 'Departures from a Position', BBC Radio 3, 14 February 1999.

51 Lingis, in *Time and Value*, eds. S. Lash, A. Quick and R. Roberts (Oxford: Blackwell, 1998), p. 19.

52 *Ibid.*, p. 19.

53 *Ibid.*, p. 23.

54 *Ibid.*, p. 24.

55 Lingis, (2000), p. 161.

56 *Time and Value*, p. 29.

57 E. Grosz, in *Encounters with Alphonso Lingis*, eds. A. E. Hooke and W. W. Fuchs (Lanham: Lexington, 2003), p. 42.

58 Derrida, *Given Time 1: Counterfeit Money* (Chicago: University of Chicago Press, 1992), p. 41.
59 Lingis, in *Time and Value*, p. 28.
60 Lingis (2000), p. 141. And conversely, elsewhere, Lingis suggests how eroticism might proceed precisely to a sense of catastrophic time: that looking at a naked body constitutes a form of caress, which may proceed to the touch, which promises a 'surprising pleasure which means that the past is cut off and the experience is totally in the present'; so 'that too is a break in time, a catastrophe'; Lingis in Hooke and Fuchs, *op. cit.*, p. 97.
61 Lingis (2000), p. 157. *Sacred Theatre* ed. Ralph Yarrow (Bristol: Intellect, 2007) promisingly initiates consideration of theatre which may specifically 'generate or open up to something which isn't definable through conventional categories', 'moments when you fall through the interstices of categories and into a kind of amazement', through 'a moment of framing' which is simultaneously 'an *aporia*, an un-or-not knowing', rather than 'a fixing (fixating?) of cultural or psychological capital' (Yarrow *et al.*, pp. 13, 16, 18). Whilst Lavery's writings on the destablizing presence of death (pp. 20–1) suggest a potential contact point with Barker's *DTOAT*, it is disappointing that the volume goes on to focus on Stoppard and Pinter rather than consider the more pertinent work of Barker, David Rudkin, Peter Barnes or Ed Thomas.

Chapter 2

1 Lyotard, *op. cit.*, p. xv.
2 'Departures from a Position'.
3 Brecht, quoted by McGrath in *A Good Night Out* (London: Methuen, 1981), p. 63.
4 For further identification and analysis of this, see Rabey, *English Drama Since 1940* .
5 *Artaud on Theatre*, ed. C. Schumacher (London: Methuen, 1989), p. 13.
6 Lingis, *Abuses* (Berkeley: University of California, 1994), p. 128.
7 Barba, in T. D'Urso and E. Barba, *Viaggi con/Voyages with Odin Teatret* (Brindisi: Editrice Alfeo, 1990), p. 202. Compare Bogart, *op. cit.*, p. 42.: 'one does not speak to a particular audience; rather you speak to a particular *part* of each individual audience member. You do not *have* your own audience; rather you address a specific component of the human experience in every audience'.
8 See Barba, *Beyond the Floating Islands* (New York: PAJ, 1986), 211ff.
9 Rustom Bharucha, *Theatre and the World* (London: Routledge, 1993), p. 7.
10 *Ibid.*, p. 41.
11 *Ibid.*, p. 10.
12 M. Mangan, *Edward Bond* (Plymouth: Northcote House, 1998), p. 3.
13 Baudrillard, *Seduction* (trans. B. Singer: Basingstoke: Macmillan, 1990), p. 81; see also p. 79.
14 Lingis, *The Community of Those who have Nothing in Common* (Bloomington: Indiana University Press, 1994), p. 10.
15 *Ibid.*, p. 124.
16 Roy Boyne, in Lash, Quick and Roberts, p. 55.
17 Martin Harries, *Forgetting Lot's Wife* (Fordham: New York, 2007), p. 93.
18 Stephen Greenblatt, *Renaissance Self-Fashioning* (Chicago: University of Chicago, 1980), p. 20.
19 *Ibid.*, pp. 20–1.

20 *Ibid.*, p. 21.
21 Iain Sinclair, *London Orbital* (London: Penguin, 2003) p. 44.
22 Louis-Ferdinand Céline, *Journey to the End of the Night*, trans. Ralph Manheim (London: Calder, 1988), p. 308.
23 Dumm, in Hooke and Fuchs, *op. cit.*, p. 168. Again, the specifics of humanity involve a keen sense of the limits of language, even as they compel its use, as at the deathbed; see Lingis, *The Community of Those who have Nothing in Common*, pp. 108, 113, which develops Heidegger's consciousness of death (*'zum Tode sein'*: 'being-towards-death'), a confrontation with one's own anxiety and limits that nevertheless encourages the subject, from that point on, to regard and live one's possibilities as specific (*eigentlich*). Lingis goes so far as to attempt to reclaim the term 'community' by proposing that this takes form 'not in elaborating a common language and reason' but in 'going to rejoin those who, fallen from the time of personal and collective history, have to go on when nothing is possible or promised' (Lingis, *Abuses*, p. 236); or, I would add, in contemplating dramatizations of these liminal human experiences, which foreshadow our own.
24 Lingis, *Dangerous Emotions*, pp. 75, 80.

Chapter 3

1 C. Lamb, *The Theatre of Howard Barker* (London: Routledge, 2004), p. 57.
2 *Ibid.*, p. 63.
3 The child's demand for help in the Second Parable, on the grounds that 'One day I'll be powerful and anyone who didn't help me will be made to suffer', particularly recalls, in order to subvert, associations with the central premises and deductions of Bond's *The Narrow Road to the Deep North* (1968) and *The Bundle* (1978).
4 See Rabey (1989), pp. 123–139, on the Nietzschean forms of breakthrough dramatised in *Victory*.
5 Lamb, *op. cit.*, p. 64.
6 See P. McCarthy, *Céline* (London: Allen Lane, 1975), for a good introductory critical biography of this writer.
7 Greenblatt (1980), p. 13.
8 J. Kristeva, *Powers of Horror*, trans. L. S. Roudiez (New York: Columbia University Press, 1982), p. 5.
9 *Ibid.*, p. 8.
10 *Ibid.*, p. 9. My favourite example of the abject in Barker's drama is the line spoken by Winterhalter in *The Ecstatic Bible*: 'CURE ME OR I WILL EMPTY MY PISS BUCKET THROUGH THE GRILLE'.
11 Brendan Kennelly, *The Book of Judas* (Newcastle-upon-Tyne: Bloodaxe, 1991), p. 10.
12 Rabey, 'The Bite of Exiled Love', in *Essays in Theatre/Études Théatrales* 13, 1 (November 1994), pp. 29–43 [42].
13 *Ibid.*, p. 31.
14 Greenblatt (1980), p. 220. See also Barker's *Downchild*, and the protagonist's climactic literal 'play on the brink of an abyss'.
15 Similarly, Apollo in *The Last Supper* is in some ways inspired by the poet Apollinaire.
16 Lamb, *op. cit.*, p. 89.
17 *Ibid.*, p. 85.

18 *Ibid.*, p. 93.
19 In Gritzner and Rabey, p. 216.
20 Doreen is haughtily oblivious to the destructive effects of her colonial and materialistic priorities; however, like other characters, her persistence brings her more complexity, as her devotion to the perfection and preservation of the object moves her towards a spiritual sense of quest and purpose.
21 Quoted in Lamb, *op. cit.*, p. 203.
22 Park's increasingly intense forms of mortification, leading to elevation on a pillar, echo those of the fourth-century Christain monk Simeon Stylites, whose retreat from society only increased his allure to those attempting to benefit from enlightenment, and spawned imitators; see Mangan (2007), p. 187.
23 See Rabey, 'For the Absent Truth Erect: Impotence and Potency in Howard Barker's Recent Drama', in *Essays in Theatre/Études Théatrales*, 10.1 (November 1991), pp. 31–37 [31].

Chapter 4

1 Kane, quoted by Saunders, *op. cit.*, p. 28.
2 Rabey, 'On Being a Shakespearian Dramatist', in *The Wye Plays* (Bristol: Intellect, 2004), p. 4.
3 Barker, *Arguments* (1997), p. 154.
4 *Ibid.* I would in fact argue that the first and second instances of Barker's instincts for, and skills in, these 'conversations' or re-visions occur much earlier in his work: in the 1971 radio play *Henry V in Two Parts*, which speculates about the imposition and maintenance of power through strategic historical fiction in the landscape of Shakespeare's *Henry V*; and *No One Was Saved* (1970), which purposefully and shockingly extends both Edward Bond's 1965 play *Saved* and the lyrical scenario of the Lennon/McCartney song *Eleanor Rigby* to challenge conventional false hopes. For further discussion of both of these, see Rabey (1989), pp. 12–13, 20–22.
5 R. Bolt, *A Man for All Seasons* (London: Methuen, 1995 edition), p. xi.
6 *Ibid.*, p. xiv.
7 *Ibid.*, p. xiii.
8 *Ibid.*, p. xvi
9 *Ibid.*, p. xiii.
10 Greenblatt (1980), p. 13.
11 *Ibid.*, p. 15.
12 *Ibid.*, p. 22.
13 *Ibid.*, p. 31.
14 Baudrillard, *op. cit.*, p. 85.
15 Lawrence Stone, 'The Rise of the Nuclear Family in Early Modern England: the Patriarchal Stage', in *The Family in History*, ed. C. E. Rosenberg (Philadelphia: University of Philadelphia Press, 1975), p. 25.
16 Greenblatt (1980), p. 44.
17 *Ibid.*, p. 53.
18 *Ibid.*, p. 51.
19 Lamb, *op. cit.*, p. 162: ' This is the case with Poussin in *Ego in Arcadia*, Chekhov in *(Uncle) Vanya* and – most explicitly – Benz in *Rome*'.
20 Greenblatt, *Shakespearean Negotiations* (Oxford: Clarendon, 1988), p. 40.
21 *Ibid.*, p. 128.

22 S. Booth, *'King Lear', 'Macbeth', Indefinition and Tragedy* (New York: Yale, 1983), p. 11.
23 Rabey (1989), pp. 171–2.
24 These figures thus amusingly reverse the function of the ghosts in Edward Bond's *Lear* (1971) – which Barker maintains he avoided reading or seeing prior to writing *Seven Lears* – in which the ghosts are nostalgic or regressive figures, drawing Bond's Lear away from engagement with the immediacies of life.
25 Shakespeare's *King Lear* contains reference to the rapacious bird, the kite; Barker makes an onstage image of the other sense of the word, the manmade paperclad frame which manifests and soars on the movements of the wind.
26 Barker, *Arguments* (1989), p. 153.
27 Rabey (2003), p. 169.
28 Cornforth, Cornforth and Rabey (1999), p. 29.
29 Cornforth, *ibid.*, p. 32.
30 Rabey, 'For the Absent Truth Erect: Impotence and Potency in Howard Barker's Recent Drama', *Essays in Theatre/Études Théâtrales* 10, 1 (November 1991), pp. 31–7 (35).
31 Cornforth, Cornforth and Rabey (1999), p. 35.
32 Cornforth, *ibid.*, p. 36.
33 See Cornforth and Rabey (1999) for more detail on our experiences of directing and performing in this production.
34 Cornforth, *ibid.*, p. 37.
35 Barker, letter to the author, 10 May 1995.
36 Cornforth, *op. cit.*, p. 37.
37 Barker, letter to the author, 10 May 1995.
38 Back cover to the Absolute Press English translation edition (Bath, 1990).
39 Barker, letter to the author, 17 August 2005.
40 Key paintings here by Watteau being 'Les De'lassements de la Guerre' and 'L'embarquement pour Cythère'.

Chapter 5

1 Kristeva, *op. cit.*, p. 8.
2 Baudrillard, *op. cit.*, pp. 128, 137.
3 Barker, Preface to *Terrible Mouth* (Universal Edition: London, 1992).
4 This effect of the doubled self is somewhat Célinian, in the terms outlined in *Journey to the End of the Night*: 'when you're weak, the best way to fortify yourself is to strip the people you fear of the last bit of prestige you're still inclined to give them. Learn to consider them as they are, worse than they are in fact and from every point of view. That will release you, set you free, protect you more than you can possibly imagine. It will give you another self. There will be two of you' (tr. Manheim: Calder: London, 1988, p. 62). Ian McDiarmid played Sleen in the BBC radio production of Barker's *Early Hours*, as well as Dancer and Goya, one performance inevitably transporting associations of the other.
5 Lamb, *op. cit.*, pp. 162–3.
6 *Ibid.*, p. 163.
7 Lili is the non-intellectual incarnation of physical perfection venerated in Céline's *Féerie pour une Autre Fois*, and Le Vig is based on that author's companion, Robert le Vigan, a volatile film actor 'given to wild outbursts of laughter and Mephistophelian poses', notorious for his power to upstage others, and frequent

    adopter of messianic postures in recollection of his having once played Christ (McCarthy, *op. cit.*, p. 127).

8 An infernal version of the classical premise of Arcadia first occurs in Barker's drama in the insistently, hellishly convivial fête in *Ten Dilemmas*.

9 The eruption of a huge crowd onto a stage otherwise populated by small scenes, and Caroline's line 'when we are exiled to some foreign city I am making hats', may be an influence on Caryl Churchill's 2000 play *Far Away*, in which restrained two-person dialogues, depicting a wartime refugee's migration to a new domestic setting and a war-effort workplace dedicated to making hats, are contrasted with the stampede, in one scene, of innumerable sacrificial victims for whom the hats are intended.

10 This name inevitably carries the literary associations of John Marston, author of numerous satires and, appropriately, the Jacobean play *The Malcontent*.

11 Anne Barton, Introduction to the Penguin edition of *The Tempest* (London, 1968), p. 14.

12 In interview, in Rabey (1989), p. 4.

13 Barton, *op. cit.*, p. 48.

14 *Ibid.*, p. 35.

15 My own 2004 production of *Ursula*, at Aberystwyth University, involved the division of the stage at this point by the lowering of a driftwood frame, recalling both picture frame and mirror, behind which Christ/Lucas stood to address Ursula and draw her towards him physically; this economically suggested an alternative reality (abruptly lost and shattered by the frame being flown back up and Lucas returning to his former prone position) as well as alluding to the play's themes of the lure and limitations of the reflected self.

16 Séan O'Callaghan, who played Lucas, reports that his delivery of Lucas's speech which begins 'I am not less beautiful than Christ nor less lonely' was helped immensely by Barker saying 'imagine you are in your hotel room playing in front of the mirror': 'That note caught the childlike quality, vanity, ego and the sensuality of Lucas rehearsing his seduction'. Compare Melanie Jessop's sense of the imaginary mirror in playing *Und*, Appendix One.

17 Placida is not literally naked here; the stage directions add '*She deceives them*'. However, in Barker's 1998 production, costume designer Lucy Weller arranged for Victoria Wicks as Placida to wear a transparent tangerine dress which partly disclosed and further dramatised her body, in accordance with Placida's declaration of shamelessness, and which slipped from her shoulder to expose a breast in appropriate demonstration of the eroticism of disarray when she moved into physical abandonment with Lucas.

18 In Barker's 1998 and my 2004 production, respectively.

19 Kristeva, *op. cit.*, p. 161.

20 Consider also the tapping of Leonora's stick, the sound of the dragged chairs, the eerie moans of Cynthia, the possible sounds of the boat on the river, defiant but isolated singing of psalms, the waves of the estuary, the clatter of the dropped sword, the smashed glass.

21 Kristeva, *op. cit.*, p. 110.

22 Sarah Rose Evans has observed how Barker and his characters subvert the convention of realistic language in theatre, creating a 'pseudo-internal narrative which is then expressed by the character eliptically', activating several possible interpretations and a struggle to decide whether we should sympathize with a character or not. Examples here are Leonora's proclaimed blindness, and Placida's epiphany 'we are the estuary': this latter ambiguous metaphor can be

interpreted as an awareness that 'the women themselves inspire fear, passion, madness and exultation in others – men such as Lucas, or women who have yet to acknowledge their own sexuality or fertility; it could be a reference to their fear of themselves, the resistance to change represented by virginity and a defence against pollution; or even that their physical and mental depths will remain mysterious and unknown to man. Indeed, subverting our expectations of sympathy, recognition and understanding in regards to the effect of women on men and the power of sex is a predilection of Barker's' (unpublished essay, Aberystwyth University, 2006; quoted by permission).

23 As summarized by Quick, in Lash, Quick and Roberts, p. 79.

Chapter 6

1 A line of Leopold's, present in Barker's 1987 first draft of *The Europeans*, but excised from the subsequently published and performed version of the text.
2 Rabey (1989), p. 170.
3 Cornforth, in Cornforth and Rabey, *op. cit.*, p. 32.
4 Booth, *Precious Nonsense* (Berkeley: University of California Press, 1998), pp. 3, 6.
5 *Ibid.*, p. 8.
6 *Ibid.*, pp. 35–6.
7 *Ibid.*, pp. 90, 100.
8 Kubiak, *Stages of Terror* (Bloomington and Indianapolis: Indiana University Press, 1991), pp. 2–3.
9 *Ibid.*, p. 5. Kubiak here quotes terms from Jonathan Dollimore.
10 Nietzsche, *The Gay Science*, #261, trans. W. Kaufmann (New York: Vintage, 1974).
11 *Ibid.*, p. 44.
12 Angel-Perez, 'Facing defacement', in Gritzner and Rabey (2006), pp. 136–49 (138).
13 Zimmermann, in Gritzner and Rabey (2006), p. 222. On a less elevated level, the familiarity of the scenario is also identified in and by the wryly romantic popular song, 'It's All in The Game' (Dawes/Sigman).
14 Hettie Judah, 'Deadly Taste of Tea for Two', *The Times* 22 June 1999, p. 43.
15 McLane, in Hooke and Fuchs (2003), p. 51.
16 Kiehl, in Gritzner and Rabey (2006), p. 209.
17 Also noted by Lamb, *op. cit.*, p. 191. Indeed, Barker's earliest radio plays, *One Afternoon…* and *Henry V in Two Parts* from 1970 and 1971 first approach the theme of a duplicitous ruler who maintains authority by laying a false trail to distract from his actual identity and actions.
18 Ace McCarron's lighting design realised this most strikingly and erotically when Turner's dress strap was deliberately slipped from her shoulder, and the lighting state contracted into a precise focus on the circumference of her exposed breast, and then to darkness.
19 Tomlin, in Rabey and Gritzner (2006), pp. 112–13.
20 Kristeva, *op. cit.*, p. 47.
21 Booth, *op. cit.*, p. 120.

Chapter 7

1 It is worth noting, in a populist age, Barker's contrary attraction to the self-isolating central figure of the intellectual who is, to use a phrase from *Albertina*,

not content to be 'trapped behind the fictions of the ordinary'; from the wry academic Savage and the surprising itinerant Gary Upright to the licensed legislator Rocklaw, the reviled irritant Machinist, the deliberately unworldly Hoik.

2 Owen, in Gritzner and Rabey, *op. cit.*, p. 195.

3 My italics, suggesting Victoria Wicks's luxurious activation of the verbs in Barker's own production of the play for radio.

4 I am here inspired by, and indebted to, Kelly V. Jones, and her discussion of Webster's *The Duchess of Malfi*, in her PhD thesis, *The Political Aesthetics of Play and Display on the Early English Stage*, Aberystwyth University, 2007.

5 Greenblatt (1980), p. 220.

6 Jones, *op. cit.*, p. 212.

7 *Ibid.*, p. 212.

8 *Ibid.*, p. 216.

9 *Ibid.*, pp. 216–7.

10 *Ibid.*, p. 217.

11 *Ibid.*, p. 213.

12 As well as Wicks, the production featured Gerrard McArthur (King), Sarah Belcher (Snow White), Julia Tarnoky, Jane Bertish, Sean O'Callaghan and Tricia Kelly.

13 Lingis, in *Time and Value*, p. 27.

14 Lingis (2000), p. 131.

15 Lingis (2005), p. 82.

16 Grosz, in Hooke and Fuchs, *op. cit.*, p. 127.

17 *Ibid.*, p. 129.

18 Angel-Perez, in Gritzner and Rabey (2006), p. 147.

19 *Ibid.*, p. 130.

20 Lingis (2000), p. 151.

21 Lamb (2005) pp.2, 63.

22 A French-language translation of the play was staged prior to Barker's own production, in Saint-Breuc, 2000.

23 Barker, in rehearsal, July 2001.

24 Barker, *Don't Exaggerate* (Calder: London, 1985), p. 23.

25 Barker, in rehearsal, July 2001.

26 Compare *DTOAT*, p. 102: 'He had to have her (he could not *not* have her...) but she was fatal to him...'.

27 Isonzo's predatory quality was reflected by one of Leipzig's design details: three white high-heeled shoes were suspended, in separate metal cages, above the rusty metal grill floor of the acting area.

28 Baudrillard, *op. cit.*, p. 100.

29 Barker, in rehearsal, July 2001.

30 *Ibid.*

31 Baudrillard, *op. cit.*, p. 125.

32 Barker, in rehearsal, July 2001.

33 Owen, in Gritzner and Rabey (2006), who offers a spectator's perspective on the production and describes the staging as resembling 'a quasi-Dadaist cabaret'.

34 Baudrillard, *op. cit.*, p. 69.

35 Baudrillard, *op. cit.*, p.53.

36 Tenna's reply 'I'm seventeen' can be coquettish: though she is certainly younger than Isonzo, she may exaggerate her youth, and he may exaggerate his age, for the purposes of heightening, the appeal of contrast between them.

37 In performance, his insistence 'I'm not blind ... I merely shut my eyes' often provoked audible gasps from the audience – as did the second shock of the revelation of my unfocussed eyes .

38 Compare the intimation of the poverty of existence in *Death, The One and the Art of Theatre*, p. 62: 'Behind every door – even the door of desire – the world is reproduced'; and Mosca's disappointment at how sex fails to provide a sublime loss of self, in *Ego in Arcadia*: 'Cunt / A corridor of self again / What hangs on the womb's wall but mirrors' (*CP3*, 312).

39 Barker, in rehearsal.

40 There is a parallel with *Und* here, when Und finally reports her dying at the sight of the machine designed to exterminate her: the confrontation is itself fatal. There is a further parallel here with the last movement of Sarah Kane's play, *Blasted*, in which Kane definitely insisted that the character Ian literally 'dies', although the play extends into a dialogue of bleak but significant reconciliation between him and the woman he has abused. Kane described how her character Ian 'dies, and he finds that the thing he has ridiculed – life after death – really does exist. And that life is worse than where he was before. It really is hell' (in Saunders, *op. cit.*, p. 64). Barker had previously featured characters who spoke posthumously, in *The Bite of the Night* and *Rome*, which may have influenced Kane.

41 Walsh draped this long rectangle of black cloth around her neck like a scarf: this concealed her frontal nakedness, but intermittently, and so seductively redramatised it.

42 Barker, in rehearsal.

43 Barker, in rehearsal.

44 *Ibid.*

45 *Ibid.*

46 *Ibid.*

47 *Ibid.*

48 *Ibid.*, p. 22.

49 Bataille, *Eroticism*, 158.

50 Lamb, *ibid.*, 63.

51 *Arguments*, p. 147.

52 Lash, in Lash, Quick and Roberts, 159.

53 *Ibid.*

54 Gritzner (2005), 97.

55 Céline, *op. cit.*, pp. 213–14.

56 Lamb, 'Appalling Solitude', in Gritzner and Rabey, *op. cit.*, p. 152.

57 Evans, *op. cit.*

58 Lamb, 'Appalling Solitude', in Gritzner and Rabey, *op. cit.*, p. 162.

59 Nightfall's speech at this juncture may even seem parodic of so-called "kitchen sink" social realism associated with most plays presented at the Royal Court Theatre, particularly Edward Bond's *Saved* (1965), the springboard for one of Barker's earliest critical subversions, *No One Was Saved* (1970).

60 Thanks to Howard Barker for description of these details, in correspondence.

61 Back cover blurb to Barker's *Collected Plays Volume Five* (London: Calder, 2001).

62 Such as that which is the principal focus of Hans-Thies Lehmann in *Postdramatic Theatre* (London: Routledge, 2006), although Lehmann's account of Gertrude Stein's landscape plays suggests some contact points with the objectives of *Found in the Ground*. As I remark in Gritzner and Rabey, *op. cit.*, p. 27, one might argue, rather, that Barker is developing a post-theatrical drama.

63 Barker, in Gritzner and Rabey, *op. cit.*, p. 31.
64 *Arguments*, p. 147.
65 Gritzner, in Gritzner and Rabey, *op. cit.*, p. 91.
66 Angel-Perez, in Gritzner and Rabey, *op. cit.*, p. 143.
67 According to Barker, the image of the pyramids should invoke associations of post-war kitsch exotica, akin to that exemplified by Vladimir Tretchikoff's 1950 painting 'Chinese Girl'.
68 One of many instances in which *Found in the Ground* can be seen to extrapolate tropes from earlier Barker plays to their extreme and ultimate form: Toonelhuis's successful rhetorical disarmament of the pistol-wielding youth recalls and subverts the ending of *Stripwell* (1975); the landscape that is a quaking bog of corpses most evidently recalls that of *The Love of a Good Man* (1978); Macedonia articulates the 'war dead' in ways which challenge comprehension, a tragic variant on the discomfitingly comic presence and effect of Murgatroyd in *Pity in History* (1984); even the motif of the pyramid invokes the image of sublimated mass endeavour, masquerade and coercion at the centre of Barker's first play, *One Afternoon...* (1970).
69 Angel-Perez, in Gritzner and Rabey, *op. cit.*, p. 141.
70 Baudrillard, *Selected Writings* (Oxford: Blackwell, 1988), p. 195.

Chapter 8

1 Barker, in Gritzner and Rabey, *op. cit.*, p. 36.
2 *A Nietzsche Reader*, selected and translated by R. J. Hollingdale (Harmondsworth: Penguin, 1977), p. 81.
3 Baudrillard, *op. cit.*, p. 91.
4 Bataille, *op. cit.*, p. 140.
5 Barker, in Gritzner and Rabey, *op. cit.*, p. 35.
6 *Defilo* in some ways subverts another Golding novel, *Pincher Martin*, in which a (literally) self-inventive wily and sensual liar is shipwrecked and harrowed into what Filo calls a 'stark and calcinated image of apology' by a punitive God.
7 Compare the description of the figure of Odysseus in Barker's unproduced filmscript, *The Blood of a Wife*: 'To be a husband I think it is necessary to be absent. Like Odysseus. Odysseus was always a husband. It distinguished him, no matter where he was, this idea of the home and the wife ... But once he returned, what kind of husband was Odysseus? It killed Odysseus to be a husband, surely? It concluded his life'.
8 'N/A' of course also has the tersely bleak modern resonance, 'not available'.
9 This phrase ominously anticipates the terminal landscape in Barker's *The Road, The House, The Road*.
10 One might also add *Defilo, A Rich Woman's Poetry, Acts Chapter One* to this list of plays in which the female lead role seems conceived for, and/or inspired by, the distinctive performative powers of Victoria Wicks.
11 Kilpatrick, 'The Myth's the Thing: Barker's Revision of Elsinore in *Gertrude – The Cry*', in *Text and Presentation* 24 (New York: McFarland, 2003). Stoppard and Müller's Hamlet plays are *Rosenkrantz and Guildenstern are Dead* and *Hamletmachine*.
12 *Nietzsche*, (Collins: Glasgow, 1978), p. 60.
13 *Ibid.*, p. 66.
14 *Ibid.*, p. 76.
15 *Ibid.*, pp. 93–4.

16 *A Nietzsche Reader*, p. 164.

17 *Ibid.*, p. 236.

18 *Ibid.*, p. 275.

19 Baudrillard, *op. cit.*, pp. 83, 88.

20 Thanks to Karoline Gritzner for assistance towards these observations.

21 Freedman, 'Frame-Up: Feminism, Psychoanalysis, Theatre', in *Performing Feminisms*, ed. Sue-Ellen Case (Baltimore, Maryland: Johns Hopkins University Press, 1990), p. 74.

22 Lingis, (2000), p.145. Lingis notes, in terms which might be designed for Barker's Gertrude, how the ethereal vision of the voluptuously feminine mesmerically breaks free from 'the world of work and reason' (pp. 146–7).

23 The most notable exception to this is Claudius's aside, in which he discloses his failure to pray to God ('My words fly up, my thoughts remain below', III.3.) – whereas Barker's Claudius increasingly defines himself in opposition to the impervious complacency of a traditional God, not least in his dedication to the secular 'God' of Gertrude's nakedness.

24 Hamlet's society of surveillance thus provides unsettling resonances for any more recent authoritarian invocations of collective "rights", expressed in adolescently conformist terms, against the unruly, because exceptional, individual.

25 Evans, *op. cit.*

26 Thanks to Karoline Gritzner for assistance towards these observations.

27 It does so in ways which exemplify philosophical propositions identified elsewhere by K. Liepe-Levinson in *Strip Show* (London: Routledge, 2002):
Like Butler, Bataille and Sontag each propose that images of death, or the *risk* of death, may conjure up an erotic infinity for the viewer or reader because death signifies an indefinable state. This death-like erotic infinity, which defies the limitations and precepts of the social world, can be represented and apprehended only in terms of its tension with the everyday – that is, through a dynamics of transgression in which social, personal and even "natural" laws are foregrounded and shattered at the same time. (pp. 147–8).

28 Bataille, *op. cit.*, p. 18.

29 *Ibid.*, p. 20.

30 See Mangan, *A Preface to Shakespeare's Tragedies* (London: Longman, 1991), pp. 120–1, and D. I. Rabey, 'Play, Satire, Self-Definition and Individuation in *Hamlet*', in *Hamlet Studies* V, nos. 1 & 2, Summer/Winter 1983, pp. 6–26.

31 Mangan, *ibid.*, p. 131.

32 *Ibid.*, p. 131.

33 Wicks wore tight-fitting dresses, transparent gowns and unfastened coats and robes which dramatized and artfully exposed her body in the scenes in which she was not completely naked, apart from Gertrude's ever-present signature high heeled shoes.

34 My 2004 production of Barker's *Ursula* in that space and configuration had proven to me the dynamism of such a staging for his work.

35 Barker/Houth, *A Style and its Origins* (London: Oberon Books, 2007), pp. 73–4.

36 Towards the end of the play Claudius recalls feeling diminished by the unsympathetic 'COLD WATCHING' to which Cascan subjects him. Consider Bogart on the theatrical gaze: 'Not only is the theatre about an audience seeing, it is also where an audience witnesses an actor seeing … The better the actor, the more differentiated the seeing. The drama is a drama of unfolding sight', *op. cit.*, p. 81.

37 Our Hamlet's delivered these lines directly to the audience: 'WOMEN ARE SO COARSE / They are / They are coarse / More so than men / Vastly more / Vastly /

I have noticed it / More coarse than men / THE POLITE ONES ARE THE WORST / Behind their downcast eyes these / SWARMING VOCABULARIES'.

38 One might even see *Fair Slaughter* as an early example of this, in relation to the secular "religion" of Communism.
39 Barker, 'The Ecstasy of the Martyr': programme essay for The Wrestling School's production of *The Seduction of Almighty God*, November 2006.
40 Kristeva, *op. cit.*, p. 8.
41 E. M. Cioran, *Tears and Saints*, (trans. Zarifopol-Johnston, Chicago: University of Chicago, 1996), p. 117.
42 This repetition, both bathetic and appalling, may appear to the modern theatre-goer as something of a subversive echo of the stage imagery of Sarah Kane's *Blasted* (1995).
43 Cioran, *op. cit.*, p. 100.
44 Artaud, *Artaud on Theatre* (London: Methuen, 1989), ed. C. Schumacher, p. 130.
45 Shaw, *Man and Superman* (London: Penguin, 1977), p. 251.
46 Cioran, *op. cit.*, p. 80.
47 G. Robinson, *A Private Mythology* (London : Associated University Presses, 1988), p. 73.

Chapter 9

1 Lingis, *Body Transformations* (London: Routledge, 2005), p. 15.
2 Lingis, *The Community of Those who Have Nothing in Common*, pp. 57–8.
3 Thanks to Ken Rabey for this observation.
4 Compare the resurgent forms of The Bishop in *Seven Lears* and the Helmsman in *Defilo*.
5 See Kristeva, *op. cit.*, p. 5; and Rabey, in Gritzner and Rabey, pp. 16–19. As a dramatic portrait of a liminal agent, preserving definition and communication at the edge of an Empire whose governors speak a separate language, *An Eloquence* is matched only by David Rudkin's 1986 stage play *The Saxon Shore*, which is set on Hadrian's Wall and similarly identifies the significant breakdown of covenant of protection for those who are instrumental in imperial control, when that control faces a threat which thoroughly challenges its terms and methods of definition and communication.
6 Compare the strategic uses of the chair, the differing gestures of authority manifested in its refusal, occupation and acceptance, in *The Twelfth Battle of Isonzo* and *Dead Hands*.
7 Summarized and considered in Rabey (1989).
8 For a thoughtful consideration of this phenomenon, including its more substantial complications, see R. Schneider, *The Explicit Body in Performance* (London: Routledge, 1997).
9 R. Girard, *Violence and the Sacred* (trans. P. Gregory, Baltimore: Johns Hopkins University Press, 1977), quoted by Kubiak, *op. cit.*, p. 17. Other Barkerian examples of the talismanic which one might further include in the category of such 'desiring objects' might be: the severed hand in *Fair Slaughter*; the eponymous edifice in *The Castle*; even the bodies of Bianca in *Women Beware Women* and Helen in *The Bite of the Night*, objectified by others (and partly by self, in the former case).
10 Barker, programme notes to The Wrestling School's production of *13 Objects*, October 2003.
11 S. Shepherd, *Theatre, Body and Pleasure* (London: Routledge, 2006), p. 77

12 In *Beyond the Pleasure Principle*, Freud relates the child's game of 'disappearance and return' to 'the instinctual renunciation' which the child makes in allowing the mother to go away without protesting; for which the child 'compensates' by 'staging' on her/his own terms 'the disappearance and return of the objects' within her/his reach'; Anna Freud (ed.), *The Essentials of Psychoanalysis* (Harmondsworth: Penguin, 1986), p. 225.

13 Bataille, *op. cit.*, p. 130.

14 Barker, in Gritzner and Rabey, p. 31.

15 *Arguments for a Theatre*, pp. 145–6.

16 Iball, in Gritzner and Rabey, pp. 70–82 (70, 74).

17 One should here be mindful of the associations of pharoahs as figures of control in and beyond death, and even of supreme deceit in Barker's first play *One Afternoon...* (1970).

18 McDiarmid, in *Players of Shakespeare 2*, eds. R. Jackson and R. Smallwood (Cambridge: Cambridge University Press, 1988), p. 47.

19 Compare Tortmann in *He Stumbled*, who manages (dramatizes?) fateful experimental encounters, apparently from beyond the grave.

20 Istvan's professed search for a barber, in which he is met only with indisposition and closure, sounds as if he had wandered into the scenario of *The Dying of Today*.

21 As James Reynolds observes (in Gritzner and Rabey, p. 65), these exchanges and the consequent collapse of logic compounds 'the surreal nature of the location both on and offstage' in *Dead Hands*, embodying 'the effect of anxiety in the audience'.

22 Barker, in correspondence to Iball, quoted in Gritzner and Rabey, pp. 80–1.

23 *Christ's Dog* awaits a full British professional production at the time of writing, though Barker directed a rehearsed reading, with Nicholas Le Prevost in the central role, at the Riverside Studios in London, to accompany the conclusion of The Wrestling School's national tour of *The Seduction of Almighty God* in December 2006.

24 The enduring fascination of the character of Casanova is reflected in, amongst other things, two television series (written by Dennis Potter, 1971, and Russell T. Davies, 2005), two films (directed by Fellini, 1976, and Hallström, 2005) and plays by Dic Edwards (*Casanova Undone*, 1992), and David Greig (*Casanova*, 2001).

25 The undercut hanging recalls Beckett's *Waiting for Godot*; the servant who disobediently precedes his master into death, rather than assisting the master's death, recalls the aptly-named Eros in Shakespeare's *Antony and Cleopatra*.

26 A 'last experience of silence' parallel to Sleen's at the conclusion of *The Early Hours of a Reviled Man*.

27 Barker's programme notes for The Wrestling School's production of *Dead Hands* (October 2004).

Chapter 10

1 With thanks to Karoline Gritzner for assistance towards these observations.

2 Jean-Luc Nancy, in Hooke and Fuchs, *op. cit.*, p. 101.

3 *Ibid.*, p. 102.

4 *Ibid.*, p. 104.

5 *Ibid.*, p. 105.

6 Barker's programme notes for The Wrestling School's production, June 2005.

7 Baudrillard, *op. cit.*, p. 69.
8 Watching the premiere production, I interpreted this regression as the effect of institutional enclosure and/or torture on Photo. Alternatively, the involution may be self-imposed, he may be 'shocked into infancy by the discovery of his incest' – Barker/Houth, *A Style and Its Origins*, p. 81.
9 Their wryly, purposefully dispassionate, developed narrative of a 'slut' mother who carelessly lets her pram be crushed beneath a lorry is strangely reminiscent of the protagonist's loss of her baby in the very early Barker play *No One Was Saved* (1970), particularly in the form of events in the 1972 film *Made* (dir: John McKenzie), for which Barker's play provided the basis.
10 Barker/Houth, *A Style and Its Origins*, p. 82; which succinctly describes the exordium of Barker's production of *The Fence* on pp. 80–1.
11 Barker's programme notes for The Wrestling School's production, June 2005.
12 Consider also the eloquence of the relinquished garment as a detail in Barker/Kaiser's set design for *Gertrude – The Cry*, a wall montage of empty white shirts, and one black one.
13 Thus this scene offers a filmic realisation of the recurrent action in *The Fence* and Tenna's aria of the beauty of the widow, seduced, in *Isonzo*.
14 Barker/Houth, *A Style and Its Origins*, p. 69.
15 *Ibid.*
16 The tension between the value invested in the art object and that in human life is a theme which can also be traced in other Barker works, particularly *Fair Slaughter*, *Pity in History*, the sketch 'Art Matters' (in *The Big One*, eds. B. Bachle and S. York, Methuen, 1984) and *I Saw Myself*.
17 Perhaps a wry echo of Kierkegaard's character-persona, Johannes the Seducer.
18 J observes the delicacy with which Barbara lifts her breast for the knife 'in three fingers' as if she 'took a young thing from a nest to show a child'.
19 F. Botting, *Gothic* (Routledge, London, 1996), p. 10.
20 This phrase was also the title of New Labour's election anthem in 1997.
21 For a good early example, see the dissection of the deliberately obfuscatory phrase 'Bombs were dropped' in *No End of Blame* (1981).
22 *A Wounded Knife* could even be played in repertoire by a company playing *Gertrude – The Cry*, with the actors of Claudius and Hamlet doubling Biro and Houth, in one or other configuration: then Gertrude/Sleev, Albert/Quittur, Cascan/Globe, Ragusa/Fashoda and Isola/Tooshay.
23 A charged statement which recalls Batter's final proclamation on the dying Stucley in *The Castle*: 'His last walk'.
24 Harries, *op. cit.*, p. 6. Harries notes how the story asks its audience to link 'extremes of vision to transgressive desire and to retrospection as they are embodied in an act of destructive viewing' and 'obliges [them] to think about mass death' (p. 9), where Lot's wife is always 'a secular figure, even a figure for secularization' (p. 14) who figures 'the coincidence of dangerous individual memory with catastrophic historical damage' (p. 16).
25 *Ibid.*, p. 15.
26 It is as if the sexual perspectives of Tortmann in *He Stumbled* and Eff's father in *Dead Hands* were more centrally explored by Barker in the figure of Lot.
27 See Rabey, 'What Do You See?' (1992).
28 Barker, *The Guardian* 5 June 2007: http://blogs.guardian.co.uk/theatre/2007/06/the_olympics_killed_my_theatre.html.
29 The title, *I Saw Myself*, directly echoes Claudius's revelatory admission when he fails to kill Albert in *Gertrude – The Cry*.

30 Thus Sleev draws level with Barker's characters Hacker and Sleen (whose name she almost echoes) in being a presence in two Barker plays.
31 This detail is a full-scale development from a premise in *The Forty*, number Twenty-Eight.
32 Baudrillard, *op. cit.*, pp. 68–9.
33 *Ibid.*, p. 69.
34 *Ibid.*, p. 63.
35 Sleev's self-blinding is one of several resonant motifs in Barker's work, often to do with an unusual (even unenviable) degree of investment in what is imagined, or sensed other than visually: compare Smith's more drastically sudden action in *Rome*, and see Barker, in *Theatre of Catastrophe*, p. 32.
36 Baudrillard, *op. cit.*, pp. 131–153 [132, 137].
37 Here Modicum is a figure redolent of the Shakespearean vicious prodigal, who, ab-jected, returns to scourge or infect his native society: Lucius in *Titus Andronicus*, Alcibiades in *Timon of Athens*, Coriolanus.
38 Baudrillard, *op. cit.*, p. 63
39 *Ibid.*, p. 64.

Chapter 11

1 Kennelly, *Now* (Tarset: Bloodaxe, 2006), p. 100.
2 Compare Mangan (2007), p. 76.
3 Thanks to Charmian Savill for assistance towards this final paragraph.

A Selection of Further Critical Reading on Barker

IN ENGLISH:

Full-length Studies:

Gritzner, Karoline, and Rabey, David Ian (eds): *Theatre of Catastrophe: New Essays on Howard Barker* (London: Oberon Books, 2006).

Lamb, Charles: *Howard Barker's Theatre of Seduction* (London: Harwood Academic Press/Routledge, 1997).

—— *The Theatre of Howard Barker* (Oxford: Routledge, 2005).

Rabey, David Ian: *Howard Barker: Politics and Desire: An Expository Study of his Drama and Poetry, 1969–1987* (London: Macmillan, 1989).

Articles and Essays:

Barnett, David: 'Howard Barker: Polemic Theatre and Dramatic Practice, Nietzsche, Metatheatre and the play *The Europeans*', in *Modern Drama* 44.4 (Winter 2001), pp. 458–75.

Barker, Howard: 'On Naturalism and its Pretensions', in *Studies in Theatre and Performance* 27: 3 (2007), pp. 289–293, doi: 10.1386/stap.27.3.289/3.

Bas, Georges: 'The Cunts, the Knobs and the Corpse: Obscenity and Horror in Howard Barker's *Victory*', in *Contemporary Theatre Review* 5 (1996), pp. 33–50.

Cornforth, Andy, and Rabey, David Ian: 'Kissing Holes for the Bullets: Consciousness in Directing and Playing Barker's *(Uncle) Vanya*', in *Performance and Consciousness* 1.4 (1999), pp. 25–45.

Gallant, Desmond: 'Brechtian Sexual Politics in the Plays of Howard Barker', in *Modern Drama*, 40 (1997), pp. 403–413.

Gritzner, Karoline: 'Catastrophic Sexualities in Howard Barker's Theatre of Transgression' in Sönser Breen M., and Peters F. (eds), *Genealogies of Identity: Interdisciplinary Readings on Sex and Sexuality* (Amsterdam and New York: Rodopi, 2006).

—— 'Adorno on Tragedy: Reading Catastrophe in Late Capitalist Culture', in *Critical Engagements* 1.2 (Autumn/Winter 2007), pp. 25–52.

Kilpatrick, David: 'The Myth's the Thing: Barker's Revision of Elsinore in *Gertrude – The Cry*', in *Text and Presentation* 24 (New York: McFarland, 2003).

Klotz, Günther: 'Howard Barker: Paradigm of Postmodernism', in *New Theatre Quarterly* 7. 25 (Feb 1991), pp. 20–26.

Megson, Chris: 'Howard Barker and the Theatre of Catastrophe' in Luckhurst, Mary (ed.), *A Companion to Modern British and Irish Drama* (Oxford: Blackwells, 2006).

Neubert, Isolde: 'The Doorman of the Century is a Transient Phenomenon: the Symbolism of Dancer in Howard Barker's *Hated Nightfall*', in Reitz, Bernhard (ed.), *Drama and Reality. Contemporary Drama in English 3*, (Trier: Wissenschaftlicher Verlag Trier, 1996), pp. 145–153.

Rabey, David Ian: 'For the Absent Truth Erect: Impotence and Potency in Howard Barker's Recent Drama', in *Essays in Theatre/Études théâtrales* 10 (1991), pp. 31–37.

—— 'What Do You See?': Howard Barker's *The Europeans*', in *Studies in Theatre Production* 6 (Dec.1992), pp. 23–34.

—— 'Howard Barker', in Demastes, W. W., (ed.), *British Playwrights, 1956–1995: A Research and Production Sourcebook* (London: Greenwood Press, 1996), pp. 28–38.

—— 'Barker: Appalling Enhancements', in Rabey, D. I., *English Drama Since 1940* (London: Longman Literature in English series, Pearson Education, 2003) pp. 182–190.

—— 'Two Against Nature: Rehearsing and Performing Howard Barker's Production of his play *The Twelfth Battle of Isonzo*', in *Theatre Research International* 30.2 (July 2005), pp. 175–189.

Sakellaridou, Elizabeth: 'A Lover's Discourse – But Whose? Inversions of the Fascist Aesthetic in Howard Barker's *Und* and Other Recent English Plays', in *European Journal of English Studies*, 7.1 (April 2003), pp. 87–108.

Saunders, Graham: 'Missing Mothers and Absent Fathers': Howard Barker's *Seven Lears* and Elaine Feinstein's *Lear's Daughters*', in *Modern Drama* 42 (1999), pp. 401–410.

Tomlin, Liz: 'The Politics of Catastrophe' in *Modern Drama* 43, no 1, (2000) pp.66–77.

—— 'Howard Barker', in Bull, John (ed.), *Dictionary of Literary Biography, Volume 233: British and Irish Dramatists Since World War II, Second Series* (New York: Buccoli Clark , 2001), pp. 9–21.

Wilcher, Robert: 'Honoring the Audience: the Theatre of Howard Barker', in Acheson, James (ed.), *British and Irish Drama Since 1960* (Basingstoke: Macmillan, 1993), pp. 176–189.

Zimmermann, Heiner: 'Howard Barker's Appropriation of Classical Tragedy', in *(Dis)Placing Classical Tragedy*, Patsalidis, Savas, and Sakellaridou, Elizabeth (eds.) (Thessaloniki: University Studio Press, 1999), pp. 359–73.

—— 'Howard Barker's Brecht or Brecht as Whipping Boy', in Reitz, Berhard, and Heiko Stahl, Heiko (eds.), *What Revels Are in Hand (Essays in Honour of Wolfgang Lippke)* (CDE-Studies 8. Trier: Wissenschaftlicher Verlag Trier, 2001. 221–26.

—— 'Howard Barker in the Nineties' in 'British Drama of the 1990s': *Anglistik & Englischunterricht* 64 (2002), pp. 181–201.

IN FRENCH:

Alternatives Théâtrales 57 (mai 1998). Numéro spécial Howard Barker, coordonné par Mike Sens.

Angel-Perez, E. (ed.). *Howard Barker et le Théâtre de la Catastrophe*, éd. Paris: Editions Théâtrales, 2006.

—— « L'espace de la catastrophe ». Éd. Geneviève Chevallier. *Cycnos* 12 (1–1995).

—— « Pour un théâtre de la barbarie: Peter Barnes et Howard Barker ». Éd. É. Angel-Perez et Nicole Boireau. *Études anglaises* 52, n° 2 (avril–juin 1999): 198–210. Rééd. in *Le Théâtre anglais contemporain (1985–2005)*. Paris: Klincksieck, 2006.

—— Préfaces aux volumes 1–5 des *Howard Barker: Œuvres choisies*. Paris: éditions Théâtrales.

—— Notice sur Howard Barker de l'*Encyclopédie Universalis*. Paris: Encyclopædia Universalis, 2003.

—— « Howard Barker: de la catastrophe à l'épiphanie » in E. Angel-Perez. *Voyages au bout du possible. Les théâtres du traumatisme*. Paris: Klincksieck, 2006.

Boireau, Nicole. « Le paysage dramatique en Angleterre: consensus et transgression ». *Alternatives Théâtrales* 61 (1999): 8–10.

—— « Dystopies » in Boireau, N., *Théâtre et société en Angleterre des années 1950 à nos jours*. Paris: PUPS, 2000.

Hirschmuller, Sarah. « Howard Barker ou la déconsécration du sens. À propos de *Maudit crépuscule.* » Éd. Jean-Marc Lantéri. *Écritures contemporaines* 5 (2002): 25–42.
Morel, Michel. « La "catastrophe" selon Barker ». Éd. Geneviève Chevallier. *Cycnos* 18, n°1 (2001): 65–76.

My thanks to Elisabeth Angel-Perez for assistance in compiling the French section of this bibliography.

Index